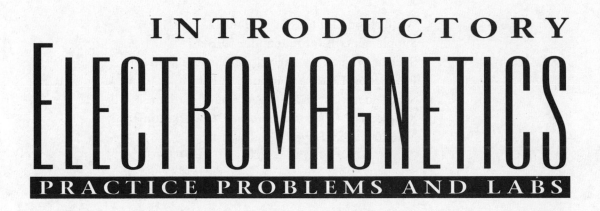

INTRODUCTORY
ELECTROMAGNETICS
PRACTICE PROBLEMS AND LABS

Zoya Popović • Branko D. Popović

PRENTICE HALL, Upper Saddle River, NJ 07458

Acquisitions Editor: Eric Frank
Supplements Editor: Jennifer DiBlasi
Special Projects Manager: Barbara A. Murray
Production Editor: Barbara A. Till
Supplement Cover Manager: Paul Gourhan
Supplement Cover Designer: PM Workshop Inc.
Manufacturing Buyer: Beth Sturla

10 9 8 7 6 5 4 3 2

ISBN 0-13-016571-9

Prentice-Hall International (UK) Limited, London
Prentice-Hall of Australia Pty. Limited, Sydney
Prentice-Hall Canada, Inc., Toronto
Prentice-Hall Hispanoamericana, S.A., Mexico
Prentice-Hall of India Private Limited, New Delhi
Pearson Education Asia Pte. Ltd., Singapore
Prentice-Hall of Japan, Inc., Tokyo
Editora Prentice-Hall do Brazil, Ltda., Rio de Janeiro

Preface to the Student

The supplement *Practice Problems and Labs* is an integral part of the textbook *Introductory Electromagnetics* by the same authors. The questions and problems from the textbook are preceded by extended chapter summaries, so the *Practice Problems and Labs* supplement can also be used as an independent textbook. References to equations in the supplement relate to the equations *in chapter summaries*.

The supplement is written with the principal intention to help you master the fundamental knowledge of electromagnetic fields. The following suggestions may help you to use this supplement efficiently:

1. Before proceeding to answering questions and solving problems, *always*

   - reread the chapter summaries, and

   - write down again all the principal conclusions and equations.

   This might seem an old-fashioned procedure, but you will be surprised how much this little effort will accelerate and improve your learning process. However, please do *not* consider the chapter summaries as substitutes for detailed explanations in your textbook.

2. While reading the texts of questions and problems, have a pencil and a piece of paper ready for your own sketches. Make absolutely sure that you understand the question or problem.

3. Most of the questions have three answers, one of which is correct (although in a few cases more than one may be acceptable). Read *very carefully* all the offered answers, and with clear arguments discard those you believe are incorrect. You also need to explain *to yourself* why you believe that the answer you have chosen is the right one. Have in mind that the learning process does not include only the learning of facts; it also includes the learning of how to avoid seemingly logical, but incorrect, answers. The offered multiple choice of answers, if approached with creativity, will help you learn the fundamental ideas of electricity and magnetism with a deep and lasting understanding.

4. Most of the problems also have multiple answers, one of which is correct. We suggest that you write down your solutions in a notebook neatly for future reference. Note also all difficulties you had in solving a problem: your false steps, dilemmas, mathematical formulas you needed for the solution but were unable to recall, etc.

5. In about 20% of questions and problems (marked by an "S" in the left margin) the choice of three answers is followed by a complete solution. We suggest that in every chapter you first study these questions and problems, as follows:

   - Try first to arrive at the correct answer without reading the solution.

   - Whether or not you have succeeded to obtain the correct answer yourself, read the solutions very carefully, several times if necessary. If you have solved the problem, see if there are any significant differences in the two solutions. If you made some serious errors, reproduce the solutions with complete understanding without referring to the *Practice Problems and Labs.*

6. To remember the facts and solution procedures properly, you will need most often to go through questions and problems *more than once.* Since you already know the solutions, and you have them in your notebook, a second or third run through the problems will be a quick one. Yet, it will fix your knowledge for a long time ahead, which a single run cannot do. We strongly recommend that you repeat all questions and problems assigned by your instructor *at least* three times before a test or the final exam.

True learning is invariably a difficult process. Miraculous recipes for "easy learning" simply do not exist. The intent of this supplement is to help you accelerate your learning so that you can master a subject in a minimum amount of time. We also hope that the process of discovering the correct answer among the offered answers will sharpen your reasoning and make your learning seem partly like a game.

We wish to express our gratitude to Dr. Branislav Notaroš of the University of Massachussetts, Dartmouth, formerly of the University of Belgrade, Yugoslavia, who provided or checked the solutions to most problems in Chapters 12–25 and Appendix 1. We thank Prof. John Dunn at the University of Colorado for very helpful comments on the Labs. We are also grateful to Professor David Rutledge at Caltech for ideas on several of the labs and many useful electromagnetic details that Z. P. learned while she was his graduate student and teaching assistant. Bruce Holland of SpectraLink (Boulder) and Steve Dunbar of Motorola Cellular Infractructure (Fort Worth) greatly contributed to the University of Colorado Electromagnetics Laboratory for the development of the experiments suggested in this book.

Z. P.
B. D. P.

Boulder, Colorado, June 1999

Contents

PART 4: TRANSMISSION LINES

PART 5: MAXWELL'S EQUATIONS AND THEIR APPLICATIONS

SIMPLE ELECTROMAGNETICS LABS

APPENDICES

Introducing Electromagnetics

1. Electromagnetics Around Us: Some Basic Concepts

- Electromagnetics deals with the theory and applications of electric and magnetic fields.

- The phenomena of electricity and magnetism were first noticed around 600 B.C. in ancient Greece. A piece of amber (in Greek, "electron") rubbed with a cloth was found to attract light objects, and pieces of a magnetic ore were found to attract small iron objects. Basically, the presence of an electric or a magnetic field is always detected by means of an electric or a magnetic force.

- There are two kinds of electric charges, which we call positive and negative charges. The charge of an electron is adopted as negative, and that of the atomic nucleus as positive. Normally, "neutral" matter contains two kinds of electric charges in equal amounts, and macroscopically does not exhibit electrical properties. If we remove a part Q of one kind of charges from a body (in practice, it can be only an extremely small percentage of the total body charge of one kind), it becomes charged with a charge $-Q$. The unit of charge is the *coulomb* (C).

- The fundamental law of the electric force is the Coulomb force between two small bodies with charges Q_1 and Q_2,

$$F_e = k_e \frac{Q_1 Q_2}{r^2}, \tag{1.1}$$

where $k_e = 9 \times 10^9 \, \mathrm{Nm^2/C^2}$, and r is the distance between the charges. The force is in newtons (N). It is attractive for unlike charges, and repulsive for like charges. The absolute value of the (negative) charge of an electron is approximately $e = 1.6 \times 10^{-19} \, \mathrm{C}$.

- Devices that are able to act as charge containers are called *capacitors*. They consist of two conducting pieces, known as *capacitor electrodes*, charged with *charges of equal magnitude, but opposite sign*. An ordered motion of a large number of electric charges is called the *electric current*. A permanent electric current can be produced by an *electric generator*. Electric generators use some other kind of energy (chemical, mechanical, thermal, the solar radiation) to separate electric charges and to obtain two charged electrodes. An electric current in substance is accompanied by heat, known as *Joule's heat*, or *Joule's losses*. The electric current intensity in a wire conductor is measured in amperes (A), equal to the number of coulombs transported through the wire cross section in one second.

• If there is a current in two nearby conductors, there is a force between them, known as the magnetic force. For two short parallel segments of wire, with their distance normal to the segments, it is of the form

$$F_m = k_m \frac{(I_1 l_1)(I_2 l_2)}{r^2},$$ (1.2)

where the force is in newtons, $k_m = 10^{-7}$ N/A^2 is a constant, l_1 and l_2 are the lengths of the segments (in meters), and I_1 and I_2 are current intensities in the segments (in amperes).

• An electric force on a charge is also obtained if a magnet, or a coil with current, is moved with respect to the charge. This is not a Coulomb force, but a new kind of electric force, responsible for the important phenomenon of *electromagnetic induction*.

• Quantitatively, the electric field is described by the *electric field strength vector*, defined trhough the electric force by the equation

$$\boldsymbol{F}_e = Q\boldsymbol{E},$$ (1.3)

where $\boldsymbol{F}_e$ is the force on charge Q situated in the electric field. The vector $\boldsymbol{E}$ is usually variable from one point to another, i.e., it is a function of coordinates, and can also be a function of time.

• The magnetic field is characterized in a similar manner by the *magnetic flux density vector*, $\boldsymbol{B}$. It is defined through the magnetic force by the equation

$$\boldsymbol{F}_m = Q\boldsymbol{v} \times \boldsymbol{B},$$ (1.4)

where $\boldsymbol{v}$ is the velocity of the charge Q moving in the magnetic field. The sign "$\times$" implies the vector, or cross, product of two vectors.

• In certain circumstances, a *combined, time-varying* electric and magnetic field can detach itself from the sources and propagate through space as an energy package, known as an *electromagnetic wave*. The speed of this wave in air is the same as the speed of light; in fact, light *is* an electromagnetic wave.

QUESTIONS

Q1.1. What is electromagnetics? — *(a) An electromagnet. (b) The subject that deals with the theory and applications of electric and magnetic fields. (c) Electricity and magnetism.*

S **Q1.2.** Think of a few examples of animals which use electricity or electromagnetic waves. What about a bat? — *(a) A person while reading. (b) A bat when flying, for observing obstacles and prey. (c) A dog while running.*

Answer. Electricity or electromagnetic waves are used by all animals, for diverse purposes. Muscles are activated by tiny electric pulses; brains operate using also such pulses; pain is transferred to the brain by electric pulses. Glow worms produce electromagnetic waves (light). Eyes are receptors of electromagnetic waves (light waves). As a rare example of animals using something else, bats do *not* use electromagnetic waves for their navigation, but ultra sound (acoustic) waves.

Q1.3. The basis of plant life is photosynthesis, i.e., synthesis (production) of life-sustaining substances by means of light. Is an electromagnetic phenomenon included? — *(a) Yes. (b) No. (c) In some cases. Explain your choice.*

Q1.4. What is the origin of the word "electricity"? — *(a) No specific origin. (b) The Roman word for force. (c) The Greek word for amber.*

Q1.5. What is the origin of the word "magnetism"? — *(a) The Roman word for iron (why?). (b) The Greek word for iron (why?). (c) The Greek name of an area in Asia Minor, "Magnesia".*

Q1.6. When did Thales of Miletus and William Gilbert make their discoveries? — *(a) Around 1200 B.C. and 1785 A.D. (b) Around 600 B.C. and 1600 A.D. (c) Arond 50 B.C. and 1784 A.D.*

S **Q1.7.** Why is it convenient to associate plus and minus signs with the two kinds of electric charges? — *(a) Because they are really positive and negative. (b) This simplifies mathematical expressions. (c) This enables compact mathematical descriptions.*

Answer. Without this convention the compact vector notation, for example in Coulomb's law, would not be possible, so the answer is (b). There is nothing "positive" or "negative" about electric charges in reality.

Q1.8. When did Coulomb perform his experiments with electric forces? — *(a) About two centuries ago. (b) About a century ago. (c) About six centuries ago.*

Q1.9. What is the definition of a capacitor? — *A system of two conducting bodies charged with (a) equal charges of opposite sign, (b) arbitrary charges, (c) equal charges of the same sign.*

Q1.10. What is electric current? — *(a) Current in an ocean of electricity. (b) An organized motion of a vast number of elemental electric charges. (c) Any motion of elemental electric charges.*

Q1.11. What are electric generators? — *(a) Devices able to separate electric charges. (b) Devices that extract positive electric charges from matter. (c) Devices which generate electrons.*

Q1.12. What common property do all electric generators have? — *(a) They all have a rotating part. (b) They all have a chemical process inside them. (c) There are forces on electric charges inside them which deposit positive and negative charges on two generator terminals.*

Q1.13. Describe in your own words the origin of Joule's losses. — *(a) Moving charges collide among themselves. (b) Friction of current with substance through which it exists. (c) Moving charges collide with neutral particles, which start to vibrate more vigorously, i.e., become warmer.*

Q1.14. What is the fundamental cause of magnetism? — *(a) Moving charges. (b) Charges at rest. (c) Both moving charges and those at rest.*

Q1.15. What is an electromagnet? — *(a) A magnet made of electric charges. (b) A body that attracts both electric charges and magnets. (c) A magnet obtained by passing a current through a coil made of wire.*

Q1.16. What did Faraday notice in 1831 when he moved a magnet around a closed wire loop? What did he expect to see? — *(a) A time-varying current in the loop during the motion of the magnet, and he expected a time-constant current if the magnet was placed near the loop. (b) A time-constant electric current in the loop, which he expected. (c) A time-constant electric current in the loop, and he expected a time-varying current.*

S **Q1.17.** Explain the concept of the electric field. — *(a) An electrified piece of land. A domain of space in which there is a force on (b) static electric charges, (c) moving electric charges.*

Answer. It is a domain of space in which there is a force on *static* electric charges.

Q1.18. Define the electric field strength vector. — *(a) The force on a charge in the field. (b) The electric force on a small charged body at a point of space, divided by the charge of the body. (c) The force on a charge in the field multiplied by the charge.*

S **Q1.19.** Explain the concept of the magnetic field. — *(a) A magnetized piece of land. A domain of space in which there is a force on (b) static electric charges, (c) moving electric charges.*

Answer. This is a natural phenomenon that is manifested by a force on *moving* electric charges.

Q1.20. Define the magnetic induction (magnetic flux density) vector. — *(a) If a small charge Q moves with a velocity v, and the magnetic force on the charge is F_m, it is defined by $F_m = Qv \times B$. (b) It is defined by $F_m = Qv \cdot B$. (c) It is defined by $F_m = Q \cdot (v \times B)$.*

Q1.21. What is an electromagnetic wave? — *(a) An electrified and magnetized water wave. (b) A combined electric and magnetic field that propagates through space as a field package. (c) Any combined electric and magnetic field.*

Q1.22. What are macroscopic quantities? — *(a) Quantities on a large scale, averaged over many atoms. (b) Quantities that can be seen only through a microscope. (c) Quantities that exist in large domains.*

PROBLEMS

S **P1.1.** How many electrons are needed to obtain one coulomb (1 C) of negative charge? Compare this number with the number of people on earth (about $5 \cdot 10^9$). — *(a) $N \simeq 6.24 \cdot 10^{18}$. (b) $N \simeq 3.24 \cdot 10^{17}$. (c) $N \simeq 1.85 \cdot 10^{19}$.*

Solution. $N = 1/(1.602 \cdot 10^{-19}) \simeq 6.24 \cdot 10^{18}$. This would be approximately obtained if every human had on his head the entire human population.

P1.2. Calculate approximately the gravitational force between two glasses of water a distance $d = 1\,\mathrm{m}$ apart, containing 2 dl (0.2 liter) of water each. — *(a) $F_{\mathrm{grav}} =\simeq 6.27 \cdot 10^{-11}\,N$, attractive. (b) $F_{\mathrm{grav}} =\simeq 2.67 \cdot 10^{-11}\,N$, attractive. (c) $F_{\mathrm{grav}} =\simeq 2.67 \cdot 10^{-12}\,N$, attractive.*

P1.3. Estimate the amount of equal negative electric charge (in coulombs) in the two glasses of water in problem P1.2 that would cancel the gravitational force. — *(a) $Q \simeq -18.2 \cdot 10^{-12}$ C. (b) $Q \simeq -17.2 \cdot 10^{-12}$ C. (c) $Q \simeq -16.2 \cdot 10^{-12}$ C.*

S **P1.4.** Two small equally charged bodies of masses $m = 1\,\mathrm{g}$ are placed one above the other at a distance $d = 10\,\mathrm{cm}$. How much negative charge would the bodies need to have so that the electric force on the upper body is equal to the gravitational force on it (i.e., so the upper body levitates)? Do you think this charge can be realized? — *(a) $Q \simeq -1.04 \cdot 10^{-7}$ C. It is easily realizable. (b) $Q \simeq -2.04 \cdot 10^{-7}$ C. It is not realizable. (c) $Q \simeq -1.04 \cdot 10^{-7}$ C. It is not easily realizable.*

Solution. $Q = -r\sqrt{4\pi\epsilon_0 mg} \simeq -1.04 \cdot 10^{-7}$ C $(g = 9.81\,\mathrm{m/s^2})$. This charge is practically impossible to realize, because the electric field around the bodies becomes so large that it ionizes the air, and the charge escapes from the bodies.

P1.5. Calculate the necessary electric field strength that would make a droplet of water of radius $a = 10\,\mu\mathrm{m}$, with an excess charge of 1000 electrons, levitate in the gravitational field of the earth. — *(a) $E \simeq 1.53 \cdot 10^5$ V/m, directed upward. (b) $E \simeq 3.65 \cdot 10^5$ V/m, directed downward. (c) $E \simeq 2.56 \cdot 10^5$ V/m, directed downward.*

P1.6. How large does the electric field intensity need to be in order to levitate a body $1\,\mathrm{kg}$ in mass and charged with -10^{-8} C? Is the answer of practical value, and why? — *(a) $E = 8.91 \cdot 10^8$ V/m, of practical value. (b) $E = 9.81 \cdot 10^8$ V/m, of no practical value. (c) $E = 9.81 \cdot 10^8$ V/m, of practical value.*

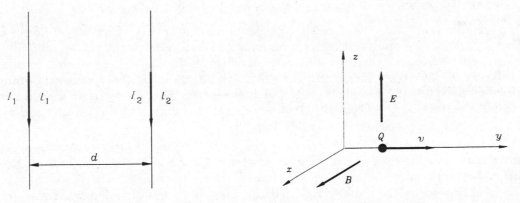

Fig. P1.8. Two parallel current elements.   **Fig. P1.9.** Charge in electric and magnetic field.

S **P1.7.** A drop of oil, $r = 2.25\,\mu\mathrm{m}$ in radius, is negatively charged and is floating above a very large, also negatively charged body. The electric field intensity of the large body happens to be $E = 7.83 \cdot 10^4$ V/m at the point where the oil drop is situated. The density of oil is $\rho_m = 0.851\,\mathrm{g/cm^3}$. (1) What is the charge of the drop equal to? (2) How large is this charge compared to the charge of an electron? (Note: the values given in this problem can realistically be achieved in the lab. Millikan used such an experiment at the beginning of the 20th century to show that charge is quantized.) — *(a) $Q \simeq 12e$ (e is the magnitude of the electron charge). (b) $Q \simeq 32e$. (c) $Q =\simeq 23e$.*

Solution. $mg = QE$, $m = 4a^3\pi\rho_m/3$, so that $Q = mg/E = 5.087 \cdot 10^{-18} \simeq 32e$ (e is the magnitude of the electron charge, $e = 1.602 \cdot 10^{-19}$ C.)

P1.8. Find the force between the two parallel wire segments in Fig. P1.8 if they are 1 mm long and 10 cm apart, and if they are parts of current loops which carry 1 A of current each. The constant k_m is equal to 10^{-7} in SI units (N/A^2). — *(a) $F_m = 10^{-10}$ N. (b) $F_m = 1.1 \cdot 10^{-11}$ N. (c) $F_m = 10^{-11}$ N.*

P1.9. A small body charged with $Q = -10^{-10}$ C finds itself in a uniform electric and magnetic field as shown in Fig. P1.9. The electric field vector and the magnetic flux density vector are E and B, respectively, everywhere around the body. If the magnitude of the electric field is $E = 100$ N/C, and the magnetic flux density magnitude is $B = 10^{-4}$ Ns/Cm, find the force on the body if it is moving with a velocity v as shown in the figure, where $v = 10$ m/s (the speed of a slow car on a mountain road). How fast would the body need to move to maintain its direction of motion? — *(a) $F \simeq 10^{-10}$ N, downward. The body maintains its direction of motion if $v = 10^6$ m/s. (b) $F \simeq 10^{-11}$ N, downward. The body maintains its direction of motion if $v = 10^5$ m/s. (c) $F \simeq 10^{-10}$ N, upward. The body maintains its direction of motion if $v = 10^6$ m/s.*

P1.10. Volta used a chemical reaction to make the first battery which could produce continuous electric current. Use the library, or any other means, to find out if electric current can be used to make chemical reactions possible. Write down about one page on the history and implications of these processes. — *Recall the effect of electrolysis and give a few examples.*

2. Circuit Theory and Electromagnetics

• The most important tool of electrical engineers is circuit theory. Circuits are electromagnetic systems, but we can analyze them approximately starting from the two Kirchhoff's laws. There are no "pure" circuit elements (resistors, capacitor, inductor, etc.), and the circuit shape and size influences the circuit properties (except in the dc case). For example, damping in a resonant circuit becomes progressively larger with the circuit size, since larger circuits radiate electromagnetic waves more efficiently. To the generator driving the circuit the radiation is just a loss of energy, i.e., it is seen as Joule's losses.

• A transmission line, e.g., a coaxial cable, is an electromagnetic system in which current and voltage propagate with a finite velocity, less than the velocity of light in a vacuum (typically, about 65 % of that velocity for a dielectric-filled coaxial cable). For air-filled transmission lines, the velocity of propagation is practically the same as the velocity of light in a vacuum.

QUESTIONS

S **Q2.1.** Why does every switch have capacitance? — *(a) When open, its two terminals represent a capacitor. (b) When open and in a circuit, its two terminals are charged with equal charges of opposite sign. (c) When closed, it represents an inifinite capacitance.*

Answer. When the switch is open, its two terminals are two bodies insulated from each other. If connected to a source, the two switch terminals will be charged with equal charges of opposite sign.

Q2.2. Try to imagine a "perfect" realistic switch (a switch with the smallest possible capacitance). How would you design a good switch? What would be its likely limitations? — *(a)*

With large, close contacts (why?). (b) With small contacts far apart (why?). (c) Any switch is a good one (why?).

Q2.3. Why does it become progressively more difficult to have an "ideal" switch as frequency increases? — *(a) The switch properties do not depend on frequency. (b) The switch leads have an inductance, which makes a closed switch resemble an open circuit as frequency increases. (c) An open switch has a capacitance, which makes it approach a short circuit as frequency increases.*

S   **Q2.4.** Why is the circuit-theory assumption that interconnecting conductors (wires) have no effect on the circuit behavior incorrect? — *(a) The shape of interconnecting conductors influences the circuit properties. (b) The size of the circuit influeneces the circuit properties. (c) Both circuit shape and size influence the circuit properties.*

Answer. Except in the dc case, the shape and size of the interconnecting conductors *always* has an influence on the circuit behaviour.

Q2.5. Imagine a resistor connected to a car battery by wires of fixed length. Does the shape of the wires influence the current in the resistor? Explain. — *(a) It does. (b) It does not. (c) In some cases (in which?).*

Q2.6. Answer question Q2.5 if the source is a (1) 60 Hz and (2) 1-GHz generator. — *(a) Pronounced at 60 Hz, negligible at 1 GHz (why?). (b) Negligible at 60 Hz, pronounced at 1 GHz (why?). (c) Pronounced at both frequencies (why?).*

S   **Q2.7.** Give at least two reasons for the failure of circuit theory when analyzing the simple circuit in Fig. Q2.7 at a frequency of 100 MHz. Assume the size of the circuit to be as in the figure. — *(a) Circuit-theory does not fail (why?). (b) Kirchhoff's laws are not valid any more (both or one of them only, and why?). (c) Capacitance between the resistor terminals, inductance of all the wires, mutual inductance between the wires, capacitance between the generator terminals (explain!).*

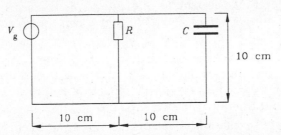

Fig. Q2.7. A simple electric circuit.

Answer. The capacitance between the terminals of the resistor, the inductance of all the wires, the mutual inductance between the wire loops, the capacitance between the generator terminals.

Q2.8. Explain why the resonant frequency of a circuit is at least to some extent dependent on the circuit shape and size. — *(a) Because they determine the circuit resistance. (b) Because they determine the circuit inductance. (c) Because they determine the circuit capacitance.*

S   **Q2.9.** Why does the damping in resonant circuits depend at least to some extent on the circuits' shape and size? Can circuit theory explain this? — *(a) Due to changed circuit* *inductance. (b) Due to changed circuit resistance. (c) Due to large currents at resonance and radiation.*

Answer. All circuits with time-varying currents radiate a certain amount of energy. This radiation increases with the size of the circuit. It also increases with the current intensity in the circuit, and so is most pronounced in resonant circuits. This radiation represents a loss of energy, and increases dumping. Circuit theory does not take radiation into account, and therefore cannot explain this part of damping.

PROBLEMS

S **P2.1.** The capacitance of a switch ranges from a fraction of a picofarad to a few picofarads. Assume that a generator of variable angular frequency ω is connected to a resistor of resistance of $1\,\text{M}\Omega$, but that the switch is open. Assuming a switch capacitance of $1\,\text{pF}$, at what frequency is the open switch reactance equal to the resistor resistance? — *(a) $f = 0.159\,MHz$. (b) $f = 1.59\,MHz$. (c) $f = 15.9\,MHz$.*

Solution. At the frequency defined by $R = 1/(\omega C)$, i.e., for $f = 1/(2\pi RC) = 0.159\,\text{MHz}$.

P2.2. A surface-mount capacitor has a 1-nH parasitic series lead inductance. Calculate and plot the frequency at which such a capacitor starts looking like an inductor, as a function of the capacitance value. — *Hint: note that this is a simple series connection of a capacitor and a coil.*

P2.3. A surface-mount resistor of resistance $R = 100\,\Omega$ has a $1\,\text{nH}$ series lead inductance. Plot the real and imaginary part of the impedance as a function of frequency. In which frequency range can this chip be used as a resistor? — *Hint: note that this is a simple series connection of a resistor and a coil.*

P2.4. The windings of a coil have a parasitic capacitance of $0.1\,\text{pF}$, which can be viewed as an equivalent series capacitance. Plot the reactance of such a $1\,\mu\text{H}$ coil as a function of frequency. — *Hint: note that this is a simple series connection of a coil and a capacitor.*

S **P2.5.** A capacitor of capacitance C is charged with a charge Q. It is then connected to an uncharged capacitor of the same capacitance C by means of conductors with practically no resistance. Find the energy contained in the capacitor before connecting it to the other capacitor, and the energy contained in the two capacitors. [The energy of a capacitor is given by $W_e = Q^2/(2C)$]. Can you explain the results using circuit-theory arguments? How can you explain the results at all? — *(a) $W_e = Q^2/(2C)$ in both cases, since there is no loss of energy. (b) $W_e' = Q^2/(2C)$ and $W_e'' = Q^2/(4C)$, the latter being smaller due to Joule's losses. (c) $W_e' = Q^2/(2C)$ and $W_e'' = Q^2/(4C)$, the latter being smaller due to radiation.*

Solution. The energy of the first capacitor alone is $Q^2/(2C)$. Since the charge, Q, remains the same after connecting the other capacitor, the energy becomes $Q^2/(4C)$, since the capacitance of the parallel connection is $2C$. So, energy is halved, in spite of the circuit having no resistance. The other half was radiated during the time-varying transient currents in the circuit. Circuit theory *cannot* explain this radiation.

P2.6. The inductance of a thin circular loop of radius R, made of wire of radius a, where $R \gg a$, is given by the approximate formula

$$L_0 \simeq \mu_0 R \left(\ln \frac{8R}{a} - 2 \right) \quad \text{(henrys)},$$

where $\mu_0 = 4\pi \, 10^{-7}$ henry/m, and R and a are in meters. A capacitor of capacitance $C = 100 \, \text{pF}$ and a coil of inductance $L = 100 \, \text{nH}$ are connected in series by wires of radii a and of the shape of a circular loop of radius R ($R \gg a$). Find: (1) the radius of the loop which results in $L_0 = L$, if $a = 0.1 \, \text{mm}$; (2) the radius of the wire which results in $L_0 = L$, if $R = 2.5 \, \text{cm}$; and (3) the resonant frequency of the circuit versus the loop radius, R, if the wire radius is $a = 0.1 \, \text{mm}$. — (a) $R = 1.56 \, cm$; $a = 1.1 \, mm$; $f = 1/[2\pi\sqrt{(L + L_0)C}]$, with $a = 0.0001 \, m$ in the expression for L_0. (b) $R = 2.56 \, cm$; $a = 1.5 \, mm$; $f = 1/[2\pi\sqrt{L_0 C}]$. (c) $R = 3.56 \, cm$; $a = 2.1 \, mm$; $f = 1/[2\pi\sqrt{LC}]$.

P2.7. The capacitance between the terminals of a resistor is $C = 0.5 \, \text{pF}$, and its resistance is $R = 10^6 \, \Omega$. Plot the real and imaginary parts of the impedance of this dominantly resistive element versus frequency, from $0 \, \text{Hz}$ to $10^7 \, \text{Hz}$. — (a) This is a plot of the impedance of R and C in series. (b) This is, approximately, a plot of the impedance of R and C in parallel. (c) This is, exactly, a plot of the impedance of R and C in parallel.

S **P2.8.** A coil is made in the form of $N = 10$ tightly packed turns of wire. Predict qualitatively the high-frequency behavior of the coil. Explain your reasoning. — (a) The same as low-frequency behavior. (b) The coil inductance becomes more pronounced at high frequencies. (c) It will behave approximately as a parallel $L - C$ circuit.

Solution. The voltage between the coil terminals is distributed along the coil. Therefore there is a voltage between any two adjacent wire turns. Since they are close, there is a capacitance between adjacent turns. In the first approximation, these capacitors add up to an equivalent capacitor parallel to the coil. As the frequency increases, this capacitance will have an increasingly pronounced effect on the coil behaviour. At a certain frequency the entire structure will behave approximately as a parallel resonant circuit (an open circuit), and at still higher frequencies the capacitance will predominate.

Part 1: Time-Invariant Electric Field

3. Coulomb's Law in Vector Form and Electric Field Strength

• Sources of an electrostatic field are stationary and time-constant electric charges. The basic relationship for such a field is the *Coulomb's law in vector form*, which gives the force due to a small charge Q_1 on another small charge Q_2 a distance r from Q_1:

$$\boldsymbol{F}_{e12} = \frac{1}{4\pi\epsilon_0} \frac{Q_1 Q_2}{r^2} \boldsymbol{u}_{r_{12}} \qquad \text{(N)}. \qquad (3.1)$$

$\boldsymbol{u}_{r_{12}}$ is the unit vector directed from charge Q_1 towards charge Q_2, and

$$\epsilon_0 = 8.854 \cdot 10^{-12} \simeq \frac{1}{36\pi \, 10^9} \qquad \text{(farads per meter } - \text{ F/m)} \qquad (3.2)$$

is known as the *permittivity of a vacuum*, or of free space. The Coulomb forces of several charges on one charge are simply added as vectors (the *principle of superposition for forces*). Electric forces around us do not exceed about $1\,\text{N}$, i.e., they are quite small.

• The electric field at a point is described by the *electric-field strength vector*, defined as

$$\boldsymbol{E} = \frac{\boldsymbol{F}_{\text{on} \, \Delta Q}}{\Delta Q} \qquad \text{(volts per meter } - \text{ V/m} = \text{N/C)}, \qquad (3.3)$$

where ΔQ is a small, "test", charge at that point. The electric-field vector due to a point charge Q is given by

$$\boldsymbol{E} = \frac{1}{4\pi\epsilon_0} \frac{Q}{r^2} \boldsymbol{u}_r \qquad \text{(V/m)}, \qquad (3.4)$$

where $\boldsymbol{u}_r$ is the unit vector *directed from the charge Q away*. The electric field strength of any number of point charges is obtained as a vector sum of individual field strengths.

• For densely packed charges inside a volume, over a surface, or along a line, the volume, surface and line charge densities are defined respectively as

$$\rho = \frac{dQ_{\text{in } dv}}{dv} \quad (\text{C/m}^3), \qquad \sigma = \frac{dQ_{\text{on } dS}}{dS} \quad (\text{C/m}^2), \qquad Q' = \frac{dQ_{\text{on } dl}}{dl} \quad (\text{C/m}). \quad (3.5)$$

These charge densities are usually functions of space coordinates. If we know these functions, the total electric field strength is obtained by integration:

$$\boldsymbol{E} = \frac{1}{4\pi\epsilon_0} \int_v \frac{\rho\, dv}{r^2} \boldsymbol{u}_r, \qquad \boldsymbol{E} = \frac{1}{4\pi\epsilon_0} \int_S \frac{\sigma\, dS}{r^2} \boldsymbol{u}_r, \qquad \boldsymbol{E} = \frac{1}{4\pi\epsilon_0} \int_L \frac{Q'\, dl}{r^2} \boldsymbol{u}_r \quad (\text{V/m}). \quad (3.6)$$

• The electrostatic field is a *vector* field. As such, it can be visualized by means of *E–field lines*, defined as lines to which $\boldsymbol{E}$ is tangential at all points. An arrow is usually added to the lines to indicate the direction of $\boldsymbol{E}$ along them.

QUESTIONS

S **Q3.1.** Discuss the statement that Eq. (3.1) shows indeed not only the magnitude but also the correct direction of the force $\boldsymbol{F}_{e12}$. Does Eq. (3.1) need an additional explanation in words? — *(a) The statement is incorrect (why?). (b) The statement is correct (why?). (c) It is correct only if both charges are of the same sign.*

Answer. The unit vector $\boldsymbol{u}_{r_{12}}$ is directed from charge Q_1 to charge Q_2. Therefore if the two charges are of the same sign, the expression on the right is positive, i.e., the force is repulsive. If the two charges are of opposite signs, this expression is negative, i.e., the force is attractive. The force is always directed along the straight line joining charges Q_1 and Q_2. Hence, no additional explanations in words is necessary.

Q3.2. Would the vector form (3.1) of Coulomb's law be possible if plus and minus signs were not associated with the two types of charges? For example, suppose that they were denoted by subscripts A and B instead of plus and minus signs. Explain your answer. (This question is intended to show how important proper conventions are for simplifying the mathematical description of physical phenomena.) — *(a) No (why?). (b) Yes (why?). (c) The other convention (subscripts A and B instead of plus and minus signs) would result in the same possibility.*

Q3.3. Is it possible to *derive* the principle of superposition of Coulomb's forces, starting from Coulomb's law? Explain. — *(a) Yes, because it is obvious. (b) No, it needs to be verified experimentally. (c) Yes, because all forces satisfy the principle of superposition.*

Q3.4. Prove that there can be no net electric force on an isolated charged body due to its charge only. — *(a) Such a force would move the body without using energy. (b) Such a force would change the shape of the body. (c) Such a force would make self-propulsion of the body possible without use of a fuel.*

S **Q3.5.** Similarly to the electric field, the gravitational field also acts "at a distance". But whereas we understand and accept that there is a downward force on an object we lift (e.g., a stone), with no visible reason, such an electric force with no visible reason is somewhat astonishing. Explain why this is so? — *(a) Electric force is different from gravitational. (b) Gravitational force is very large. (c) We encounter the gravitational force all the time and are used to it.*

Answer. We have been noticing the effect of the gravitational force ever since we were born. We never *expect* something else (i.e., to see a body going up when we drop it), and therefore we are accustomed to consider this to be natural. We do not have such a constant experience with electric forces.

Q3.6. Of five equal conducting balls one is charged with a charge Q, and the other four are not charged. Find all possible charges the balls can obtain by touching one another, assuming that the two balls are allowed to touch only once, and that while two balls are touching, the influence of the other three can be neglected. — *(a) Practically any charge can be obtained. (b) Some of the charges that can be obtained are $Q/2$, $Q/4$, $Q/8$, and $Q/16$. (c) Some of the charges that can be obtained are $2Q$ and $3Q$.*

Q3.7. If an electrified body (e.g., a plastic ruler rubbed against a wool cloth) is brought near small pieces of a thin aluminum foil, you will see (make the experiment!) that the body first attracts, but after the contact repels, the small pieces. Explain! — *(a) The foil is charged with charge of opposite sign, and upon contact gets the charge of the same sign. (b) The charge on the foil is zero, but the field of the body induces equal opposite charges on the foil, attracts the closer charge, and upon contact transfers some charge of the same sign to the foil. (c) Attraction is due to the gravitational force, repulsion to the charge transferred to the foil upon contact.*

Q3.8. Imagine that you electrified a body, e.g., by rubbing it against another body. How could you determine the sign of the charge on the body? Try to perform the experiment. — *(a) Since the electrified body is attracted by the other body, its charge is positive. (b) There is no way to determine the sign of the charge. (c) We can determine the sign of the charge if we have a reference body with of a charge of known sign.*

Q3.9. You have two identical small metal balls. How can you obtain identical charges on them? — *(a) Charge both by touching a large charged body. (b) Press the balls against each other and charge them by touching a small charged body. (c) Charge one of them, move both far from other bodies, and touch them.*

Q3.10. Two small balls carry charges of unknown signs and magnitudes. Experiment shows that there is no electric force on a third charged ball placed at the midpoint between the first two. What can you conclude about the charges on the first two balls? — *(a) The charges on the balls are zero. (b) The charges are equal, of opposite sign. (c) The charges are equal in both magnitude and sign.*

Q3.11. An uncharged small ball is introduced into the electric field of a point charge. Is there a force on the ball? Explain. — *(a) There is no force. (b) The force is repulsive. (c) The force is attractive.*

S **Q3.12.** Is it correct to write the following: (1) Q ($Q > 0$); (2) $-Q$ ($Q < 0$); (3) Q ($Q < 0$); and (4) $-Q$ ($Q > 0$)? Explain. — *(a) All expressions are correct. (b) Only the first and the third are correct. (c) Only the second and the fourth are correct.*

Answer. All are correct. The charge Q is defined to have an algebraic sign. So, for example, (2) is analogous to, e.g., $-(-3) = 3$ in algebra.

Q3.13. To measure the electric field strength at a distance r from a small charge Q, a test charge ΔQ ($\Delta Q \ll Q$) in the form of a sphere of radius $a = r/2$ is centered at that point. Discuss the correctness of the measurement. — *(a) It is correct, because we determine the force on a charge. (b) It is not correct, because $\Delta Q \ll Q$. (c) It is not correct, because the radius of the test charge is too large.*

Q3.14. What would the form of the expression in Eq. (3.4) be if we adopt that u_r is *towards* the charge. What form does Eq. (3.4) take if we do not associate a sign with the charge Q? — *(a) A minus sign in front of the right-hand side. Nothing would change. (b) Everything remains the same. (c) A minus sign in front of the right-hand side, and explanation in words concerning the direction of force.*

Q3.15. Is ρ in Eq. (3.5) a function of coordinates, in general? — *(a) It is always constant. (b) Yes, e.g., it can be of the form $\rho = \rho(x, y, z)$. (c) It changes only across the boundary of a charged cloud.*

S **Q3.16.** Assuming ρ in Eq. (3.6) to be known, explain in detail how you would numerically evaluate the *vector* integral to obtain $\mathbf{E}$. — *(a) Divide v into small domains Δv, consider the charges $\rho \Delta v$ inside them as point charges, and find the total field. (b) As the field of a point charge at the center of the volume. (c) It is not possible to evaluate the integral numerically, because it is a vector integral.*

Answer. It is necessary to divide the domain with nonzero ρ into small volumes, to consider the charges inside these volumes, ρdv, as point charges, and to add vectorially all the elemental vectors $d\mathbf{E}$ to obtain the total vector $\mathbf{E}$.

Q3.17. Repeat question Q3.16 for a surface distribution of charges over a surface S, and for a line distribution of charges along a line L. — *The same three answers as for the preceding question, relating to a surface and a line, instead to a domain.*

Q3.18. Why are the formulas in Eqs. (3.6) only of limited practical value? — *(a) We cannot evaluate the integrals, not even numerically. (b) We rarely know the distribution of charge. (c) They are of considerable practical value, not of limited value.*

PROBLEMS

S **P3.1.** What would be the charge of a copper cube, $1\,cm$ on the side, if one electron were removed from all the atoms on the cube surface? A cubic meter of copper has about $8.4 \cdot 10^{28}$ atoms. — *(a) $Q = 1.849\,mC$. (b) $Q = 2.325\,mC$. (c) $Q = 1.389\,mC$.*

Solution. One "atom" occupies a small cube of side $a = 1/\sqrt[3]{8.4 \cdot 10^{28}} = 2.28 \cdot 10^{-10}$ m. So there are $N = 4.386 \cdot 10^9$ atoms/m, or $n = 1.924 \cdot 10^{15}$ atoms/cm^2. There are six sides of the cube, so on its surface there are $M = 1.154 \cdot 10^{16}$ atoms. If one electron per atom is removed from all these atoms, the cube becomes charged with a positive charge $Q = Me = 1.849\,$mC.

P3.2. Evaluate the force that would exist between two cubes as described in problem P3.1 when they are (1) $d = 1\,$m, and (2) $d = 1\,$km apart. — *(a) $F_1 = 2.23 \cdot 10^4$ N, $F_2 = 21.6\,mN$. (b) $F_1 = 3.07 \cdot 10^4$ N, $F_2 = 30.7\,mN$. (c) $F_1 = 5.35 \cdot 10^4$ N, $F_2 = 45.6\,mN$.*

P3.3. Three small charged bodies arranged along a straight line are at distances a, b, and $(a + b)$ apart. Determine the conditions that the charges on the bodies have to satisfy so that the electric forces on all three are zero. — *(a)* $Q_1 = Q_2(a+b)^2/b^2 = -Q_3 a^2/b^2$. *(b)* $Q_1 = -Q_2(a+2b)^2/b^2 = Q_3 a^2/b^2$. *(c)* $Q_1 = -Q_2(a+b)^2/b^2 = Q_3 a^2/b^2$.

P3.4. Assume that the earth is electrified by a charge $2Q$, and the moon by a charge Q. How large does Q have to be so that the repulsive electric force between the earth and the moon be equal to the attractive gravitational force? The masses of the earth and the moon are $m_E = 5.983 \cdot 10^{24}$ kg and $m_M = 7.347 \cdot 10^{22}$ kg. The gravitational constant is $\gamma = 6.67 \cdot 10^{-11}$ N $\cdot$ m^2/kg^2. — *(a)* $Q = 5.25 \cdot 10^{12}$ C. *(b)* $Q = 3.25 \cdot 10^{14}$ C. *(c)* $Q = 4.04 \cdot 10^{13}$ C.

P3.5. Evaluate the specific charge of the electron (the ratio of charge and mass). Estimate the charge of the book you are reading if it had the same specific charge. What would the force between two such charged books be if they were at a distance of 10 m? — *Let the mass of the book be $m = 0.4$ kg.* *(a)* $F = 4.4 \cdot 10^{29}$ N. *(b)* $F = 5.1 \cdot 10^{25}$ N. *(c)* $F = 2.3 \cdot 10^{18}$ N.

P3.6. A given charge Q is divided between two small bodies, so that one is charged with a charge Q', and the other has the rest. Determine the ratio Q/Q' resulting in the greatest electric force between them, assuming the distance between them is fixed. — *(a)* $Q'/Q = 0.25$. *(b)* $Q'/Q = 0.5$. *(c)* $Q'/Q = 0.4$.

S **P3.7.** Three small charged bodies of charge Q are placed at three vertices of an equilateral triangle with sides of length a. What is the direction and magnitude of the electric force on each of them if $a = 3$ cm, and $Q = 1.8 \cdot 10^{-10}$ C? — *(a)* $F = 5.6 \cdot 10^{-7}$ N. *(b)* $F = 3.7 \cdot 10^{-8}$ N. *(c)* $F = 4.3 \cdot 10^{-9}$ N.

Solution. The force is directed along the line from the triangle center point towards the considered triangle vertex. The magnitude of the force is $F = Q^2 \cos(\pi/6)/(2\pi\epsilon_0 a^2) = 5.6 \cdot 10^{-7}$ N.

P3.8. A charge Q exists at all vertices of a cube with sides of length a. Determine the direction and magnitude of the electric force on one of the charges. — *(a)* $F = Q^2/(4\pi\epsilon_0 a^2)$. *(b)* $F = Q^2(\sqrt{6}/2 + 1/3)/(4\pi\epsilon_0 a^2)$. *(c)* $F = Q^2(\sqrt{3} + \sqrt{6}/2 + 1/3)/(4\pi\epsilon_0 a^2)$.

S **P3.9.** Two identical, small, conducting balls with centers that are d apart have charges Q_1 and Q_2. The balls are brought into contact and returned to their original positions. Determine the electric force in both cases, if charges Q_1 and Q_2 are (1) of the same sign, and (2) of opposite signs. — *(a)* $F_1 = Q_1 Q_2/(4\pi\epsilon_0 d^2)$, repulsive or attractive, $F_2 = (Q_1 + Q_2)^2/(16\pi\epsilon_0 d^2)$, repulsive. *(b)* $F_1 = Q_1 Q_2/(4\pi\epsilon_0 d^2)$, repulsive, $F_2 = (Q_1 + Q_2)^2/(4\pi\epsilon_0 d^2)$, attractive. *(c)* $F_1 = Q_1 Q_2/(4\pi\epsilon_0 d^2)$, attractive, $F_2 = (Q_1 + Q_2)^2/(8\pi\epsilon_0 d^2)$, repulsive.*

Solution. The magnitude of force before the contact is $F = Q_1 Q_2/(4\pi\epsilon_0 d^2)$, and the force can be either repulsive or attractive, depending on the sign of charges Q_1 and Q_2. After the contact, their charge is the same, equal to $Q = (Q_1 + Q_2)/2$. The magnitude of force is now $F = (Q_1 + Q_2)^2/(16\pi\epsilon_0 d^2)$, and the force is repulsive in all cases.

P3.10. Evaluate the velocity of an electron orbiting around the nucleus of a hydrogen atom along an approximately circular orbit of radius $a = 0.528 \cdot 10^{-10}$ m. How many revolutions does the electron make in one second? — *(a)* $v = 3.32 \cdot 10^6$ m/s, $n = 8.7 \cdot 10^{15}$ revolutions/s. *(b)* $v = 2.19 \cdot 10^6$ m/s, $n = 6.6 \cdot 10^{15}$ revolutions/s. *(c)* $v = 5.29 \cdot 10^6$ m/s, $n = 11.8 \cdot 10^{15}$ revolutions/s.*

S **P3.11.** Two small balls of mass m each have a charge Q and are suspended at a common point by separate thin, light, conducting filaments of length l. Assuming the charges are located approximately at the centers of the balls, find the angle α between the filaments. Suppose that α is small. (Such a system can be used as a primitive device for measuring charge, and is called an *electroscope.*) — *Let the geometry of the system with charged balls be as in Fig. P3.11. (a)* $\tan \alpha = F_g/F_e$. *(b)* $\tan \alpha = F_e/F_g$. *(c)* $\tan \alpha = 2F_e/F_g$.

Solution. Let the geometry of the system with charged balls be as in Fig. P3.23. There are two forces acting on both balls, the gravitational force $F_g = mg$, directed downward, and the repulsive electric (Coulomb) force, $F_e = Q^2/[4\pi\epsilon_0(2l\sin\alpha)^2]$, directed horizontally. The vector sum of these two forces must be along the filament. Let the filament make an angle α with a vertical axis (see Fig. P3.23). The angle α is obtained from the equation $\tan \alpha = F_e/F_g$.

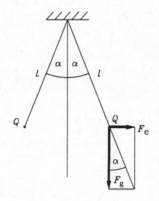

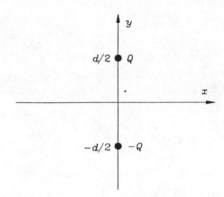

Fig. P3.11. Two suspended charged balls.   **Fig. P3.14.** Two equal charges of opposite sign.

P3.12. A small body with a charge $Q = 1.8 \cdot 10^{-10}$ C is situated at a point A in the electric field. The electric force on the body is of intensity $F = 5.4 \cdot 10^{-4}$ N. Evaluate the magnitude of the electric field strength vector at that point. — *(a)* $E = 4 \cdot 10^6$ *V/m. (b)* $E = 5 \cdot 10^6$ *V/m. (c)* $E = 3 \cdot 10^6$ *V/m.*

P3.13. A point charge Q ($Q > 0$) is located at the point $(0, d/2)$, and a charge $-Q$ at the point $(0, -d/2)$, of a rectangular coordinate system. Determine and plot the magnitude of the total electric field strength vector at any point in the xy plane. — *The solution is of the form*

$$E = E_1 + E_2 = \frac{Q}{C\pi\epsilon_0}\left(\frac{r_1}{r_1^3} - \frac{r_2}{r_2^3}\right),$$

where $r_1 = xu_x + (y - d/D)u_y$, $r_2 = xu_x + (y + d/D)u_y$, $r_1 = \sqrt{x^2 + (y - d/D)^2}$ *and* $r_2 = \sqrt{x^2 + (y + d/D)^2}$. *One of the following pairs of constants C and D gives the right answer: (a)* $C = 4, D = 2$, *(a)* $C = 8, D = 4$, *(a)* $C = 2, D = 2$.

S **P3.14.** An electric dipole consists of two equal and opposite point charges Q and $-Q$ which are a distance d apart, Fig. P3.14. (1) Find the electric field vector along the x axis in the figure. (2) Find the electric field vector along the y axis. (3) How does the electric field strength behave at distances $x \gg d$ and $y \gg d$ away from the dipole? How does this behaviour compare to that of the field of a single point charge? — *Hint: use the result from the preceding problem and consider the specified points.*

Solution. The general result is given in the answer to P3.13. If we set in that answer $y = 0$ (the points along the x axis), we obtain

$$E = E_1 + E_2 = -\frac{Qd}{4\pi\epsilon_0}\left(\frac{u_y}{[x^2 + (d/2)^2]^{3/2}}\right),$$

since $r_1 - r_2 = (xu_x - d/2u_y) - (xu_x + d/2u_y) = -du_y$.

If $x \gg d$, $d/2$ can be neglected with respect to x, so we have

$$E = -\frac{Qdu_y}{4\pi\epsilon_0 x^3}.$$

Similarly, if we set that $x = 0$ (the points along the y axis), we get

$$E = E_1 + E_2 = \frac{Q}{4\pi\epsilon_0}\left(\frac{r_1}{r_1^3} - \frac{r_2}{r_2^3}\right) = \frac{Qu_y}{4\pi\epsilon_0}\left(\frac{1}{r_1^3} - \frac{1}{r_2^3}\right),$$

where $r_1 = (y - d/2)u_y$, $r_2 = (y + d/2)u_y$, $r_1 = y - d/2$ and $r_2 = y + d/2$.

If $y \gg d$, $d/2$ in r_1 and r_2 can be neglected with respect to y, so that $r_1 = r_2 = y$, and we have

$$E = \frac{Qu_y}{4\pi\epsilon_0}\left[\frac{1}{(y - d/2)^2} - \frac{1}{(y + d/2)^2}\right] \simeq \frac{Qu_y}{4\pi\epsilon_0 y^2}\left[\left(1 + \frac{d}{y}\right) - \left(1 - \frac{d}{y}\right)\right] = \frac{Qdu_y}{2\pi\epsilon_0 y^3}.$$

P3.15. Find the x and y components of the electric field vector at an arbitrary point in the field of the electric dipole from problem P3.14, assuming that the distance of the observation point from the dipole center is much greater than d. Plot your results. — *Hint: use the result from the answer to problem P3.13, and find the projection of the vector E onto the x and y axes.*

P3.16. A thin, straight rod $a = 10$ cm long is uniformly charged along its length with a total charge $Q = 2 \cdot 10^{-10}$ C. The rod extends from the point $(-a/2, 0)$ to the point $(a/2, 0)$ in an xy rectangular coordinate system. Evaluate the electric field strength vector at points $A(0, a/4)$ and $B(3a/4, 0)$. — (a) $E_A = E_y = 1286$ V/m, $E_B = E_x = 575.2$ V/m. (b) $E_A = E_y = 1635$ V/m, $E_B = E_x = 667.3$ V/m. (c) $E_A = E_x = 1286$ V/m, $E_B = E_y = 575.2$ V/m.

P3.17. Solve problem P3.16 approximately, by dividing the rod into n segments. Compare the results with the exact solution for $n = 1, 2, 3, 4, 5, 6, 10$, and 20. — *Hint: divide the rod into n segments of equal length $b = a/n$. Assume that the charges on these segments, equal to $\Delta Q = Q/n$, are in the form of point charges located at their centers.*

*P3.18. An L-shaped rod with sides $a = 10$ cm extends from the origin of an xy rectangular system to the point $(a, 0)$, and from the origin to the point $(0, a)$. The rod is charged uniformly along its length with a total charge $Q = 2.6 \cdot 10^{-9}$ C. Evaluate the electric field strength vector at points $A(a,a)$ and $B(3a/2,0)$. — (a) $E_{Ax} = E_{Ay} = 2328$ V/m, $E_{Bx} = 1866$ V/m, $E_{By} = -290$ V/m. (b) $E_{Ax} = E_{Ay} = 1168$ V/m, $E_{Bx} = 1989.9$ V/m, $E_{By} = -130.8$ V/m. (c) $E_{Ax} = E_{Ay} = 876$ V/m, $E_{Bx} = 1234$ V/m, $E_{By} = -235$ V/m.

*P3.19. Solve problem P3.18 approximately, by dividing the L-shaped rod into $2n$ segments. Compare the results with the exact solution for $n = 2, 3, 4, 5, 6$, and 20. — *Hint: use similar procedure as suggested for P3.17.*

S **P3.20.** A thin ring of radius a is uniformly charged along its length with a total charge Q. Determine the electric field strength along the ring axis. — (a) $E_z(z) = Qz/[2\pi\epsilon_0(z^2+2a^2)^{3/2}]$. (b) $E_z(z) = Qz/[8\pi\epsilon_0(z^2 + a^2)^{3/2}]$. (c) $E_z(z) = Qz/[4\pi\epsilon_0(z^2 + a^2)^{3/2}]$.

Solution. Let the z axis pass through the ring center and be normal to it, and let the origin, $z = 0$, be at the ring center. The electric field strength vector has only the E_z component, obtained by summing the contributions of all the elements of the ring. The result is

$$E_z(z) = \frac{Qz}{4\pi\epsilon_0(z^2 + a^2)^{3/2}}.$$

P3.21. A thin circular disk of radius a is charged uniformly over its surface with a total charge Q. Determine the electric field strength along the disk axis normal to its plane and plot your results. What do you expect the expression for the electric field to become at large distances from the disk? What do you expect the expression to become if the radius of the disk increases indefinitely, and the surface charge density is kept constant? — *Hint: divide the disk into elemental rings of radii r ($0 < r < a$) and widths dr. Note that the charge dQ on such a ring is $2\pi r dr Q/(a^2\pi)$. The problem is thus reduced to integrating the field of the form given in the solution to the preceding problem, with Q substituted by dQ, and a by r, from $r = 0$ to $r = a$.*

P3.22. Calculate the electric field along the axis of the disk in problem P3.21, if the charge is not distributed uniformly but increases linearly along the disk radius, and it is zero at the disk center. Plot your result and compare it to those for problem P3.21. — *The procedure is the same as in the preceding problem, except that charge density is a function of r.*

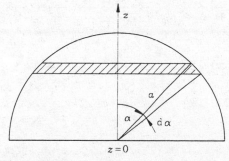

Fig. P3.27. A uniformly charged shell.

S **P3.23.** A dielectric cube with sides of length a is charged over its volume with a charge density $\rho(x) = \rho_0 x/a$, where x is the normal distance from one side of the cube. Determine the charge of the cube. — (a) $Q = 3\rho_0 a^3/2$. (b) $Q = \rho_0 a^3/3$. (c) $Q = \rho_0 a^3/2$.

Solution.

$$Q = \int_0^a \rho_0 \frac{x}{a} a^2 \, dx = \frac{1}{2}\rho_0 a^3.$$

P3.24. The volume charge density in a spherical charged cloud of radius a is $\rho(r) = \rho_0(a-r)/a$, where r is the distance from the cloud center, and ρ_0 is a constant. Determine the charge of the cloud. — (a) $Q = \rho_0 a^3 \pi/3$. (b) $Q = \rho_0 a^3 \pi/6$. (c) $Q = 3\rho_0 a^3 \pi/2$.

P3.25. Determine and plot the electric field strength as a function of the distance r from a straight, very long, thin charged filament with a charge Q' per unit length. — *Assume that the origin of the x axis is at the filament, that the axis is normal to it, and that the z axis is along the filament.* (a) $E_x(x) = Q'/(4\pi\epsilon_0 x)$. (b) $E_x(x) = Q'/(2\pi\epsilon_0 x)$. (c) $E_x(x) = 3Q'/(2\pi\epsilon_0 x)$.

P3.26. A wire in the form of a semicircle of radius a is charged with a total charge Q. Assuming the charge to be uniformly distributed along the wire, determine the electric field strength vector at the center of the semicircle. — *Let the z axis be as in Fig. P3.27.* (a) $E_z = -Q/(\pi^2\epsilon_0 a^2)$. (b) $E_z = -Q/(3\pi^2\epsilon_0 a^2)$. (c) $E_z = -Q/(2\pi^2\epsilon_0 a^2)$.

S **P3.27.** A hemispherical shell of radius a is charged uniformly over its surface by a total charge Q. Determine the electric field strength at the center of the sphere, one-half of which is the shell. — *Let the z axis be as in Fig. P3.27. Use the result of problem P3.20, and integrate the contributions to the total E_z field of narrow circular strips of width $a\,d\alpha$, radius $r = a\sin\alpha$, and the total charge $dQ = 2\pi a \sin\alpha\sigma a\,d\alpha$, where $\sigma = Q/(2a^2\pi)$.* (a) $E_z = Q/(\pi\epsilon_0 a^2)$. (b) $E_z = Q/(8\pi\epsilon_0 a^2)$. (c) $E_z = Q/(4\pi\epsilon_0 a^2)$.

Solution. Let the z axis be normal to the shell contour, with the origin at the shell (sphere) center (Fig. P3.27). We can then use the result of problem P3.20, and integrate the contributions of small circular strips of radius $r = a\sin\alpha$ and total charge $dQ = 2\pi a \sin\alpha\sigma a\,d\alpha$, where $\sigma = Q/(2a^2\pi)$. The distance to the observation point is also variable, $z = a\cos\alpha$, where α is the angle between the z axis and the radius towards the strip under consideration. So we obtain

$$E_z = \frac{1}{4\pi\epsilon_0} \int_{\alpha=0}^{\pi/2} \frac{(2\pi a \sin\alpha\sigma\,d\alpha)(a\cos\alpha)}{[(a\cos\alpha)^2 + (a\sin\alpha)^2]^{3/2}}.$$

Since the numerator under the integral sign is a^3, the integral reduces to that of $2\sin\alpha\cos\alpha = \sin 2\alpha$, and the result is

$$E_z = \frac{Q}{8\pi\epsilon_0 a^2}.$$

4. The Electric Scalar Potential

- The electric potential is a scalar function for the description of the electrostatic field. It is equal to the work done by the electric field in moving a small charge from an arbitrary point A in the field to a "reference" point, R, per unit charge:

$$V_A = \int_A^R \boldsymbol{E} \cdot \mathrm{d}\boldsymbol{l} \qquad \text{(volts} - \text{V)}. \qquad (4.1)$$

The reference point is a fixed point that can be any point in the field, or outside the field. Most frequently it is adopted to be very far, theoretically at infinity. It cannot be adopted at infinity only if there are charges extending to infinity (e.g., in the case of an infinitely long line charge). By adopting a different reference point the potential at all points is changed by the same amount.

- Definition (4.1), combined with the law of conservation of energy, yields the following fundamental property of the electrostatic field:

$$\oint_C \boldsymbol{E} \cdot \mathrm{d}\boldsymbol{l} = 0. \qquad (4.2)$$

The closed contour C can be completely, but also only partly, in the field.

- The basic expression for the potential is the potential at distance r from a point charge Q, with respect to the reference point at infinity:

$$V(r) = \frac{Q}{4\pi\epsilon_0 r} \quad \text{(reference point at infinity)} \qquad \text{(V)}. \qquad (4.3)$$

From this expression, the potential of a given distribution of volume, surface, or line charges at a point P of the field, with respect to the reference point at infinity, is given respectively by

$$V_P = \frac{1}{4\pi\epsilon_0} \int_v \frac{\rho \, \mathrm{d}v}{r}, \qquad V_P = \frac{1}{4\pi\epsilon_0} \int_S \frac{\sigma \, \mathrm{d}S}{r}, \qquad V_P = \frac{1}{4\pi\epsilon_0} \int_L \frac{Q' \, \mathrm{d}l}{r} \qquad \text{(V)}. \quad (4.4)$$

- The *potential difference*, or *voltage*, in the electrostatic field is defined as

$$V_{AB} = V_A - V_B = \int_A^B \boldsymbol{E} \cdot \mathrm{d}\boldsymbol{l} \qquad \text{(V)}. \qquad (4.5)$$

The voltage is a broader concept than the potential difference. They are the same *only in the electrostatic field.*

• The electric field strength is obtained from the electric scalar potential as $E = -\text{grad}\,V = -\nabla V$. In the rectangular coordinate system this relationship is of the form

$$E = -\text{grad}\,V = -\nabla V = -\left(\frac{\partial V}{\partial x}u_x + \frac{\partial V}{\partial y}u_y + \frac{\partial V}{\partial z}u_z\right) \quad (\text{V/m}), \quad (4.8)$$

where u_x, u_y and u_z are the base unit vectors of the three coordinate axes, and ∇ is the "nabla" or "del" operator.

• A surface in an electrostatic field having the same potential at all points is referred to a an *equipotential surface*.

QUESTIONS

Q4.1. Consider a uniform electric field of electric field strength E, and two planes normal to vector E, that are a distance d apart. What is the work done by the field in moving a test charge ΔQ from one plane to another? Can the work be negative? Does it depend on the location of the two points on the planes? Explain. — *(a) $W = -\Delta Q E d$, negative, does not depend on the location of the two points. (b) $W = \Delta Q E d$, can be positive or negative, does not depend on the location of the two points. (c) $W = \Delta Q E d$, positive, depends on the location of the two points.*

S **Q4.2.** Is it possible to have an electrostatic field with circular closed field lines, with the vector E in the same direction along the entire lines? Explain. — *(a) In some cases (which ones?). (b) No. (c) Yes.*

Answer. It is not possible. This would imply that the line integral of the vector E around such a circle be nonzero.

Q4.3. Is it possible to have an electrostatic field with parallel lines, but of different magnitude of vector E in the direction normal to the lines? Explain. — *(a) Yes. (b) No. (c) Under certain circumstances (which circumstances?).*

S **Q4.4.** If the potential of the earth were taken to be $100,000\,\text{V}$ (instead of the usual $0\,\text{V}$), would it be dangerous to walk around? What influence would this have on the potential at various points, and what on the difference of the potential at two points? — *(a) Yes. The potential and potential difference will also increase. (b) No. The potential is thereby increased, but the potential difference remains the same. (c) No, although both the potential and the potential difference will increase.*

Answer. The potential is defined within an additive constant. What matters is the potential difference, which is not affected when a constant is added at *all* points in the field. So, by adopting the potential of the earth to be $100,000\,\text{V}$, the potential of all points has to be increased by the same amount, the potential difference between any two points remaining the same.

Q4.5. If we know $E(x,y,z)$, is the electric scalar potential $V(x,y,z)$ determined uniquely? Explain. — *(a) No. (b) Yes. (c) In some cases (which cases?).*

Q4.6. Eq. (4.2) is satisfied by the electric field of a point charge. Does the expression for the electric field of a point charge *follow* from Eq. (4.2)? — *(a) Yes (explain). (b) Only for a positive point charge. (c) No (why?).*

Q4.7. Why does eq.(4.2) represent the law of conservation of energy in the electrostatic field? — *(a) Because it was proved by using the law of conservation of energy. (b) It does not represent this law. (c) Because the definition of E implies the law of conservation of energy.*

Q4.8. What is the potential of the reference point? — *(a) It depends on its position in the field. (b) Infinite. (c) Zero.*

S **Q4.9.** As we approach a point charge Q $(Q > 0)$, the potential tends to infinity. Explain. — *(a) This is true if we approach any charged body. (b) Because r tends to zero. (c) The integral of $\mathrm{d}r/r^2$ has infinite value if integrated up to $r = 0$.*

Answer. A point charge is a fictitious charge. The electric field of the charge increases as $1/r^2$ as we approach the charge, and the integral of $\mathrm{d}r/r^2$ has infinite (undetermined) value if integrated up to $r = 0$.

Q4.10. How much energy do you transfer to the electric field of a point charge when you move the reference point from a point at a distance r_R from the charge to a point at infinity? — *(a) Zero energy. (b) Q times the potential at r_R. (c) Infinite energy.*

Q4.11. Why do we usually adopt the reference point at infinity? — *(a) This is the most natural thing to do. (b) To simplify the expressions for potential. (c) Because the potential at infinity is zero.*

Q4.12. Is the potential of a positively charged body always positive, and that of a negatively charged body always negative? Give examples that illustrate your conclusions. — *(a) No (explain). (b) Yes (explain). (c) Only in special cases it is not (give an example).*

Q4.13. Why are the expressions for the potential in Eqs. (4.4) valid for a reference point at infinity? — *(a) There is no reference point in the expression. (b) They are based on Eq. (4.3). (c) Because it is impossible to adopt any other reference point in such cases.*

Q4.14. Does it make sense to speak about voltage between a point in the field and the reference point? If it does, what is this voltage? — *(a) No. (b) Yes, it is equal to the difference of potential at the reference point and that point. (c) Yes, it is equal to the potential at that point.*

S **Q4.15.** A charge ΔQ is moved from a point where the potential is V_1 to a point where the potential is V_2. What is the work done by the electric forces? What is the work done by the forces acting against the electric forces? — *(a) $\Delta Q(V_1 - V_2)$, $-\Delta Q(V_1 - V_2)$. (b) $-\Delta Q(V_1 - V_2)$, $\Delta Q(V_1 - V_2)$. (c) $\Delta Q(V_1 - V_2)$, $\Delta Q(V_1 - V_2)$.*

Answer. The work done by the electric forces is $Q(V_1 - V_2)$. The work done by the forces acting against the electric forces is the negative of this.

Q4.16. A charge ΔQ $(\Delta Q < 0)$ is moved from a point at potential V_1 to a point at potential V_2. What is the work done by the electric forces? — *(a) $-\Delta Q(V_1 - V_2)$. (b) $|\Delta Q|(V_1 - V_2)$. (c) $\Delta Q(V_1 - V_2)$.*

Q4.17. Is $V_{AB} = -V_{BA}$? Explain. — *(a) It is, because dl in the two calculations differs in sign. (b) It is, because $BA = -AB$. (b) It is not, because dl in the two cases is the same.*

Q4.18. Why is the vector E at a point directed towards the adjacent equipotential surface of *lower* potential? — *(a) Because $E = -\mathrm{grad}\, V$. (b) Because there is a negative charge* *on the equipotential surface of lower potential. (c) Vector E is directed towards the adjacent equipotential surface of higher potential.*

Q4.19. Why do we have $E = -\text{grad}\,V$, and not $E = +\text{grad}\,V$? — *(a) It follows from the definition of potential. (b) It follows from Coulomb's law. (c) It is a convention, and can be either way.*

Q4.20. A cloud of positive and negative ions is situated in an electrostatic field. Which ions will tend to move towards the points of higher potential, and which towards the points of lower potential? — *(a) Positive to higher, negative to lower. (b) Positive to lower, negative to higher. (c) Depends on the field configuration.*

Q4.21. Suppose that $V = 0$ at a point. Does it mean that $E = 0$ at that point? Explain. — *(a) Yes, because if at a point $V = 0$, grad $V = 0$ also. (b) No, because E is not determined by the value of V at a single point. (c) This depends on the potential at other points.*

S **Q4.22.** Assume we know E at a point. Is this sufficient to determine the potential V at that point? Conversely, if we know V at that point, can we determine E? — *(a) Yes, no. (b) No, no. (c) No, yes.*

Answer. No. To determine V, we need to know E along any *line* from that point to the reference point. Knowing V at a point does not allow us to find E. To determine the *component* of E in a direction x at a point, we need to know the potential at two close points along this direction.

Q4.23. The potential in a region of space is constant. What is the magnitude and direction of the electric field strength vector in the region? — *(a) E is also constant. (b) $E = 0$. (c) E is nonzero, directed towards the points of lower potential.*

Q4.24. Prove that E is normal to equipotential surfaces. — *(a) The tangential component of E at any equipotential surface is zero. (b) The potential at all points of an equipotential surface is the same. (c) Potential of equipotential surfaces from one point to another of the surface is not constant, but the tangential component of E on the surface is zero.*

PROBLEMS

S **P4.1.** Two point charges, $Q_1 = -3 \cdot 10^{-9}$ C and $Q_2 = 1.5 \cdot 10^{-9}$ C, are $r = 5$ cm apart. Find the potential at the point which lies on the line joining the two charges and halfway between them. Find the zero-potential point(s) lying on the straight line which joins the two charges. — *(a) $V_{\text{midpoint}} = 539.3\,V$; $V = 0$ at points $x_1 = 10\,cm$ and $x_2 = -10/3\,cm$, where the x axis is directed from Q_1 to Q_2, and the origin is at Q_1. (b) $V_{\text{midpoint}} = 663.3\,V$; $x_1 = 8\,cm$, $x_2 = 6\,cm$. (b) $V_{\text{midpoint}} = 685.1\,V$; $x_1 = 7\,cm$, $x_2 = -6\,cm$.*

Solution. The potential at the midpoint between the charges is

$$V = \frac{Q_1}{4\pi\epsilon_0 (r/2)} + \frac{Q_2}{4\pi\epsilon_0 (r/2)} = \frac{Q_1 + Q_2}{4\pi\epsilon_0 (r/2)} = -539.3\,\text{V}.$$

The points where the potential is zero is obtained from the equation

$$V = \frac{Q_1}{4\pi\epsilon_0\, x} + \frac{Q_2}{4\pi\epsilon_0 (x - r)} = 0,$$

where x is the coordinate along the axis directed from Q_1 towards Q_2, with $x = 0$ at Q_1. So we have

$$x_{1/2} = \frac{Q_1 r}{Q_1 \pm Q_2}.$$

The distances from charge Q_1 of the two points are $x_1 = 2r = 10 \, \text{cm}$, and $x_2 = -2r/3 = -10/3 \, \text{cm}$. The two points are on the two sides of charge Q_1.

P4.2. Two small bodies, with charges Q ($Q > 0$) and $-Q$, are a distance d apart. Determine the potential at all points with respect to the reference point at infinity. Is there a zero-potential equipotential surface? How much work do the electric forces do if the distance is increased to $2d$? — *Hint: the total potential is the sum of the potentials due to the two charges.*

P4.3. A ring of radius a is charged with a total charge Q. Determine the potential along its axis normal to the ring plane with reference to the ring center. — *Let the z axis be along the ring axis, with the origin at the ring center. (a)* $V(z) = Q/[2\pi\epsilon_0\sqrt{a^2 + z^2}]$. *(b)* $V(z) = Q/[8\pi\epsilon_0\sqrt{a^2 + z^2}]$. *(c)* $V(z) = Q/[4\pi\epsilon_0\sqrt{a^2 + z^2}]$.

S **P4.4.** A soap bubble of radius R and very small wall thickness a is at a potential V with respect to the reference point at infinity. Determine the potential of a spherical drop obtained when the bubble explodes, assuming all the soap in the bubble is contained in the drop. — *If $V = 100 \, V$, $R = 1 \, cm$ and $a = 3.3 \, \mu m$, (a)* $V_{\text{drop}} = 850 \, V$, *(b)* $V_{\text{drop}} = 500 \, V$, *(c)* $V_{\text{drop}} = 1000 \, V$.

Solution. The charge on the bubble is obtained from the expression for the potential of a uniformly charged sphere,

$$V = \frac{Q}{4\pi\epsilon_0 R},$$

from which $Q = 4\pi\epsilon_0 RV$. The same charge is on the drop, the potential of which is obtained from the above equation if we substitute R by the (as yet unknown) radius of the drop.

The radius of the drop, r, is obtained by assuming that the entire volume of the soap bubble wall transformed into a spherical drop,

$$\frac{4}{3}r^3\pi = 4R^2\pi a,$$

from which $r = \sqrt[3]{3R^2 a}$.

As a specific example, if $V = 100 \, \text{V}$, $R = 1 \, \text{cm}$ and $a = 3.3 \, \mu\text{m}$, we find for the potential of the drop $V_{\text{drop}} = 1000 \, \text{V}$.

P4.5. A volume of a liquid conductor is sprayed into N equal spherical drops. Then, by some appropriate method, each drop is given a potential V with respect to the reference point at infinity. Finally, all these small drops are combined into a large spherical drop. Determine the potential of the large drop. — *If $N = 10^6$ and $V = 10 \, V$, (a)* $V_{\text{large drop}} = 10,000 \, V$, *(b)* $V_{\text{large drop}} = 100,000 \, V$, *(c)* $V_{\text{large drop}} = 1,000 \, V$.

P4.6. Two small conducting spheres of radii a and b are connected by a very thin, flexible conductor of length d. The total charge of the system is Q. Assuming that d is much larger than a and b, determine the force F that acts on the wire so as to extend it. Charges may be considered to be located on the two spheres only, and to be distributed uniformly over their surfaces. (Hint: when connected by the conducting wire, the spheres will be at the same potential — see Chapter 6.) — (a) $F = abQ^2/[4\pi\epsilon_0 d^2(a+b)^2]$. (b) $F = 3abQ^2/[4\pi\epsilon_0 d^2(a+b)^2]$. (c) $F = abQ^2/[8\pi\epsilon_0 d^2(a+b)^2]$.

P4.7. Two small conducting balls of radii a and b are charged with charges Q_a and Q_b, and are at a distance d ($d \gg a,b$) apart. Suppose that the balls are connected with a thin conducting wire. What will the direction of flow of positive charges through the wire be? Discuss the question for various values of Q_a, Q_b, a, and b. (Hint: when connected by the conducting wire, the balls will be at the same potential — see Chapter 6.) — *If $Q_a/a > Q_b/b$, the flow will be (a) from the ball of radius a towards the other ball, (b) conversely, (c) there will be no flow of charges.*

*P4.8. The source of an electrostatic field is a volume charge distribution of finite charge density ρ, distributed in a finite region of space. Prove that the electric scalar potential has a finite value at all points, including the points inside the charge distribution. — *Hint: consider a point P inside the charged region and a small sphere of radius a enclosing this point. Let a be sufficiently small so that the charge density inside the small sphere can be considered constant. Prove that the potential at the sphere center is finite.*

*P4.9. Prove that the electric scalar potential due to a surface charge distribution of density σ over a surface S is finite at all points, including the points of S. — *Hint: consider a small circle of radius a on the charged surface and prove that the potential at its center due to the charge on the circle is finite.*

P4.10. The reference point for the potential is changed from point R to point R'. Prove that the potential of all points in an electric field changes by the voltage between R and R'. — *Hint: in determining the potential with respect to R', adopt paths that pass through R.*

P4.11. Four small bodies with equal charges $Q = 0.5 \cdot 10^{-9}$ C are located at the vertices of a square with sides $a = 2$ cm. Determine the potential at the center of the square, and the voltage between the square center and a midpoint of a square side. What is the work of electric forces if one of the charges is moved to a very distant point? — (a) $V_0 - V_1 = 29.7$ V. (b) $V_0 - V_1 = -92.9$ V. (c) $V_0 - V_1 = -29.7$ V.

P4.12. An insulating disk of radius $a = 5$ cm is charged by friction uniformly over its surface with a total charge of $Q = -10^{-8}$ C. Find the expression for the potential of the points which lie on the axis of the disk perpendicular to its surface. Plot your result. What are the numerical values for the potential at the center of the disk, and at a distance $z = a$ from the center, measured along the axis? What is the voltage between these two points equal to? — (a) $V(0) = -4595$ V, $V(a) = -2489$ V. (b) $V(0) = -3595$ V, $V(a) = -1489$ V. (c) $V(0) = -595$ V, $V(a) = -489$ V.

P4.13. The volume charge density inside a spherical surface of radius a is such that the electric field vector inside the sphere is pointing towards the center of the sphere, and varies with radial position as $E(r) = E_0 r/a$ (E_0 is a constant). Find the voltage between the center and the surface of the sphere. — *(a) $V = -3E_0 a/2$. (b) $V = -E_0 a/2$. (c) $V = -E_0 a/4$. Why do we have the minus sign?*

P4.14. Two large parallel equipotential plates at potentials $V_1 = -10\,\text{V}$ and $V_2 = 55\,\text{V}$ are a distance $d = 2\,\text{cm}$ apart. Determine the electric field strength between the plates. — *(a) $E = 5.25\,kV/m$. (b) $E = 4.25\,kV/m$. (c) $E = 3.25\,kV/m$.*

S **P4.15.** Determine the potential along the line joining two small bodies carrying equal charges Q. Plot your result. Starting from that expression, prove that the electric field strength at the midpoint between the bodies is zero. — *Hint: start from the expression for the potential of a point charge.*

Solution. Let the x axis have the origin at the midpoint between the charges, and be directed towards one of the charges. Let the distance between the two charges be d. The potential on the x axis between the charges is then

$$V(x) = \frac{Q}{4\pi\epsilon_0} \left(\frac{1}{d/2 - x} + \frac{1}{d/2 + x} \right).$$

The electric field strength between the charges is x-directed, of magnitude

$$E(x) = -\frac{dV(x)}{dx} = \frac{Q}{4\pi\epsilon_0} \left[-\frac{1}{(d/2 - x)^2} + \frac{1}{(d/2 + x)^2} \right].$$

Evidently, $E(0) = 0$, i.e., the electric field strength at the midpoint between the two charges is zero.

P4.16. Two small bodies with charges $Q_1 = 10^{-10}\,\text{C}$ and $Q_2 = -Q_1$ are a distance $d = 9\,\text{cm}$ apart. Determine the potential along the line joining the two charges, and from that expression, determine the electric field strength along the line. Plot your results. — *Hint: the solution is given by that of the preceding problem, if one of the charges is assumed to be $-Q$, instead of Q.*

P4.17 From the expression for potential found in problem P4.3, find the electric field strength vector along the ring axis. (See problem P3.20.) — *Possible solutions are given in problem P3.20.*

P4.18. From the general expression for the potential along the axis of the disk from problem P4.12, determine the electric field strength along the disk axis. (See problem P3.21.) — *Possible solutions are given in problem P3.21.*

5. Gauss' Law

• Gauss' law is an integral relationship between vector E on an entire closed surface, and the total charge enclosed by the surface. In a vacuum (free space) it is of the form

$$\oint_S \boldsymbol{E} \cdot \mathrm{d}\boldsymbol{S} = \frac{Q_{\text{total in S}}}{\epsilon_0} \qquad (\text{V} \cdot \text{m}). \qquad\qquad (5.1)$$

S is *any* closed surface, $\mathrm{d}\boldsymbol{S}$ is a small vector surface element locally normal to the surface and (by convention) directed outward, and $\boldsymbol{E}$ is the electric field strength vector at $\mathrm{d}\boldsymbol{S}$. The expression on the left-hand side of Eq. (5.1) is known as the flux of $\boldsymbol{E}$ through S.

- Gauss' law is a direct consequence of Coulomb's law (more precisely, of the term $1/r^2$) and can be extended to charges in any medium.

QUESTIONS

Q5.1. Prove that in a uniform electric field the flux of the electric field strength vector through any closed surface is zero. — *Hint: divide the entire field into narrow tubes along the lines of vector E and determine the total flux as the sum of fluxes through these tubes.*

Q5.2. Can the closed surface in Gauss' law be infinitesimally small in the mathematical sense? Is the answer different for the case of a vacuum and some other material? Explain. — *Yes. (b) No. (c) Only inside matter.*

Q5.3. Assume we know that the vector E satisfies Gauss' law in Eq. (5.1), but we do not know the expression for the vector E of a point charge. Can this expression be *derived* from Gauss' law? — *(a) Yes, with no further assumptions. (b) No. (c) Yes, if we assume that E is radial and symmetrical with respect to the point charge.*

S   **Q5.4.** The center of a small spherical body of radius r, uniformly charged over its surface with a charge Q, coincides with the center of one side of a cube of edge length a ($a > 2r$). What is the flux of the electric field strength vector through the cube? — *(a) Q/ϵ_0. (b) $Q/(2\epsilon_0)$. (c) Zero.*

Answer. Since only one half of the small body is enclosed by the cube, the flux equals $Q/(2\epsilon_0)$.

Q5.5. A dielectric cube of edge length a is charged by friction uniformly over its surface, with a surface charge density σ. What is the flux of the electric field strength vector through a slightly smaller and slightly larger imaginary cube? Do the answers look logical? Explain. — *(a) Zero, zero. (b) Half the charge on the cube in both cases. (c) Zero, the charge on the cube.*

Q5.6. Is it possible to apply Gauss' law to a large surface enclosing a domain with a number of holes? If you think it is possible, explain how it should be done. — *(a) Yes, considering the large surface to be bounded from inside by the holes. (b) Yes, not taking the holes into account. (c) No, because the flux through such a complex surface cannot be defined.*

S   **Q5.7.** Inside an imaginary closed surface S the total charge is zero. Does this mean that at all points of S the vector E is zero? Explain. — *(a) Yes. (b) No. (c) In some cases (give an example).*

Answer. No. Imagine, for example, charges Q and $-Q$ enclosed by S.

Q5.8. A spherical rubber balloon is charged by friction uniformly over its surface. How does the electric field inside and outside the balloon change if it is periodically inflated and deflated to change its radius? — *(a) It does not change anywhere. (b) It changes only inside the balloon. (c) When the balloon is inflated, the field at points just outside of it drops to zero; when it is deflated, the field at points just inside of it jumps from zero to a nonzero value.*

Q5.9. Assume that the flux of the electric field strength vector through a surface enclosing a point A is the same for any size and shape of the surface. What does this tell us about the charge at A or in its vicinity? — *(a) There is a charge of nonzero volume charge density at A. (b) There is a point charge at A. (c) There is a charge of nonzero surface charge density at A.*

Q5.10. The electric field strength is zero at all points of a closed surface S. What is the charge enclosed by S? — *(a) Zero. (b) Can have any value. (c) Some positive charge.*

Q5.11. An electric dipole (two equal charges of opposite signs) is located at the center of a sphere of radius greater than half the distance between the charges. What is the flux of vector E through the sphere? — *(a) Equal to the positive dipole charge, Q, over ϵ_0. (b) Zero. (c) $-Q/\epsilon_0$.*

S **Q5.12.** Would it be possible to apply Gauss' law for the determination of the electric field for charged planes with nonuniform charge distribution? Explain. — *(a) Yes. (b) For some charge distributions (give an example). (c) No.*

Answer. No. In such a case we do not know the direction of the vector E.

Q5.13. What would be the form of Gauss' law if the unit vector normal to a closed surface were adopted to point into the surface, instead of out of the surface? — *(a) It would remain the same. (b) The sign on the right-hand side would be negative. (c) It cannot be formulated in that case at all.*

Q5.14. Gauss' law is a consequence of the factor $1/r^2$ in the expression for the electric field strength of a point charge (i.e., in Coulomb's law). At what step in the derivation of Gauss' law is this the condition for Gauss' law to be valid? — *Hint: read carefully the derivation of Gauss' law and find where the dependence on r was removed.*

Q5.15. Try to derive Gauss' law for a hypothetical electric field where the field strength of a point charge is proportional to $1/r^k$, where $k \neq 2$. — *Hint: see the answer to the preceding question.*

PROBLEMS

P5.1. The flux of the electric field strength vector through a closed surface is 100 Vm. How large is the charge inside the surface? — *(a) $Q = 4.427 \cdot 10^{-10}$ C. (b) $Q = 17.608 \cdot 10^{-10}$ C. (c) $Q = 8.854 \cdot 10^{-10}$ C.*

S **P5.2.** A point charge $Q = 2 \cdot 10^{-11}$ C is located at the center of a cube. Determine the flux of vector E through one side of the cube using Gauss' law. — *(a) 0.3767 Vm. (b) 0.4854 Vm. (c) 0.7534 Vm.*

Solution. Since the charge is at the cube center, the flux of vector E through each of the six cube sides is the same. Since the total flux is Q/ϵ_0, the flux through one side only is 1/6 of this, i.e., $Q/(6\epsilon_0) = 0.3767$ Vm.

S **P5.3.** A point charge $Q = -3 \cdot 10^{-12}$ C is $d = 5$ cm away from a circular surface S of radius $a = 3$ cm as shown in Fig. P5.3. Determine the flux of vector E through S. — *Hint: make use of Gauss' law to simplify the problem.* (a) -0.02414 Vm. (b) 0.03215 Vm. (c) -0.03215 Vm.

Solution. The flux of vector E through the circular surface is difficult to determine directly, since the magnitude of vector E and the angle it makes with the surface vary from point to point.

Imagine a spherical cap of radius $R = \sqrt{d^2 + a^2}$ on the circular surface. Since there are no charges in the domain limited by the circular surface and the cap, the flux through this closed surface is zero, i.e., the flux through the circular surface and the cap is the same. The latter is much simpler to determine, however, since vector E has the same magnitude over the cap, and is normal to it. So the flux equals the product of the cap area and the magnitude of vector E, i.e.,

$$(\psi_E)_{\text{cap}} = \frac{Q}{4\pi\epsilon_0 R^2} 2\pi R(R - d) = \frac{Q(R - d)}{2\epsilon_0 R},$$

since $h = (R - d)$ is the cap height, and the area of the cap is $2\pi Rh$. So we obtain

$$(\psi_E)_{\text{cap}} = (\psi_E)_{\text{circ.surface}} = -0.02414 \text{ Vm.}$$

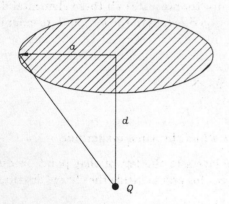

Fig. P5.3. A circular surface near a point charge.

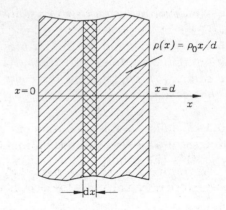

Fig. P5.7. A nonuniformly charged plate.

P5.4. Determine the flux of vector E through a hemispherical surface of radius $a = 5$ cm, if the field is uniform, with $E = 15$ mV/m, and if vector E makes an angle $\alpha = 30°$ with the hemisphere axis. Use Gauss' law. — *Hint: make use of Gauss' law to simplify the problem.* (a) $3.22 \cdot 10^{-4}$ Vm. (b) $2.12 \cdot 10^{-4}$ Vm. (c) $1.02 \cdot 10^{-4}$ Vm.

P5.5. Three parallel thin large charged plates have surface charge densities $-\sigma$, 2σ and $-\sigma$. Find the electric field everywhere for all combinations of the relative sheet positions and $\sigma = 10^{-6}$ C/m^2. Do the results depend on the distances between the plates? Determine the equipotential surfaces in all cases, and the potential difference between pairs of plates, if the distance between them is 2 cm. — *Hint: there are only two possibilities of plate combinations, (1) as in the text of the problem, and (2) 2σ, $-\sigma$ and $-\sigma$. In both cases, E is normal to the plates.* (a) $V_{\text{middle}} - V_{\text{side}} = 1129$ V in case 1, zero in case 2. (b) $V_{\text{middle}} - V_{\text{side}} = 1129$ V in both cases. (c) $V_{\text{middle}} - V_{\text{side}} = 0$ in case 1, and 2158 V in case 2.

P5.6. A very large flat plate of thickness d is uniformly charged with volume charge density ρ. Find the electric field strength at all points. Determine the potential difference between the two boundary planes, and between the plane of symmetry of the plate and a boundary plane. — *Let the x axis be normal to the plate, with the origin at the middle of the plate. (a) $V(d/2) - V(-d/2) = \rho d^2/(2\epsilon_0)$, $V(d/2) - V(0) = \rho d^2/(4\epsilon_0)$. (b) $V(d/2) - V(-d/2) = \rho d^2/(4\epsilon_0)$, $V(d/2) - V(0) = \rho d^2/(8\epsilon_0)$. (c) $V(d/2) - V(-d/2) = 0$, $V(d/2) - V(0) = \rho d^2/(4\epsilon_0)$.*

S **P5.7.** The volume charge density of a thick very large plate varies as $\rho = \rho_0 x/d$ through the plate, where x is the distance from one if its boundary planes. Find the electric field strength vector everywhere. Plot your result. How large is the potential difference between the two boundary surfaces of the plate? — *Hint: divide the plate into a large number of thin layers of thicknesses dx, as in Fig. P5.7, and assume that $\rho_0 > 0$. (a) Outside the plate $|E| = \rho_0 d/(4\epsilon_0)$, inside $E(x) = \rho_0(2x^2 - d^2)/(4\epsilon_0 d)$. (b) Outside the plate $|E| = \rho_0 d/(8\epsilon_0)$, inside $E(x) = \rho_0(2x^2 - d^2)/(4\epsilon_0 d)$. (c) Outside the plate $|E| = \rho_0 d/(4\epsilon_0)$, inside $E(x) = \rho_0(2x^2 - d^2)/(8\epsilon_0 d)$.*

Solution. Since the charge distribution is not symmetrical, we cannot find the solution as in the preceding problem. Instead, we divide the plate into a large number of thin layers of thickness dx, as in Fig. P5.7. Such a layer at a coordinate x can be considered as a plate with a surface charge of density $d\sigma = \rho(x)dx = \rho_0(x/d)dx$. The field of this charge is uniform, and the magnitude of the electric field vector is $dE = d\sigma/(2\epsilon_0)$.

To obtain the total field at a point X, we need to add up (to integrate) all these elemental fields. To the left of the plate (for $x < 0$), and if $\rho_0 > 0$, the vector $\mathbf{E}$ is directed to the left, and its magnitude is

$$E = \int_0^d \frac{\rho_0 x}{2\epsilon_0 d} dx = \frac{\rho_0 d}{4\epsilon_0}.$$

The electric field vector (directed from left to right) for $x > d$ had the same magnitude.

For a point $0 \le X \le d$, note that (for $\rho_0 > 0$) all the layers to the left of that point produce a $+x$-directed field, and those to the right a $-x$-directed field. So, for points inside the charge distribution

$$E(X) = \int_0^X \frac{\rho_0 x}{2\epsilon_0 d} dx - \int_X^d \frac{\rho_0 x}{2\epsilon_0 d} dx = \frac{\rho_0}{4\epsilon_0 d}(2X^2 - d^2) \quad (0 \le X \le d).$$

P5.8. Two concentric spherical surfaces, of radii a and $b > a$, are uniformly charged with the same amounts of charge Q, but of opposite signs. Find the electric field strength at all points and present your expressions graphically. — *This is known as a spherical capacitor. Let r be the distance from the capacitor center. (a) $E = 0$ for $r < a$ and $r > b$, $E(r) = Q/(8\pi\epsilon_0 r^2)$ for $(a < r < b)$. (b) $E = 0$ for $r < a$ and $r > b$, $E(r) = Q/(6\pi\epsilon_0 r^2)$ for $a < r < b$. (c) $E = 0$ for $r < a$ and $r > b$, $E(r) = Q/(4\pi\epsilon_0 r^2)$ for $a < r < b$.*

P5.9. The spherical surfaces from the previous problem do not have the same charge, but are charged with $Q_{\text{inner}} = 10^{-10}$ C and $Q_{\text{outer}} = -5 \cdot 10^{-11}$ C. The radii of the spheres are $a = 3$ cm and $b = 5$ cm. Find the electric field strength and potential at all points and present your expressions graphically. — *The correct result for the electric field strength is: $E = 0$ for*

$r < a$, $E(r) = Q_{\text{inner}}/(4\pi\epsilon_0 r^2)$ for $a < r < b$, and $E(r) = (Q_{\text{inner}} + Q_{\text{outer}})/(4\pi\epsilon_0 r^2)$ for $r > b$.

P5.10. A spherical cloud of radius a has a uniform volume charge of density $\rho = -10^{-5}\,\text{C/m}^3$. Find the electric field strength and potential at all points and present your expressions graphically. — *(a) Inside the cloud, $E(r) = \rho r/(6\epsilon_0)$, outside it as that of a point charge $Q = 4a^3\pi\rho/3$. (b) Inside the cloud, $E(r) = \rho r/(3\epsilon_0)$, outside it as that of a point charge $Q = 4a^3\pi\rho/3$. (c) Inside the cloud, $E(r) = \rho r/(4\epsilon_0)$, outside it as that of a point charge $Q = 4a^3\pi\rho/3$.*

P5.11. A spherical cloud shell has a uniform volume charge of density $\rho = 10^{-3}\,\text{C/m}^3$, an inner radius $a = 2\,\text{cm}$, and an outer radius $b = 4\,\text{cm}$. Find the electric field strength and potential at all points and present your expressions graphically. — *The correct result for the electric field strength is: inside the cavity $E = 0$, inside the shell $E(r) = (r^3 - a^3)\rho/(3\epsilon_0 r^2)$, and outside the shell $E(r) = (b^3 - a^3)\pi\rho/(9\epsilon_0 r^2)$.*

S **P5.12.** The volume charge density of a spherical charged cloud is not constant, but varies with the distance from the cloud center as $\rho(r) = \rho_0 r/a$. Determine the electric field strength and potential at all points. Represent your results graphically. — *(a) Inside the cloud $E(r) = \rho_0 r^2/(2\epsilon_0 a)$, outside $E(r) = \rho_0 a^3/(4\epsilon_0 r^2)$. (b) Inside the cloud $E(r) = \rho_0 r^2/(4\epsilon_0 a)$, outside $E(r) = \rho_0 a^3/(4\epsilon_0 r^2)$. (c) Inside the cloud $E(r) = \rho_0 r^2/(4\epsilon_0 a)$, outside $E(r) = \rho_0 a^3/(8\epsilon_0 r^2)$.*

Solution. Due to symmetry, the lines of the electric field vector are radial, and the magnitude of the vector $\mathbf{E}$ is only a function of the distance r from the cloud center.

Imagine a spherical shell of radius r and thickness dr, $r < a$. The charge inside this shell is

$$dQ = \rho_0 \frac{r}{a} 4r^2 \pi dr.$$

The charge enclosed by a sphere or radius R ($R \le a$) is thus

$$Q(R) = \int_0^R \rho_0 \frac{r}{a} 4r^2 \pi dr = \frac{\rho_0 R^4 \pi}{a}.$$

For $R > a$, the sphere encloses the entire charge of the cloud, obtained if in the above expression we set $R = a$.

The electric field inside the cloud is obtained from the expression $4\pi R^2 E(R) = Q(R)/\epsilon_0$, or

$$E(R) = \frac{\rho_0 R^2}{4\epsilon_0 a} \quad (R < a),$$

and the field outside the cloud is given by

$$E(R) = \frac{\rho_0 a^3}{4\epsilon_0 R^2} \quad (R > a).$$

From these expressions, we obtain for the potential

$$V(R) = \frac{\rho_0(a^3 - R^3)}{12\epsilon_0 a} + \frac{\rho_0 a^2}{4\epsilon_0} \quad (R < a),$$

$$V(R) = \frac{\rho_0 a^3}{4\epsilon_0 R} \quad (R > a).$$

P5.13. Find the expression for the electric field strength and potential between and outside two long coaxial cylinders of radii a and b $(b > a)$, carrying charges Q' and $-Q'$ per unit length. (This structure is known as a coaxial cable, or coaxial line.) Plot your results. Determine the voltage between the two cylinders. — *(a) $E = 0$ inside the inner cylinder and outside the outer cylinder, $E(r) = Q'/(4\pi\epsilon_0 r)$ between the cylinders. (b) $E = 0$ inside the inner cylinder and outside the outer cylinder, $E(r) = Q'/(8\pi\epsilon_0 r)$ between the cylinders. (c) $E = 0$ inside the inner cylinder and outside the outer cylinder, $E(r) = Q'/(2\pi\epsilon_0 r)$ between the cylinders. Note that for determining the potential the reference point in this case cannot be at infinity.*

P5.14. Repeat problem P5.13 assuming that the two cylinders carry unequal charges per unit length, when these charges are (1) of the same sign, and (2) of opposite signs. Plot your results and compare to problem P5.13. — *Let Q'_a be the charge per unit length of the inner cylinder, Q'_b the charge per unit length of the outer cylinder, and $r = a$ the reference surface for the potential. Then the correct result is: $E = 0$ for $r < a$, $E(r) = Q'_a/(2\pi\epsilon_0 r)$ for $a < r < b$, $E(r) = (Q'_a + Q'_b)/(2\pi\epsilon_0 r)$ for $r > b$; $V(r) = [Q'_a \ln(b/r)]/(2\pi\epsilon_0)$ for $a < r < b$, $V(r) = [(Q'_a + Q'_b)\ln(b/r)]/(2\pi\epsilon_0 r)$ for $r > b$.*

S **P5.15.** A very long cylindrical cloud of radius a has a constant volume charge density ρ. Determine the electric field strength and potential at all points. Present your results graphically. Is it possible in this case to adopt the reference point at infinity? Explain. — *If the reference point for potential is at the cloud axis, the correct result is: $E(r) = (\rho r)/(2\epsilon_0)$ and $V(r) = -(\rho r^2)/(4\epsilon_0)$ for $r \leq a$, $E(r) = (\rho a^2)/(2\epsilon_0 r)$ and $V(r) = -(\rho a^2)/(4\epsilon_0)$ for $r \geq a$.*

Solution. The vector $\boldsymbol{E}$ is radial, and its magnitude depends only on the distance r from the cloud axis. The intensity $E(r)$ can be obtained by the application of Gauss' law to a right cylinder of height h and radius r, coaxial with the cloud. The flux of vector $\boldsymbol{E}$ through the two bases is zero, and through the curved surface it equals $2\pi r h E(r)$.

Noting that for $r < a$ only a part of the cloud charge is enclosed, and for $r > a$ the total charge is enclosed, we obtain

$$E(r) = \frac{\rho r}{2\epsilon_0} \quad (r \leq a),$$

$$E(r) = \frac{\rho a^2}{2\epsilon_0 r} \quad (r \geq a).$$

The potential with respect to the reference point at the cloud axis is

$$V(r) = -\frac{\rho r^2}{4\epsilon_0} \quad (r \leq a),$$

$$V(r) = -\frac{\rho a^2}{4\epsilon_0} - \frac{\rho a^2 \ln(r/a)}{2\epsilon_0} \quad (r \geq a).$$

Why are the expressions for $E(r)$ in this case valid also if $r = a$? Is it possible in this case to adopt the reference point at infinity? Explain.

P5.16. Repeat problem P5.15 assuming that the charge density is not constant, but varies with distance r from the cloud axis as $\rho(r) = \rho_0 r/a$. — *If the reference point for potential is at the charge axis, the correct result is:* $E(r) = (\rho_0 r^2)/(3\epsilon_0 a)$, $V(r) = -(\rho_0 r^3)/(9\epsilon_0 a)$ *for* $r \leq a$, *and* $E(r) = (\rho_0 a^2)/(3\epsilon_0 r)$, $V(r) = -(\rho_0 a^2)/(9\epsilon_0) - [\rho_0 a^2 \ln(r/a)]/(3\epsilon_0)$ *for* $r \geq a$.

P5.17. Repeat problem P5.15 assuming that the cloud has a coaxial cavity of radius b ($b < a$) with no charges. — *With the same reference point for potential, the correct answer is:* $E(r) = 0$, $V(r) = 0$ *for* $r \leq b$, $E(r) = [\rho(r^2 - b^2)]/(2\epsilon_0 r)$, $V(r) = -\{\rho[(r^2 - b^2)/2 - b^2 \ln(r/b)]\}/(2\epsilon_0)$ *for* $b \leq r \leq a$, *and* $E(r) = [\rho(a^2 - b^2)]/(2\epsilon_0 r)$, $V(r) = -\{\rho[(a^2 - b^2)/2 - b^2 \ln(a/b)]\}/(2\epsilon_0) - [\rho(a^2 - b^2) \ln(r/a)]/(2\epsilon_0)$ *for* $r \geq a$.

*P5.18. Prove that the electric scalar potential cannot have a maximum or a minimum value, except at points occupied by positive and negative charges, respectively. — *Hint: suppose that at a point M the potential has a maximum or minimum value with respect to neighboring points and apply Gauss' law.*

*P5.19. Prove *Earnshow's theorem:* A stationary system of charges cannot be in a stable equilibrium without external nonelectric forces. — *Hint: use the conclusion from problem P5.18.*

*P5.20. Prove that the average potential of any sphere S is equal to the potential at its center, if the charge density inside the sphere is zero at every point. — *Hint: (1) Write the integral expressing the average potential on S. (2) Introduce the (constant) radius of the sphere, r, under the integral sign, and note that $\mathrm{d}S/r^2$ is independent of r. (3) Perform partial differentiation with respect to r, note that $\partial V/\partial r = -E_r$, and apply Gauss' law.*

6. Conductors in the Electrostatic Field

• Conductors are substances with a relatively large proportion of freely movable electric charges. As a consequence of this property, the following conclusions are valid:

$$\text{In electrostatics, } E = 0 \text{ inside conductors,} \tag{6.1}$$

$$\text{In electrostatics, a conductor has charges only on its surface,} \tag{6.2}$$

$$\text{In electrostatics, } E_{\text{tangential}} = 0 \text{ on conductor surfaces,} \tag{6.3}$$

In electrostatics, the surface and volume of a conductor are equipotential, (6.4)

$$\text{On surfaces of conducting bodies } E_{\text{normal}} = \frac{\sigma}{\epsilon_0} \qquad \text{(V/m)}. \qquad (6.5)$$

In the last equation, σ is the surface charge density at a point of the conductor surface, and E_{normal} is the normal component of vector E on the conductor surface at that point, just above the conductor surface.

• The largest surface charge density (and thus also the electric field strength) over conductors is at sharp points. For example, this is why a lightning rod initiates a lightning discharge at its tip, and thus protects an object.

• If a conductor with zero total charge is introduced into an electrostatic field, positive and negative charges over its surface are separated, a process known as *electrostatic induction*. A charge-free cavity in a conducting body is totally shielded from the electrostatic field (so-called *Faraday's cage*). By connecting a conducting charged body to the ground, the body is completely discharged.

• Charged bodies above a conducting plane induce surface charges on the plane. The field of these induced charges above the plane is the same as that of "images" of charged bodies in the plane. Images of bodies are positioned symmetrically with respect to the plane, and have charges of opposite sign.

QUESTIONS

Q6.1. Prove that all points of a conducting body situated in an electrostatic field are at the same potential. — *Hint: use the property in Eq. (6.1).*

S **Q6.2.** Two thin aluminum foils of area S are pressed onto each other and introduced into an electrostatic field, normal to the vector E. The foils are then separated while *in the field*, and moved separately out of the field. What is the charge of the foils? — *(a) Zero. (b) $\pm\epsilon_0 ES$. (c) $\pm ES$.*

Answer. The charge induced on the two foils when they were pressed was $\pm\epsilon_0 ES$. This is the charge they retain when separated in the field, and then moved out of the field.

Q6.3. An uncharged conducting body has four cavities. In every cavity there is a point charge, $-Q_1$, $-Q_2$, $-Q_3$, and $-Q_4$. What is the induced charge on the surfaces of the cavities? What is the charge over the outer surface of the body? — *(a) The induced charges on the four cavities are Q_1, Q_2, Q_3, and Q_4, respectively. The charge over the outer body surface is $Q_1 + Q_2 + Q_3 + Q_4$. (b) $-Q_1$, $-Q_2$, $-Q_3$, $-Q_4$, viz. $-(Q_1 + Q_2 + Q_3 + Q_4)$. (c) Q_1, Q_2, Q_3, Q_4, viz. $-(Q_1 + Q_2 + Q_3 + Q_4)$.*

Q6.4. We know that there is no electrostatic field inside a conductor. Assume that we succeeded in producing an electric field which is tangential to a conducting body just above its surface. Is this physically possible? If you think that it is not, which law you think would be violated in that case? — *(a) Yes. (b) No, because the law of conservation of energy would be violated (why?). (c) No, Coulomb's law would be violated (why?).*

Q6.5. If *uncharged* pieces of aluminum foil are brought close to an electrified metal body, you will notice that they will be attracted, and then some of them repelled. Explain. — *(a) Attracted by gravitational attraction, repelled when charged with charges of the same sign. (b) Attracted because induced charges of opposite sign are closer to the body, repelled when charged with charges of the same sign. (c) There will be no attraction at all.*

Q6.6. If an uncharged body (e.g., your finger) is brought near a small charged body, you will notice that the body is attracted by the uncharged body (your finger). Explain. — *(a) The small charged body will induce charges of opposite sign in the uncharged body close to the charged body. (b) Gravitational attraction. (c) This effect cannot be noticed.*

Q6.7. A very thin short conducting filament is hanging from a large conducting sphere. If the sphere is charged with a charge Q, is the charge on the filament greater or less than that which remains on the sphere? Explain. — *(a) Greater, because it is sharp. (b) Approximately the same. (c) Much smaller, because the filament is thin, but the charge density at its tip is large.*

Q6.8. An uncharged conducting flat plate is brought into a uniform electrostatic field. In which position of the plate will its influence on the field distribution be minimal, and in which maximal? — *(a) Maximal if parallel to the field lines, minimal if normal to them (why?). (b) Opposite to the preceding answer (why?). (c) The influence of the plate is nearly the same in all positions (why?).*

S **Q6.9.** Assume that the room in which you are sitting is completely covered by thin aluminum foil. To signal to a friend outside the room, you move a charge around the room. Is your friend going to receive your signal? Explain. — *(a) Yes. (b) Only if the foil is thin. (c) No.*

Answer. No. The room is a perfect shield for electrostatic fields. (Only if the charge is moved very fast some field will penetrate the foil. This is explained in the propagation of electromagnetic waves through conducting media.)

Q6.10. Assume that in question Q6.9 your friend would like to signal you by moving a charge. Would you receive his signal? Explain. — *The same answers as for the preceding question.*

S **Q6.11.** A small charged conducting body is brought to a large uncharged conducting body and connected to it. What will happen to the charge on the small body? Is this the same as if a charged conducting body is connected to the ground? — *(a) It will be shared with the large body in proportion to the body size, so the small body will practically be discharged. (b) The small body will remain charged as before. (c) The charge will be shared in about equal amounts.*

Answer. No, for the same reason as in the answer to the preceding problem.

S **Q6.12.** A point charge Q is brought through a small hole into a thin uncharged metallic spherical shell of radius R, and fixed at a point which is a distance d $(d < R)$ from its center. What is the electric field strength outside the shell? — *(a) The same as at that of the point charge at its location. (b) The same as that of the point charge at the shell center. (c) Zero.*

Answer. The point charge will induce a *nonuniform* charge distribution on the inner shell surface, of total amount $-Q$. The same amount of charge of opposite sign, i.e., $+Q$, will be distributed on the outer shell surface. Since there is no electric field in the shell conducting wall, there is no electrostatic coupling between the outer charge and the inner charges. Therefore, the outer charge is distributed uniformly over the outer surface. The field outside the shell is therefore the same as that of a charge Q at the center of the shell.

Q6.13. A very thin metal foil is introduced exactly on a part of the equipotential surface in an electrostatic field. Is there any change in the field? Are there any induced charges on the foil surfaces? Explain. — *(a) No change, no induced charges. (b) No change, there are induced charges. (c) Field will be changed, there are induced charges.*

Q6.14. A closed equipotential surface enclosing a total charge Q is completely covered with very thin metal foil. Is there any change in the field inside and outside the foil? What is the induced charge on the inner surface of the foil, and what on the outer surface? — *(a) No change, no induced charge. (b) Field changed, induced charges $-Q$ and Q. (c) Field unchanged, induced charges $-Q$ and Q.*

Q6.15. A thin wire segment is introduced in the field and placed so that it lies completely on an equipotential surface. Is there any change in the field? Are there any induced charges on the wire surface? Explain. — *(a) No change, no induced charges. (b) No change, there are induced charges. (c) Field will be changed, there are induced charges.*

Q6.16. Repeat question Q6.15 assuming that the wire segment is made to follow a part of the line of vector E. — *(a) No change, no induced charges. (b) No change, there are induced charges. (c) Field will be changed, there are induced charges.*

Q6.17. Describe what happens as an airplane, charged negatively by friction with a charge $-Q$, is landing and finally touches down. — *Hint: use image theory and follow the image as the airplane approaches the ground.*

PROBLEMS

P6.1. A small conducting sphere of radius $a = 0.5$ cm is charged with a charge $Q = 2.3 \cdot 10^{-10}$ C, and is at a distance $d = 10$ m from a large uncharged conducting sphere of radius $b = 0.5$ m. The small sphere is then brought into contact with the large sphere, and moved back into its original position. Determine approximately the charges and potentials of the small and the large spheres in the final state. Take into account that $a \ll b$. — *(a) $V_{\text{large sphere}} \simeq V_{\text{small sphere}} \simeq 3.49$ V, $Q_{\text{small sphere}} \simeq 1.5 \cdot 10^{-12}$ C. (b) $V_{\text{large sphere}} \simeq V_{\text{small sphere}} \simeq 4.49$ V; $Q_{\text{small sphere}} \simeq 2.5 \cdot 10^{-12}$ C. (c) $V_{\text{large sphere}} \simeq V_{\text{small sphere}} \simeq 2.49$ V, $Q_{\text{small sphere}} \simeq 0.5 \cdot 10^{-12}$ C.*

P6.2. A large charged conducting sphere of radius $a = 0.4$ m is charged with a charge $Q = -10^{-9}$ C. A small uncharged conducting sphere of radius $b = 1$ cm is brought into contact with the large sphere, and then taken to a very distant point. Determine approximate charges and potentials of the large and small spheres in the end state, as well as the potential of the large sphere in the beginning. — *(a) The potential of the large sphere in the beginning is $V_0 = -22.47$ V. In the end state, $Q_{\text{small sphere}} \simeq -25$ pC, and the potential of both is approximately V_0. (b) $V_0 = -11.23$ V, $Q_{\text{small sphere}} \simeq -15$ pC, and the potential of both is approximately V_0. (c) $V_0 = -31.23$ V, $Q_{\text{small sphere}} \simeq -5$ pC, and the potential of both is approximately V_0.*

S **P6.3.** Two conducting spheres of equal radii $a = 2$ cm are far away from each other, and carry charges $Q_1 = -4 \cdot 10^{-9}$ C and $Q_2 = 2 \cdot 10^{-9}$ C. The spheres are brought to each other, touched, and moved back to their positions. Determine the charges of the spheres in the final state, as well as the potentials of the spheres in the initial and final states. — *(a) $V_{1 \text{ init}} = -198$ V and $V_{2 \text{ init}} = 89$ V. $Q_{1 \text{ fin}} = Q_{2 \text{ fin}} = -10^{-9}$ C. $V_{1 \text{ fin}} = V_{2 \text{ fin}} = -449$ V. (b) $V_{1 \text{ init}} = -1798$ V and $V_{2 \text{ init}} = 899$ V. $Q_{1 \text{ fin}} = Q_{2 \text{ fin}} = -10^{-9}$ C. $V_{1 \text{ fin}} = V_{2 \text{ fin}} = -449$ V. (c) $V_{1 \text{ init}} = -2798$ V and $V_{2 \text{ init}} = 2899$ V. $Q_{1 \text{ fin}} = Q_{2 \text{ fin}} = -10^{-9}$ C. $V_{1 \text{ fin}} = V_{2 \text{ fin}} = -449$ V.*

Solution. Since the spheres are far apart, their potentials in the initial positions are the same as if they were isolated,

$$V_{1/2} = \frac{Q_{1/2}}{4\pi\epsilon_0 a},$$

from which $V_1 = -1798$ V and $V_2 = 899$ V.

When the two spheres are touched, their charge becomes equal to half the sum of their charges, i.e., to $Q = -10^{-9}$ C. When moved back to the initial positions their potentials are equal, and are obtained from the above equation with $Q_{1/2}$ substituted by Q, i.e., $V = -449$ V.

S **P6.4.** The electric field strength at a point A on the surface of a very thin charged conducting shell is E. Determine the electric field strength in the middle of a small round hole made in the shell and centered at point A. — *(a) Zero. (b) E. (c) $E/2$.*

Solution. The electric field at point A before the hole is drilled is due partly to charges in the immediate neighborhood of the point, and partly to all distant charges. The two components of the vector E just above the surface make the total vector E, and just below the surface they must add to zero (that point is inside the conducting shell wall).

The field of distant charges, E_{dist}, is practically the same just above and just below the conductor surface. The field of local charges, E_{local}, is similar to that of a plane charged sheet, and is in opposite directions on the two sides of the conductor surface. So above the conductor surface we have

$$E_{\text{dist}} + E_{\text{local}} = E,$$

and just below the surface

$$E_{\text{dist}} - E_{\text{local}} = 0.$$

So, $E_{\text{dist}} = E_{\text{local}} = E/2$. If we drill a small hole in the shell, at the center of the hole, at point A, the field will be only that due to distant charges, i.e., it will be equal to $E/2$.

P6.5. Inside a spherical conducting shell of radius b is a conducting sphere of radius a $(a < b)$, charged with a charge Q_a. What is the potential of the shell: (1) if it is uncharged? (2) if it is charged with a charge Q_b? Does the potential depend on the position of the sphere inside the shell? Will it change if we move the sphere into contact with the inner surface of the shell? — *(a) $V_1 = Q_a/(2\pi\epsilon_0 b)$, $V_2 = (Q_a + Q_b)/(2\pi\epsilon_0 b)$. (b) $V_1 = Q_a/(4\pi\epsilon_0 b)$, $V_2 = (Q_a + Q_b)/(8\pi\epsilon_0 b)$. (c) $V_1 = Q_a/(4\pi\epsilon_0 b)$, $V_2 = (Q_a + Q_b)/(4\pi\epsilon_0 b)$.*

P6.6. Suppose that the shell in problem P6.5 is connected by a thin conducting wire to the reference point of the potential. Determine its charge, and determine the electrostatic potential function outside the shell. — *(a) $-Q_a$. The potential outside is zero. (b) $-Q_a + Q_b$. The* *potential is as that of a point charge $-Q_a + Q_b$. (c) Q_b. The potential is as that of a point charge Q_b.*

P6.7. A conducting sphere of radius a carries a charge Q_1. Concentric with the sphere there is a spherical shell of inner radius b ($b > a$) and outer radius c, carrying a charge Q_2. Determine the electric field intensity and the electric scalar potential at every point of the system. Plot the dependence of E and V on the distance r from the common center. — *Hint: note that, due to electrostatic induction, there will be an induced charge $-Q_1$ on the inner surface of the shell, so that on the outer surface there is a charge $Q_1 + Q_2$.*

P6.8. Twenty small charged bodies each carrying a charge $Q = 10^{-10}$ C are brought into an uncharged metallic shell of radius $R = 5$ cm. Evaluate the potential of the shell and the electric field strength on its surface. — *(a) $V(R) = 259.5\,V$. (b) $V(R) = 59.5\,V$. (c) $V(R) = 359.5\,V$.*

P6.9. How large an electric charge must be brought into the shell from problem P6.8 to achieve a field of 30 kV/cm at its surface? (This is approximately the greatest electric field strength in air; for larger fields, the air ionizes and becomes a conductor, or breaks down.) — *(a) $Q_{max} = \pm 3.338 \mu C$. (b) $Q_{max} = \pm 5.338 \mu C$. (c) $Q_{max} = \pm 4.338 \mu C$.*

S **P6.10.** A metal shell with a small hole is connected to ground with a conducting wire. A small charged body with a charge Q ($Q > 0$) is periodically brought through the hole into the shell without touching it, then taken out of it, and so on. Determine the charge which passes through the conducting wire from the shell to ground. — *(a) Oscillating between $-Q$ and Q. (b) Oscillating between 0 and Q. (c) Oscillating between $-Q$ and 0.*

Solution. When the charge is brought into the shell, a charge $-Q$ is induced on its inner surface, and a charge $+Q$ on its outer surface. Since the shell is connected to ground, this charge will flow through the wire into the ground. When we take the charge out of the shell, the charge Q will flow from the ground through the wire to the shell. So, there will be an alternating current in the wire.

P6.11. Three coaxial conducting hollow cylinders have radii $a = 0.5$ cm, $b = 1$ cm and $c = 2$ cm, and equal lengths $d = 10$ m. The middle cylinder is charged with a charge $Q = 1.5 \cdot 10^{-10}$ C, and the other two are uncharged. Determine the voltages between the middle cylinder and the other two. Neglect effects at the ends of the cylinders. — *(a) $V_b - V_a = -0.231\,V$, $V_b - V_c = 0.187\,V$. (b) $V_b - V_a = 0$, $V_b - V_c = 0.187\,V$. (c) $V_b - V_a = 0.345\,V$, $V_b - V_c = -0.223\,V$.*

P6.12. A charged conducting sphere of radius $b = 1$ cm and with a charge $Q = 2 \cdot 10^{-12}$ C is located at the center of an uncharged conducting spherical shell of outer radius $a = 10$ cm. The inner sphere is moved to touch the shell, and returned to its initial position. Calculate the potential of the spheres in the initial and end states for the following values of the wall thickness of the large sphere: $d = 0$ (i.e., vanishingly small), $d = 1$ cm, and $d = 5$ cm. — *Hint: before the two spheres are touch, the induced charge on the inner shell surface is $-Q$, and on the outer surface $+Q$. When the spheres touch, the charge on the inner sphere and that on the inner surface of the shell cancel out, so that only the charge on the outer surface of the shell remains.*

S **P6.13.** A line charge Q' is at a height h above a large flat conducting surface. Determine the electric field strength along the conducting surface in the direction normal to the line charge. — *Hint: recall that the field of charges induced on the conducting surface is the same as that of the line charge image, Fig. P6.13.*

Solution. The field of the charges induced on the conducting surface is the same as the field of the "image" of the line charge in the surface. So, above the surface, we have the same field as that due to the line charge Q' and a line charge $-Q'$ a distance $2h$ from the first.

The electric field on the conducting surface is therefore obtained as a vector sum of the fields due to these two line charges. Referring to Fig. P6.13, the total field is normal to the surface (as it should be), and is of intensity

$$E_{\text{total}}(x) = \frac{2Q'\cos\alpha}{2\pi\epsilon_0 R} = \frac{Q'h}{\pi\epsilon_0(h^2 + x^2)}.$$

What is the density of surface charges induced over the conducting surface?

P6.14. A point charge Q is at a point $(a, b, 0)$ of a rectangular coordinate system. The half-planes $(x \geq 0, y = 0)$ and $(x = 0, y \geq 0)$ are conducting. Determine the electric field at a point $(x, y, 0)$, where $x > 0$ and $y > 0$. — *Hint: referring to Fig. P6.14, it is seen that boundary conditions on the two conducting half planes are satisfied if they are removed, and the three indicated "image" charges introduced instead. Why is the third (positive) image necessary?*

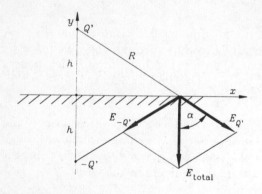

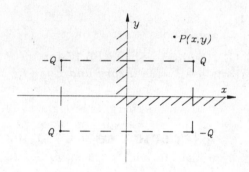

Fig. P6.13. A line charge above flat surface.   **Fig. P6.14.** A charge in a conducting corner.

P6.15. Repeat problem P6.14 for a line charge parallel to the z axis. — *Hint: replace the induced charges on the conducting corner by three line charges, the same as in Fig. P6.14.*

P6.16. A thunderstorm cloud can be represented as an electric dipole with $\pm 10\,\text{C}$ of charge. The bottom part of the cloud is at $h_1 = 5\,\text{km}$ above the ground, and the top is $h_2 = 8\,\text{km}$ above the ground (Fig. P6.16). The soil is wet and can be assumed to be a good conductor. (1) Find the potential and the electric field at the surface of the earth right under the cloud. (2) Find the surface charge density at points A and B on the surface (Fig. P6.16), for $x = 5\,\text{km}$. — *Hint: replace the charges induced on the ground by the image of the dipole, to obtain a total of four point charges which produce the field and potential.*

S **P6.17.** Find the induced charge distribution $\sigma(r)$ on the ground when a point charge $-Q$ is placed at a height h above ground, assuming the ground is an infinite flat conductor. Plot your results. — *Hint: use image theory and Eq. (6.5).*

Solution. The density of induced surface charge is obtained from the relation $\sigma = \epsilon_0 E_{\text{total}}$. The total electric field is simply the sum of the vertical components of the field due to the original charge and its image. So we obtain

$$\sigma(x) = \epsilon_0 E_{\text{total}} = -\frac{2Qh}{(h^2 + x^2)^{3/2}},$$

where x is the coordinate axis on the ground, with the origin below the charge. Is the minus sign correct, and why?

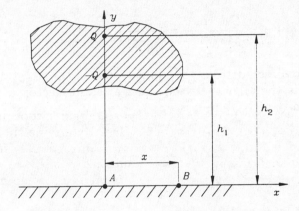

Fig. P6.16. A thunderstorm cloud.

P6.18. Repeat problem P6.17 for the case of a dipole such as the one shown in Fig. P6.16. — *Hint: use image theory and Eq. (6.5).*

7. Dielectrics in the Electrostatic Field

• Dielectrics are substances which do not have freely movable electric charges inside them. When in an electric field, their atoms and molecules become tiny electric dipoles aligned with the electric field. This happens mostly in two ways. In one type of substances, in the absence of the electric field molecules are not dipoles (*nonpolar* molecules), but when in the field they become dipoles aligned with the field lines. In the other type of substances, molecules are permanent dipoles (*polar molecules*); when in the field, they align partly with the field, due to torques of electric forces on them. In both cases, a vast number of oriented elemental dipoles results. This process is known as *polarization*, and the dielectric is said to be *polarized*. An individual dipole is characterized by its *dipole moment*, $p = Qd$, where Q is the charge of the positive dipole end, and d is the vector distance from the negative to the positive dipole charge.

• A polarized dielectric is described at all points by the *polarization vector*, P. It is defined as

$$P = \frac{\sum_{\text{d}v} p}{\text{d}v} = Np \qquad (\text{C/m}^2), \tag{7.1}$$

where $\text{d}v$ is a small volume enclosing the point at which we determine P, and N is the number of dipoles per unit volume at that point. So a small volume $\text{d}v$ of a polarized dielectric can be replaced by a single dipole of moment

$$d\boldsymbol{p} = \boldsymbol{P}\,dv \qquad (\text{C}\cdot\text{m}). \tag{7.2}$$

• If we adopt the z axis to be along the dipole axis, in the direction of vector $\boldsymbol{d}$ and centered at the midpoint between the dipole charges, the potential due to the dipole is given by

$$V_P = \frac{Qd\,\cos\theta}{4\pi\epsilon_0 r^2} = \frac{\boldsymbol{p}\cdot\boldsymbol{u}}{4\pi\epsilon_0 r^2} \qquad (\text{V}), \tag{7.3}$$

where $\boldsymbol{u}$ is the unit vector directed from the dipole towards the point at which the potential is being evaluated, and θ is the angle between the z axis and vector $\boldsymbol{u}$. So the potential due to a polarized dielectric body of volume v is

$$V = \frac{1}{4\pi\epsilon_0}\int_v \frac{\boldsymbol{P}\cdot\boldsymbol{u}}{r^2}dv \qquad (\text{V}). \tag{7.4}$$

• Experiments show that for most substances

$$\boldsymbol{P} = \chi_e\epsilon_0\,\boldsymbol{E} \qquad (\boldsymbol{P} - \text{C/m}^2,\ \chi_e - \text{dimensionless}). \tag{7.5}$$

The constant $\chi_e \geq 0$ is the *electric susceptibility*. If it is the same at all points, the dielectric is *homogeneous*, else it is *inhomogeneous*. If Eq. (7.5) holds, the dielectric is *linear*, otherwise it is *nonlinear*.

• A polarized dielectric can always be replaced by an equivalent volume and surface charge distributions in a vacuum, of densities

$$\rho_\text{p} = -\text{div}\,\boldsymbol{P} = -\nabla\cdot\boldsymbol{P} \qquad (\text{C/m}^3), \tag{7.6}$$

and

$$\sigma_\text{p} = \boldsymbol{n}\cdot(\boldsymbol{P}_2 - \boldsymbol{P}_1) \qquad (\text{C/m}^2), \tag{7.7}$$

where unit vector $\boldsymbol{n}$ is directed into dielectric 1. For a homogeneous dielectric $\rho_\text{p} = 0$. The divergence in Eq. (7.6) in a rectangular coordinate system has the form,

$$\text{div}\,\boldsymbol{P} = \frac{\partial P_x}{\partial x} + \frac{\partial P_y}{\partial y} + \frac{\partial P_z}{\partial z} = \nabla\cdot\boldsymbol{P} \qquad (\text{C/m}^3), \tag{7.8}$$

where P_x, P_y and P_z are scalar rectangular components of the vector $\boldsymbol{P}$.

• The *electric displacement vector*, $\boldsymbol{D}$, is defined as

$$\boldsymbol{D} = \epsilon_0\boldsymbol{E} + \boldsymbol{P} \qquad (\text{C/m}^2), \tag{7.9}$$

and the generalized Gauss' law, valid for any media, reads

$$\oint_S \boldsymbol{D} \cdot \mathrm{d}\boldsymbol{S} = Q_{\text{free in S}} \qquad \text{(C)}. \qquad (7.10)$$

- If the dielectric is linear, vector $\boldsymbol{D}$ can be expressed as

$$\boldsymbol{D} = \epsilon_0(1 + \chi_e)\boldsymbol{E} = \epsilon_0\epsilon_r\boldsymbol{E} = \epsilon\boldsymbol{E} \qquad \text{(C/m}^2\text{)}, \qquad (7.11)$$

where

$$\epsilon_r = (1 + \chi_e) \qquad \text{(dimensionless)}. \qquad (7.12)$$

is the *relative permittivity* of the dielectric, and

$$\epsilon = \epsilon_r\epsilon_0 \qquad \text{(F/m)}. \qquad (7.13)$$

as the *permittivity* of the dielectric.

- At two close points on two sides of a boundary surface between two different media, vectors $\boldsymbol{E}$ and $\boldsymbol{D}$ must satisfy the *boundary conditions*,

$$E_{1\text{ tangential}} = E_{2\text{ tangential}}, \qquad (7.14)$$

$$D_{1\text{ normal}} - D_{2\text{ normal}} = \sigma, \qquad (7.15)$$

where the reference direction is into dielectric 1. Specifically, if medium 2 is a conductor, this becomes

$$D_n = \sigma \quad \text{(on boundary of dielectric and conductor)}. \qquad (7.16)$$

- The *differential form* of generalized Gauss' law is a partial differential equation obtained from its integral form in Eq. (7.10):

$$\operatorname{div}\boldsymbol{D} = \rho. \qquad (7.17)$$

- The electric scalar potential satisfies a partial differential equation of the second order, known as the *Poisson equation*. In the case of homogeneous dielectrics it is of the form

$$\operatorname{div}(\operatorname{grad}V) = \nabla \cdot (\nabla V) = -\frac{\rho}{\epsilon} \quad \text{(Poisson's equation)}. \qquad (7.18)$$

In the practically most frequent case when $\rho = 0$ it becomes *Laplace's equation*,

$$\text{div}(\text{grad}V) = \nabla \cdot (\nabla V) = \nabla^2 V = 0 \quad \text{(Laplace's equation).} \tag{7.19}$$

The operator $\text{div}(\text{grad}) = \nabla \cdot \nabla = \nabla^2$ is known as *Laplace's operator*, or the *Laplacian*. In a rectangular coordinate system

$$\nabla^2 = \frac{\partial^2}{\partial x^2} + \frac{\partial^2}{\partial y^2} + \frac{\partial^2}{\partial z^2} \quad \text{(Laplace's operator).} \tag{7.20}$$

• In addition to ϵ_r, every dielectric has two other important properties. The first is its *dielectric strength* (the largest magnitude of the electric field which can exist in a dielectric without damaging it — for larger field values *dielectric breakdown* occurs). For air, it is about $30\,\text{kV/cm}$ ($3 \cdot 10^6\,\text{V/m}$). The other property are losses in time-varying fields, known as the *polarization losses*.

QUESTIONS

S **Q7.1.** At a point of a polarized dielectric there are N dipoles per unit volume. Each dipole is of moment p. What is the polarization vector at that point? — *(a)* $P = N^3 p$. *(b)* $P = N^2 p$. *(c)* $P = Np$.

Answer. From the definition of the polarization vector, $P = Np$.

Q7.2. A body is made of a linear, homogeneous dielectric. Explain what this means. — *(a) At all points of the body $D = \epsilon(x, y, z)E$, where $\epsilon(x, y, z)$ varies from one point to another. (b) $D = \epsilon(x, y, z)E$, where $\epsilon(x, y, z)$ is the same at all points of the body. (c) $D = \epsilon(E)E$, where $\epsilon(E)$ depends on the magnitude of E.*

S **Q7.3.** What is the difference between an inhomogeneous linear dielectric, and a homogeneous nonlinear dielectric? — *(a) $D = \epsilon(x, y, z)E$ in the first case, $D = \epsilon(E)E$ in the second. (b) $D = \epsilon E$ in the first case, $D = \epsilon(E)E$ in the second. (c) $D = \epsilon(x, y, z)E$ in the first case, $D = \epsilon E$ in the second.*

Answer. An inhomogeneous linear dielectric is characterized by a permittivity which is a function of coordinates, e.g., $\epsilon(x, y, z) = xyz\epsilon_0/(abc)$.

A homogeneous nonlinear dielectric is a dielectric having the same properties at all points, but these properties are a function of the electric field strength. For example, if at all points of a dielectric $D = \epsilon EE/E_0$, where ϵ and E_0 are constants, the dielectric is homogeneous, but nonlinear.

Q7.4. Why is $\chi_e = 0$ for a vacuum? — *(a) Because in a vacuum $D = 0$. (b) Because in a vacuum $P = 0$. (c) Because in a vacuum $\epsilon = 0$.*

Q7.5. Are there substances for which $\chi_e < 0$? Explain. — *(a) Yes. (b) No. (c) They are very rare.*

Q7.6. An atom acquires a dipole moment proportional to the electric field strength E of the external field, $p = \alpha E$ (α is often referred to as the *polarizability*). Determine the electric force on the atom if it is introduced into a *uniform* electric field of intensity E. — *(a) Zero. (b) Towards the region of higher field. (c) Towards the region of smaller field.*

Q7.7. Answer question Q7.6 for the case in which the atom is introduced into the field of a point charge Q. Determine only the direction of the force, not its magnitude. — *(a) Zero. (b) Towards the charge. (c) Away from the charge.*

Q7.8. A small body — either dielectric or conducting — is introduced into a nonuniform electric field. In which direction (qualitatively) does the force act on the body? — *(a) There is no force. (b) Towards the region of higher field. (c) Towards the region of smaller field.*

Q7.9. Two point charges are placed near a piece of dielectric. Explain why Coulomb's law cannot be used to determine the *total* force on the two charges. — *(a) The polarized dielectric also exerts a force. (b) Coulomb's law is the correct way to find that force. (c) There is also a gravitational force due to the dielectric.*

Q7.10. A small charged body is placed near a large dielectric body. Will there be a force acting between the two bodies? Explain. — *(a) There fill be no force. (b) The force will be attractive. (c) The force will be repulsive.*

S **Q7.11.** A closed surface S situated in a vacuum encloses a total charge Q and a polarized dielectric body. Using a sound physical argument, prove that in this case also the flux of the electric field strength vector E through S is Q/ϵ_0. — *(a) In the dielectric there are no charges. (b) The flux is not equal to Q/ϵ_0. (a) The total polarization charge in a dielectric is always zero.*

Answer. A polarized dielectric body can be considered from the electrostatic point of view as a vast ensemble of dipoles situated in a vacuum. The dipoles are displaced equal positive and negative charges, and their total charge is zero. Therefore, *provided that S encloses the entire dielectric body,* the total charge enclosed by S is still Q.

Q7.12. Arbitrary pieces of dielectrics and conductors carrying a total charge Q are introduced through an opening in a hollow, uncharged metal shell. The opening is then closed. Using a physical argument and Gauss' law for a vacuum, prove that the charge appearing on the outer surface of the shell is exactly equal to Q. — *Hint: apply Gauss' law to a closed surface entirely inside the shell wall.*

Q7.13. A positive point charge is placed in air near the interface of air and a liquid dielectric. Will the interface be deformed? If you think it will be deformed, then will it raise or sink? What if the charge is negative? — *(a) It will not be deformed. (b) It will raise, irrespective of the charge sign. (c) It will raise if the charge is positive, sink if negative.*

Q7.14. Explain in your own words why equations

$$\epsilon_0 \oint_{\Delta S} E \cdot dS = Q_{\text{p in } \Delta S}$$

and

$$Q_{\text{p in } \Delta S} = -\oint_{\Delta S} P \cdot dS = -\chi_e \epsilon_0 \oint_{\Delta S} E \cdot dS$$

imply that the flux of E through a closed surface ΔS is zero. — *(a) They result in the equation of the form $x = -ax$, $a > 0$. (b) This is self evident. (c) The flux of E through any small closed surface is zero.*

Q7.15. Electric dipoles are arranged along a line (possibly curved) so that the negative charge of one dipole coincides with the positive charge of the next. Describe the electric field of this arrangement of dipoles. — *(a) As that of a continuous line charge. (b) As that of an uncharged line. (c) As that of two equal point charges, of opposite signs, located at the ends of the line.*

Q7.16. Write Eq. (7.7) for the interface of a dielectric and a vacuum. For case (1) assume the dielectric to be medium 1, and for case (2) medium 2. — *Possible results for case 1: (a) $\sigma_{\mathrm{P}} = (\epsilon - \epsilon_0)n \cdot E$. (b) $\sigma_{\mathrm{P}} = (\epsilon + \epsilon_0)n \cdot E$. (c) $\sigma_{\mathrm{P}} = -(\epsilon - \epsilon_0)n \cdot E$.*

Q7.17. Is there a pressure of electrostatic forces acting on a boundary surface between two different dielectrics situated in an electrostatic field? Explain. — *(a) There is no pressure. (b) There is a pressure on surface polarization charges on the boundary. (c) There is a pressure on volume polarization charges on the two sides of the boundary.*

Q7.18. Prove that the total polarization charge in any piece of a dielectric material is zero. — *Hint: recall the definition of polarization and polarization charges.*

S **Q7.19.** A point charge Q is placed inside a spherical metal shell, a distance d from its center. In addition, the shell is filled with an inhomogeneous dielectric. Determine the electric field strength outside the shell. — *(a) There is no field outside the shell. (b) The field is as that of a charge Q at its actual location. (c) The field is as that of a charge Q at the shell center.*

Answer. The dielectric, although it will be polarized, has a zero total charge. So there will be an induced charge $-Q$ on the shell inner surface, distributed nonuniformly, the distribution depending on the position of Q and of the dielectric bodies. A charge $+Q$ will appear on the outer surface of the shell, and will be distributed uniformly, since there is no electric field of the inside charges outside the shell. The electric field strength outside the shell is therefore the same as that of a point charge Q at the center of the shell.

Q7.20. Does the equation

$$\oint_S E \cdot dS = \frac{Q_{\text{free in } S} + Q_{\text{polarization in } S}}{\epsilon_0}$$

mean exactly the same as the equation

$$\oint_S (\epsilon_0 E + P) \cdot dS = Q_{\text{free in S}} ?$$

Explain. — *(a) It does not. (b) Only in some cases. (c) It does.*

Q7.21. Can the relative permittivity of a dielectric be less than one, or negative? Explain. — *(a) It can be less than one, but not negative. (b) It can be negative. (c) It cannot be less than one.*

S **Q7.22.** Can you find an analogy between properly connecting sleeves to a jacket, and using boundary conditions in solving electrostatic field problems? Describe. — *(a) None at all. (b) Yes. (c) A vague one.*

Answer. In both cases a rule is used which follows from certain laws. The "boundary conditions" for the sleeves are that they must be connected to the jacket so that they hang without wrinkles. The boundary conditions in the electrostatic fields follow from the fundamental field equations.

S **Q7.23.** Prove that a charged conductor situated in an inhomogeneous but linear dielectric has a potential proportional to its charge. — *Hint: consider the polarized dielectric as a large number of dipoles situated in a vacuum.*

Answer. A charge Q will be distributed over a conductor so that, together with the field of the dipoles of the polarized dielectric, it makes the conductor surface equipotential, e.g., at a potential V. Let the electric field strength in this case be $\boldsymbol{E}$ (a function of coordinates).

Let now the charge be kQ, with $k \neq 1$. The conductor surface can remain equipotential only if the charge over the conductor is distributed in the same manner, in which case the dipoles will also be polarized in the same manner. The field at all points will be $k\boldsymbol{E}$, and the potential of the body kV.

Q7.24. Discuss question Q7.23 for a case in which the dielectric is not linear. — *(a) The proof is the same. (b) The proof is too complicated. (c) In this case the statement is not correct.*

Q7.25. What is the unit of dielectric strength of a dielectric? — *(a) Newton. (b) Volt per meter. (c) Coulomb.*

Q7.26. Explain how $30\,\mathrm{kV/cm}$ is the same as $3 \cdot 10^6\,\mathrm{V/m}$. — *Hint: express the kV in volts, and cm in meters.*

Q7.27. Are polarization losses in a dielectric the same as resistive Joule's losses? Explain. — *(a) No. (b) Yes. (c) Only at high frequencies.*

PROBLEMS

P7.1. Using the relation $\boldsymbol{E} = -\nabla V$, determine the spherical components E_r, E_θ, and E_ϕ of the electric field strength of the electric dipole in Fig. P7.1. — *Hint: refer to Fig. P7.1 and use the expression for gradient in a spherical coordinate system.*

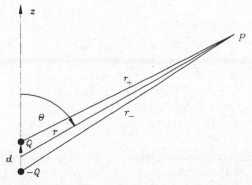

Fig. P7.1. An electric dipole. It is assumed that $r \gg d$.

S **P7.2.** Determine the electric force on a dipole of moment p located at a distance r from a point charge Q_0, if the angle between p and the direction from the charge is arbitrary. — *Hint: note that the two dipole charges are at slightly different distances from the point charge.*

Solution. Let the dipole center be at a distance r from the point charge, and let the dipole moment, $p = Qd$, make an angle α with the unit vector u_r (directed from the point charge toward the dipole). Then, if $Q_0 > 0$, the negative charge of the dipole is at a distance $(r - d\cos\alpha/2)$ from the point charge, and the positive charge of the dipole at a distance $(r + d\cos\alpha/2)$ from it. If we adopt the reference direction of the total force on the dipole to be in the direction of u_r, the magnitude of the force (that can be both positive and negative) is

$$F = -\frac{Q_0 Q}{4\pi\epsilon_0 (r - d\cos\alpha/2)^2} + \frac{Q_0 Q}{4\pi\epsilon_0 (r + d\cos\alpha/2)^2}.$$

Since $r \gg d/2$, this can be written as

$$F = -\frac{Q_0 Q}{4\pi\epsilon_0 r^2}\left[\left(1 - \frac{d\cos\alpha}{2r}\right)^{-2} - \left(1 + \frac{d\cos\alpha}{2r}\right)^{-2}\right]$$

$$\simeq -\frac{Q_0 Q}{4\pi\epsilon_0 r^2}\left[\left(1 + \frac{d\cos\alpha}{r}\right) - \left(1 - \frac{d\cos\alpha}{r}\right)\right],$$

from which

$$F = -\frac{Q_0 Q d\cos\alpha}{2\pi\epsilon_0 r^3}, \quad \text{or} \quad \boldsymbol{F} = -\frac{Q_0 \boldsymbol{p}\cdot\boldsymbol{u}_r}{2\pi\epsilon_0 r^3}.$$

P7.3. An atom acquires a dipole moment proportional to the electric field strength E of the external field, $p = \alpha E$. Determine the force on the dipole if it is introduced into the field of a point charge Q at a distance r from the charge. — *Hint: use the solution to the preceding problem.*

P7.4. A homogeneous dielectric sphere is polarized uniformly over its volume. The polarization vector is P. Determine the distribution of the polarization charges inside and on the surface of the sphere. — *Let u_n be the unit vector normal to the sphere, directed outward, and α the angle between vectors P and u_n.* (a) $\rho_p = 0$, $\sigma_p = P\cos\alpha$, (b) $\rho_p = 0$, $\sigma_p = P\sin\alpha$, (c) $\rho_p = P$, $\sigma_p = 0$.

S **P7.5.** A thin circular dielectric disk of radius a and thickness d is permanently polarized with a dipole moment per unit volume P, parallel to the axis of the disk which is normal to its plane faces. Determine the electric field strength and the electric scalar potential along the disk axis. Plot your results. — *Hint: note that the polarized disk is equivalent to two layers of polarization charges on the two disk faces, Fig. P7.5.*

Solution. The electric field of the dipoles in the polarized disk is the same as that of two layers of polarization charges on the two disk plane faces, of densities (see Fig. P7.5)

$$\sigma_p = \boldsymbol{P}\cdot\boldsymbol{u}_n = \pm P.$$

We can calculate the potential at point M shown in Fig. P7.5 as that due to two uniformly charged circular disks (see problem P4.12). So we have

$$V_M = \frac{\sigma_\mathrm{p}}{2\epsilon_0}\left(\sqrt{z^2 + a^2} - z\right) - \frac{\sigma_\mathrm{p}}{2\epsilon_0}\left[\sqrt{(z+d)^2 + a^2} - (z+d)\right]$$

$$= \frac{\sigma_\mathrm{p}}{2\epsilon_0}\left[d + \sqrt{z^2 + a^2} - \sqrt{(z+d)^2 + a^2}\right].$$

This expression is valid for any thickness d of the disk. If $d \ll z$,

$$\sqrt{(z+d)^2 + a^2} \simeq \sqrt{z^2 + a^2 + 2zd} = \sqrt{z^2 + a^2}\sqrt{1 + \frac{2zd}{z^2 + a^2}}$$

$$\simeq \sqrt{z^2 + a^2}\left(1 + \frac{zd}{z^2 + a^2}\right),$$

so that, in that case, the expression for the potential becomes

$$V_M = \frac{\sigma_\mathrm{p}d}{2\epsilon_0}\left(1 - \frac{z}{\sqrt{z^2 + a^2}}\right).$$

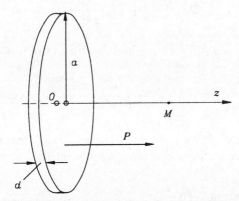

Fig. P7.5. A polarized dielectric disk.

P7.6. Determine the density of volume polarization charges inside a linear but inhomogeneous dielectric of permittivity $\epsilon(x, y, z)$, at a point where the electric field strength is E. There is no volume distribution of free charges inside the dielectric. — (a) $\rho_\mathrm{p} = -(\epsilon_0/\epsilon)E$. (b) $\rho_\mathrm{p} = -(\epsilon_0/\epsilon)E \cdot \nabla\epsilon$. (c) $\rho_\mathrm{p} = -E \cdot \nabla\epsilon$.

P7.7. The permittivity of an infinite dielectric medium is given as the following function of the distance r from the center of symmetry: $\epsilon(r) = \epsilon_0(1 + a/r)$. A small conducting sphere of radius R, carrying a charge Q, is centered at $r = 0$. Determine and plot the electric field strength and the electric scalar potential as functions of r. Determine the volume density of polarization charges. — *(a)* $E(r) = Q/[8\pi\epsilon_0 r(a+r)]$, $r > R$. *(b)* $E(r) = Q/[4\pi\epsilon_0 r(a+2r)]$, $r > R$. *(c)* $E(r) = Q/[4\pi\epsilon_0 r(a+r)]$, $r > R$.

S **P7.8.** A conducting sphere of radius a carries a charge Q. Exactly one half of the sphere is pressed into a dielectric half-space of permittivity ϵ. Air is above the dielectric. Determine the free and polarization surface charge density on the sphere and in the dielectric. — *Hint: note that, due to symmetry, the lines of vectors D and E are radial, and apply the generalized Gauss' law to find these vectors.*

Solution. Due to symmetry, the lines of vectors D and E are radial. Consider a point at the interface air/dielectric, a distance r $(r > a)$ from the sphere center. The boundary condition for the tangential component of vector E requires that E be the same at two close points on the two sides of the interface.

The electric displacement vectors at these two points are therefore different, and are given by

$$D_0(r) = \epsilon_0 E(r), \qquad D(r) = \epsilon E(r).$$

Let us apply the generalized Gauss' law in integral form to a sphere of radius r, centered at the center of the sphere:

$$\oint_S \mathbf{D} \cdot d\mathbf{S} = \oint_S D(r)dS = D_0(r)2\pi r^2 + D(r)2\pi r^2 = Q,$$

or

$$D_0(r) + D(r) = \frac{Q}{2\pi r^2}.$$

Substituting the above expressions for $D_0(r)$ and $D(r)$ into the last equation, we get

$$E(r) = \frac{Q}{2\pi r^2 (\epsilon + \epsilon_0)}.$$

The surface density of free charges on the sphere surface is different for the part of the surface in contact with air, and the part that is in contact with the dielectric:

$$\sigma_0 = D_0(a) = \epsilon_0 E(a) = \frac{\epsilon_0 Q}{2\pi a^2(\epsilon + \epsilon_0)}, \qquad \sigma = D(a) = \epsilon E(a) = \frac{\epsilon Q}{2\pi a^2(\epsilon + \epsilon_0)}.$$

(Of course, the total charge over the sphere is Q.)

The surface polarization charge exists only on the surface of the dielectric adjacent to the sphere. There, its density is given by

$$\sigma_{\mathrm{p}} = -P(a) = -(\epsilon - \epsilon_0)E(a).$$

Thus, the sum of σ on the part of the sphere covered with the dielectric is equal to the free charge surface density on the part of the sphere covered with air, i.e., the total surface charge density over the sphere is constant:

$$\sigma + \sigma_\mathrm{p} = \epsilon E(a) - (\epsilon - \epsilon_0)E(a) = \epsilon_0 E(a) = \sigma_0.$$

P7.9. Repeat problem P7.8 for a circular cylinder of radius a with charge Q' per unit length. — *Hint: the same as in the preceding problem.*

P7.10. A small spherical charged body with a charge $Q = -1.9 \cdot 10^{-9}$ C is located at the center of a spherical dielectric body of radius a and relative permittivity $\epsilon_\mathrm{r} = 3$. Determine the vectors E, P, and D at all points, volume and surface density of polarization charges, and the potential at all points. Is it possible to determine the field and potential outside the dielectric body without solving for the field inside the body? Explain. — *Hint: note that, due to symmetry, the lines of vectors D and E are radial, and apply the generalized Gauss' law to find these vectors.*

P7.11. What is E equal to in a needlelike air cavity inside a homogeneous dielectric of permittivity ϵ if the cavity is parallel to the electric field vector E_d inside the dielectric (Fig. P7.11)? — *(a) Zero. (b) $E = E_\mathrm{d}$. (c) $E = (\epsilon/\epsilon_0)E_\mathrm{d}$.*

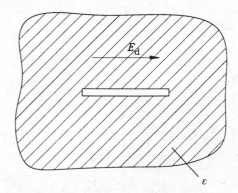

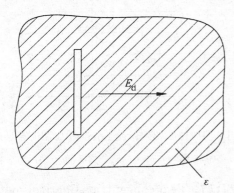

Fig. P7.11. A needlelike cavity.   Fig. P7.12. A disklike cavity.

P7.12. What is E equal to in a disklike air cavity with faces normal to the electric field vector E_d inside a homogeneous dielectric of permittivity ϵ (Fig. P7.12)? — *(a) $E = \epsilon E_\mathrm{d}/\epsilon_0$. (a) $E = -\epsilon E_\mathrm{d}/\epsilon_0$. (a) $E = \epsilon_0 E_\mathrm{d}/\epsilon$.*

S **P7.13.** At a point of the boundary surface between dielectrics of permittivities ϵ_1 and ϵ_2, the electric field strength vector in medium 1 makes an angle α_1 with the normal to the boundary, and that in medium 2 an angle α_2. Prove that $\tan\alpha_1/\tan\alpha_2 = \epsilon_1/\epsilon_2$. — *Hint: start from boundary conditions in Eqs. (7.14) and (7.15), with $\sigma = 0$.*

Solution. Recall that $E_{1\,\mathrm{tang}} = E_{2\,\mathrm{tang}}$, and $D_{1\,\mathrm{norm}} = D_{2\,\mathrm{norm}}$, or $\epsilon_1 E_{1\,\mathrm{norm}} = \epsilon_2 E_{2\,\mathrm{norm}}$. Thus

$$\frac{\tan\alpha_1}{\tan\alpha_2} = \frac{E_{1\,\mathrm{tang}}/E_{1\,\mathrm{norm}}}{E_{2\,\mathrm{tang}}/E_{2\,\mathrm{norm}}} = \frac{\epsilon_1}{\epsilon_2}\frac{E_{1\,\mathrm{tang}}/(\epsilon_1 E_{1\,\mathrm{norm}})}{E_{2\,\mathrm{tang}}/(\epsilon_2 E_{2\,\mathrm{norm}})} = \frac{\epsilon_1}{\epsilon_2}.$$

This is frequently referred to as the law of refraction of lines of vector E.

P7.14. A dielectric slab of permittivity $\epsilon = 2\epsilon_0$ is situated in a vacuum in an external uniform electric field E so that the field lines are perpendicular to the faces of the slab (Fig. P7.14). Sketch the lines of the resulting vectors E and D. — *Hint: have in mind the boundary conditions in Eqs. (7.14) and (7.15).*

P7.15. Repeat problem P7.14 assuming that the dielectric slab is at an angle of 45 degrees with respect to the lines of the external electric field (Fig. P7.15). — *Hint: have in mind the boundary conditions in Eqs. (7.14) and (7.15).*

P7.16. One of two very large parallel metal plates is at a zero potential, and the other at a potential V. Starting from Laplace's equation, determine the potential, and hence the electric field strength, at all points. — *Hint: the problem is one-dimensional. Assume the x axis to be normal to the plates, with the origin at the zero-potential plate, and solve the one-dimensional Laplace's equation.*

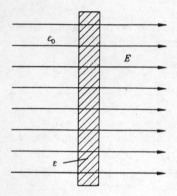

Fig. P7.14. E-lines normal to dielectric slab.   **Fig. P7.15.** E-lines oblique to dielectric slab.

S **P7.17.** Two concentric spherical metal shells, of radii a and b ($b > a$), are at potentials V (the inner shell) and zero. Starting from Laplace's equation in spherical coordinates, determine the potential, and hence the electric field strength, at all points. Plot your results. — *Hint: if we adopt the spherical coordinate system with the origin at the center of the spheres, the potential is a function of one coordinate, r, only. Solve the Laplace's equation in spherical coordinate system with this in mind.*

Solution. If we adopt the spherical coordinate system with the origin at the center of the spheres, the potential is again a function of one coordinate, r, only. So Laplace's equation is of the form (see Appendix 1)

$$\frac{1}{r^2}\frac{d}{dr}\left[r^2\frac{dV(r)}{dr}\right] = 0.$$

The first integration yields

$$r^2\frac{dV(r)}{dr} = A, \quad \text{or} \quad \frac{dV(r)}{dr} = \frac{A}{r^2},$$

where A is a constant. The solution of the last equation is

$$V(r) = -\frac{A}{r} + B,$$

where B is another constant. The two constants are determined from the conditions that $V(a) = V$, and $V(b) = 0$. We thus obtain two equations,

$$V(a) = V = -\frac{A}{a} + B, \qquad V(b) = 0 = -\frac{A}{b} + B.$$

From the second equation $B = A/b$, and then from the first equation we have

$$V = -\frac{A}{a} + \frac{A}{b}, \quad \text{so that} \quad A = -V\frac{ab}{b-a}.$$

The final potential function is thus

$$V(r) = \frac{V}{b-a}\left(\frac{ab}{r} - a\right).$$

S **P7.18.** The charge density at all points between two large parallel flat metal sheets is ρ_0. The sheets are d apart. One of the sheets is at a zero potential, and the other at a potential V. Find the potential at all points between the plates starting from Poisson's equation. Plot your results. *Hint: the problem is one-dimensional. Adopt the x axis to be normal to the sheets, with the origin at the sheet at potential zero, and solve the one-dimensional Poisson equation.*

Solution. The problem is one-dimensional. Let the x axis be normal to the sheets, with the origin at the sheet at potential zero. The Poisson equation for the problem is then of the form

$$\frac{\mathrm{d}^2 V(x)}{\mathrm{d}x^2} = -\frac{\rho_0}{\epsilon_0}.$$

The solution of this equation is

$$V(x) = A + Bx - \frac{\rho_0 x^2}{2\epsilon_0}.$$

The constants A and B are determined from the conditions that $V(0) = 0$, and $V(d) = V$, i.e., that $0 = A$, and

$$V = Bd - \frac{\rho_0 d^2}{2\epsilon_0},$$

which gives the value of B,

$$B = \frac{V}{d} + \frac{\rho_0 d}{2\epsilon_0}.$$

Thus

$$V(x) = V\frac{x}{d} + \frac{\rho_0 x d}{2\epsilon_0} - \frac{\rho_0 x^2}{2\epsilon_0} = V\frac{x}{d} + \frac{\rho_0 x}{2\epsilon_0}(d - x).$$

P7.19. Repeat problem P7.18 if the charge density between the plates is $\rho(x) = \rho_0 x/d$, x being a coordinate normal to the plates, with the origin at the zero-potential plate. Plot your results and compare to problem P7.18. — *Hint: adopt the same coordinate system as in the previous problem, and solve the one-dimensional Poisson equation.*

P7.20. Repeat problem P7.19 if the origin is at the plane of symmetry of the system. — *Hint: the general solution is the same, but integration constants are determined in a slightly different way.*

P7.21. Two long coaxial cylindrical thin metal tubes of radii a and b ($b > a$) are at potential zero (the outer tube) and V. Starting from Laplace's equation in cylindrical coordinates, determine the potential between the cylinders, and hence the electric field strength. — *Hint: adopt the radial coordinate, r, to be normal to the axis of the cylinders, with the origin on the axis. The potential is then a function of r only, and the Laplace equation in cylindrical coordinate system becomes a one-dimensional equation.*

P7.22. Prove that if V_1 and V_2 are solutions of Laplace's equation, their product is not generally a solution of that equation. — *Hint: consider the expression $\nabla^2(V_1 V_2)$ and prove that it is not zero if $\nabla^2 V_1 = 0$ and $\nabla^2 V_2 = 0$.*

S **P7.23.** The radii of conductors of a coaxial cable with air dielectric are a and b ($b > a$). Determine the maximum value of the potential difference between the conductors for which a complete breakdown of the air dielectric does not occur. The dielectric strength of air is E_0. — *Hint: note that the maximal electric field is on the surface of the inner cable conductor. Prove that, for given b and voltage between the conductors, the minimal electric field on the inner cable conductor is obtained if the ratio $b/a = \mathrm{e} = 2.71828\ldots$.*

Solution. The maximal electric field strength inside the cable is on its inner conductor. If the voltage between the conductors is V, this electric field strength is given by

$$E(a) = \frac{Q'}{2\pi\epsilon_0 a} = \frac{C'V}{2\pi\epsilon_0 a} = \frac{V}{a\ln(b/a)},$$

since the capacitance per unit length of the air-filled coaxial line is $C' = 2\pi\epsilon_0 / \ln(b/a)$.

Let $a = b/x$, and the above expression becomes

$$E(x) = \frac{Vx}{b\ln x}.$$

The minimal $E(x)$ is obtained if

$$\frac{dE(x)}{dx} = \frac{V}{b\ln x} - \frac{V}{b\ln^2 x} = 0.$$

This equation is satisfied only if $x = e = 2.7182818\ldots$, the basis of natural logarithms. Thus, if $a = b/e$, we have the minimal field on the inner conductor for fixed b and given voltage. Both for $a < b/e$ and $a > b/e$ the field on the inner conductor for a given voltage between the conductors will be larger.

Having this in mind, assume first that $a < b/e$. If we increase the voltage so that the electric field strength on the inner conductor is greater than the dielectric strength of air, a corona (ionized, i.e. conducting, air) will be formed around the inner conductor. This is equivalent to increasing the inner conductor radius. If this radius remains less than b/e, the ionization of the air will stop. We can increase the voltage as long as this condition is satisfied. Once the corona radius becomes equal to b/e, further increase of the voltage will increase the radius of the corona, but the electric field strength on its surface will also increase. Consequently, spark discharge will follow.

Thus, if $a < b/e$, the largest voltage is determined by the equation

$$E(a) = E_0 = \frac{V_{max}}{(b/e)\ln[b/(b/e)]} = \frac{V_{max}}{b/e} \quad (a < b/e).$$

If $a > b/e$ and we increase the voltage, once we reach the field E_0 on the inner surface conductor, a spark discharge will follow. So, in that case the greatest possible voltage is defined by the equation

$$E(a) = E_0 = \frac{V_{max}}{a\ln(b/a)} \quad (a > b/e).$$

8. Capacitance and Related Concepts

• *Capacitors* consist of two metal bodies, known as the capacitor *electrodes*, charged with equal charges of opposite sign. Capacitors are characterized by their *capacitance*. If one of the electrodes is at infinity (i.e., very far away), the capacitance, C, of the other electrode, which can be considered as isolated, is defined by the equation

$$Q = CV, \quad \text{or} \quad C = \frac{Q}{V} \quad \text{(farads - F)}, \tag{8.1}$$

where V is the potential of the body when charged with a charge Q. For a capacitor with two close electrodes, the capacitance, C, is defined by

$$Q = C(V_Q - V_{-Q}), \quad \text{or} \quad C = \frac{Q}{V_Q - V_{-Q}} \quad \text{(F)}. \tag{8.2}$$

V_Q is the potential of the electrode with charge Q, and V_{-Q} of that with charge $-Q$. Usually, the capacitance does not depend on the charge on the electrodes, nor on the voltage between them (*linear capacitors*). For *nonlinear capacitors* this is not true (e.g., a varactor diode).

• The equivalent capacitance of n capacitors connected is parallel is obtained as

$$C_{\text{equivalent}} = C_1 + C_2 + \ldots + C_n \qquad (\text{F}), \qquad\qquad (8.3)$$

and that of n capacitors connected in series is given by

$$\frac{1}{C_{\text{equivalent}}} = \frac{1}{C_1} + \frac{1}{C_2} + \ldots + \frac{1}{C_n} \qquad \left(\frac{1}{\text{F}}\right). \qquad\qquad (8.4)$$

• A parallel-plate capacitor consists of two parallel metal plates of areas S, a small distance d apart. If the dielectric in the capacitor is homogeneous, of permittivity ϵ, its capacitance is

$$C = \epsilon \frac{S}{d} \qquad (\text{F}). \qquad\qquad (8.5)$$

For a coaxial cable of conductor radii a and b ($a < b$), filled with a homogeneous dielectric of permittivity ϵ, the capacitance per unit length is given by

$$C' = \frac{2\pi\epsilon}{\ln\frac{b}{a}} \qquad (\text{F/m}). \qquad\qquad (8.6)$$

• Electrostatic coupling in multibody systems is described by several quantities, derivable from one another. For a system consisting of a reference body (e.g., the ground) and n bodies with charges $Q_1, \ldots, Q_n$, at corresponding potentials $V_1, \ldots, V_n$, the *coefficients of potential* are defined by n linear equations

$$V_i = a_{i1}Q_1 + a_{i2}Q_2 + \ldots + a_{in}Q_n, \qquad i = 1, 2, \ldots, n. \qquad\qquad (8.7)$$

By solving these equations for Q_i, we obtain the definition of the *coefficients of electrostatic induction, c_{ij},*

$$Q_i = c_{i1}V_1 + c_{i2}V_2 + \ldots + c_{in}V_n, \qquad i = 1, 2, \ldots, n, \qquad\qquad (8.8)$$

and of the *coefficients of capacitance, C_{ij},*

$$Q_i = C_{i1}(V_i - V_1) + C_{i2}(V_i - V_2) + \ldots + C_{ii}V_i + \ldots + C_{in}(V_i - V_n), \ i = 1, 2, \ldots, n. \quad (8.9)$$

The coefficients c_{ij} and C_{ij} are interrelated in a simple manner,

$$C_{ij} = -c_{ij} \ \text{if} \ i \neq j, \ \text{and} \ C_{ii} = c_{i1} + \ldots + c_{in}. \qquad\qquad (8.10)$$

QUESTIONS

S **Q8.1.** A conducting body is situated in a vacuum. Prove that the potential of the body is proportional to its charge. — *Hint: note that the surface charge density at any point of the conductor surface is proportional to the total charge on the body.*

Answer. The surface charge density at any point of the conductor surface is proportional to the total charge on the body. (If this were not so, it would not be possible that the electric field strength be zero inside the conductor for any charge on the body.) Therefore the electric field strength at any point, and therefore also the potential, are proportional to the charge on the body.

Q8.2. Repeat the preceding question if the body is situated in a linear (1) homogeneous and (2) inhomogeneous dielectric. — *(c) For a homogeneous dielectric the statement is true, but not for an inhomogeneous dielectric. (b) The statement is true in both cases. (c) The statement is true only in the second case.*

Q8.3. Two conducting bodies with charges Q and $-Q$ are situated in a homogeneous linear dielectric. Prove that the potential difference between them is proportional to Q. Does the conclusion remain true if the dielectric is inhomogeneous (but still linear)? — *Hint: the answer is based on correct answers to two preceding questions.*

Q8.4. Two conducting bodies with charges Q and $-Q$ are situated in a homogeneous, but nonlinear, dielectric. Is the potential difference between them proportional to Q? — *(a) Yes, because the dielectric is homogeneous. (b) No, because the dielectric is nonlinear. (c) No, because the dielectric is both homogeneous and nonlinear.*

Q8.5. The capacitance of a diode is a function of the voltage between its terminals. Is this a linear or nonlinear capacitor? — *(a) Depends on the value of the voltage. (b) Linear. (c) Nonlinear.*

S **Q8.6.** Prove in your own words that a parallel connection of capacitors is indeed just a single unconventional capacitor. — *Hint: recall that a capacitor is defined as a system of two conducting bodies with equal charges of opposite sign, irrespective of the shape of the bodies, and draw a few capacitors connected in parallel.*

Answer. A parallel connection of capacitors can be visualized as, for example, two hands with small plates at the fingertips placed close to one another. The two hands with the small plates are two conducting bodies. So, this is an ordinary capacitor, except that its electrodes are of a strange shape.

Q8.7. Four metal spheres of radii R are centered at the corners of a square of side length $a = 3R$. Two pairs of the spheres are considered to be the electrodes of two capacitors, and are connected "in series". Is it possible to calculate the equivalent capacitance exactly using Eq. (8.3)? Explain. — *(a) Yes, it does not depend on the positions of the electrodes. (c) No, because the electrodes are not charged with equal charges of opposite sign. (c) Yes, because the formula is valid always.*

Q8.8. A parallel-plate capacitor is connected to a source of voltage V. A dielectric slab is periodically introduced between the capacitor electrodes and taken out. Explain what happens with the capacitor charge. — *(a) Nothing. (b) When the slab is in the capacitor, it is increased. (c) When the slab is in the capacitor, it is decreased.*

S **Q8.9.** Explain in your own words why the capacitance of a capacitor filled with a dielectric is larger than the capacitance of the same capacitor without the dielectric. — *(a) The dielectric* *induces free charges on the electrodes. (b) The dielectric attracts free charges on the electrodes. (c) The polarization charges on the electrode/dielectric interfaces are of opposite sign compared to those on the electrodes, which reduces the potential difference between the electrodes for the same free charge Q and −Q on them.*

Answer. The polarization charges on the electrode/dielectric interfaces are of opposite sign to those on the electrodes. Therefore, for the same free charge Q and $-Q$ on the electrodes, the electric field between the electrodes, and therefore also the potential difference between them, is less than when the dielectric is not present. The capacitance, defined as $C = Q/V$, is larger when the dielectric is present.

Q8.10. A negligibly thin metal foil is introduced between and parallel to the plates of a parallel-plate capacitor, parallel to the plates. Is there any change in the capacitor capacitance? Can it be regarded as a series connection of two capacitors? Explain. — *(a) The capacitance is unchanged, it can be regarded as a series connection of capacitors (which ones?). (b) The capacitance is changed, it cannot be regarded as a series connection. (c) The capacitance is changed, but can be regarded as a series connection.*

Q8.11. Repeat question Q8.10 assuming that the foil is not parallel to the plates. — *The choice of answers is the same as for the preceding question.*

Q8.12. Repeat question Q8.10 assuming that the thickness of the foil is not small. — *(a) The capacitance is increased, can be regarded as a series connection of capacitors (which ones?). (b) The capacitance is decreased, can be regarded as a series connection. (c) The capacitance is not changed, but can be regarded as a series connection.*

Q8.13. A metal foil of thickness a is introduced between the plates of a parallel-plate capacitor that are a distance d $(d > a)$ apart, parallel to the plates. If the area of the foil and the capacitor plates is S, what is the capacitance of the capacitor without, and with, the foil? — *Assume that the dielectric is air. (a) $C_{\text{without}} = \epsilon_0 S/d$, $C_{\text{with}} = \epsilon_0 S/(d + a)$. (b) $C_{\text{without}} = \epsilon_0 S/d$, $C_{\text{with}} = \epsilon_0 S/(d - a)$. (c) $C_{\text{without}} = \epsilon_0 S/d = C_{\text{with}}$.*

S **Q8.14.** Describe the procedure for measuring the coefficients of potential, a_{ij}, in Eq. (8.7). — *Hint: assume one body is charged at a time, while all others are uncharged, and use Eq. (8.7).*

Answer. The definition of the coefficients of potential, a_{ij}, for n bodies at potentials $V_1, V_2, \ldots, V_n$ and with charges $Q_1, Q_2, \ldots, Q_n$, is

$$V_i = a_{i1}Q_1 + a_{i2}Q_2 + \ldots + a_{in}Q_n, \qquad i = 1, 2, \ldots, n.$$

From this definition, these coefficients can be measured as follows. Let us charge only body no.1 with a charge Q_1, and leave all the other bodies uncharged ($Q_2 = Q_3 = \ldots = Q_n = 0$). If we measure the potentials of all the bodies in that case, we can evaluate all the coefficients of potential a_{i1}, $i = 1, 2, \ldots, n$. By repeating this procedure with bodies $2, 3, \ldots, n$, we similarly obtain the other coefficients.

Q8.15. Describe the procedure for measuring the coefficients of electrostatic induction, c_{ij}, in Eq. (8.8). — *Hint: assume all the bodies to be connected to the ground except body 1, which is connected to a source of voltage V_1, etc., and use Eqs. (8.8).*

Q8.16. Describe the procedure for measuring the coefficients of capacitance, C_{ij}, in Eq. (8.9). — *Hint: measurement of C_{ij} is the same as of c_{ij}, and to measure C_{ii} connect all bodies to the same source, of voltage V_i.*

PROBLEMS

P8.1. Two large parallel metal plates of area S are a distance d apart, have equal charges of opposite sign, Q and $-Q$, and the dielectric between the plates is homogeneous. Using Gauss' law, prove that the field between the plates is uniform. Calculate the capacitance of the capacitor per unit area of the plates. — *(a)* $C_{\text{per unit area}} = \epsilon S/d$. *(b)* $C_{\text{per unit area}} = \epsilon/d$. *(c)* $C_{\text{per unit area}} = \epsilon S^2/d$.

S **P8.2.** The permittivity between the plates of a parallel-plate capacitor varies as $\epsilon(x) = \epsilon_0(2 + x/d)$, where x is the distance from one of the plates, and d the distance between the plates. If the area of the plates is S, calculate the capacitance of the capacitor. Determine the volume and surface polarization charges if the plate at $x = 0$ is charged with a charge Q ($Q > 0$), and the other with $-Q$. — *Hint: note that there are both volume and surface polarization charges. (a)* $C = \epsilon_0 S/[d\ln(3/2)]$. *(b)* $C = \epsilon_0 S/d$. *(c)* $C = \dot{\epsilon}_0 S/(3d)$.

Solution. As in the solution to the preceding problem, $D = Q/S$ in the entire dielectric of the capacitor. Since the permittivity is not constant, in this case the magnitude of vector $\boldsymbol{E}$ is not constant. So the potential difference is not simply Ed, but instead

$$
V_Q - V_{-Q} = \int_0^d \frac{D}{\epsilon(x)}\mathrm{d}x = \frac{Q}{\epsilon_0 S}\int_0^d \frac{\mathrm{d}x}{2 + x/d} = \frac{Qd}{\epsilon_0 S}\ln(3/2).
$$

The capacitance of the capacitor is thus

$$
C = \frac{Q}{V_Q - V_{-Q}} = \frac{\epsilon_0 S}{d\ln(3/2)}.
$$

It is assumed that the plate charged with a charge $Q > 0$ is at $x = 0$. The density of surface polarization charge at that plate is given by

$$
\sigma_{0\mathrm{p}} = -P(0) = -[\epsilon(0) - \epsilon_0]\frac{Q}{\epsilon(0)S},
$$

and on the other plate

$$
\sigma_{1\mathrm{p}} = P(d) = [\epsilon(d) - \epsilon_0]\frac{Q}{\epsilon(d)S}.
$$

The polarization charge density inside the dielectric is

$$
\rho_{\mathrm{p}} = -\mathrm{div}\boldsymbol{P}(x) = -\frac{\mathrm{d}}{\mathrm{d}x}\left([\epsilon(x) - \epsilon_0]\frac{Q}{\epsilon(x)S}\right).
$$

P8.3. A parallel plate capacitor with plates of area $S = 100\,\text{cm}^2$ has a two-layer dielectric, as in Fig. P8.3. One layer, of thickness $d_1 = 1\,\text{cm}$, has a relative permittivity $\epsilon_r = 3$, and a dielectric strength five times that of air. The other layer is air, of thickness $d_0 = 0.5\,\text{cm}$. How large a voltage will produce breakdown of the air layer, and how large does the voltage need to be to cause breakdown of the entire capacitor? — *(a)* $V_{\text{breakdown capacitor}} = 50{,}000\ V$. *(b)* $V_{\text{breakdown capacitor}} = 100{,}000\ V$. *(c)* $V_{\text{breakdown capacitor}} = 150{,}000\ V$.

S **P8.4.** A capacitor with an air dielectric was connected briefly to a source of voltage V. After the source had been disconnected, the capacitor was filled with transformer oil. Evaluate the new voltage between the capacitor terminals. — *(a)* $V_1 = V/\epsilon_r$. *(b)* $V_1 = \epsilon_r V$. *(c)* $V_1 = V$.

Solution. Let the capacitance of the capacitor with air dielectric be C_0. Then the capacitance of the capacitor with transformer oil is $C_1 = \epsilon_r C_0$, where ϵ_r is relative permittivity of the transformer oil.

The capacitor was charged with a charge $Q = C_0 V$. This charge remained when it was filled with the transformer oil. So the new voltage between the capacitor terminals is $V_1 = Q/C_1 = V/\epsilon_r$. How is it possible that the voltage is less than V? Explain.

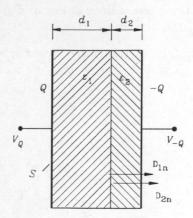   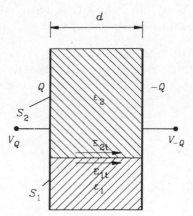

Fig. P8.3. A capacitor with two dielectrics.   **Fig. P8.8.** A capacitor with two dielectrics.

P8.5. A capacitor of capacitance C, with a liquid dielectric of relative permittivity ϵ_r, is connected to a source of voltage V. The source is then disconnected and the dielectric drained from the capacitor. Determine the new voltage between the capacitor electrodes. — *(a)* V/ϵ_r. *(b)* V. *(c)* $\epsilon_r V$.

P8.6. Two conducting bodies with charges Q and $-Q$ are situated in a linear, but inhomogeneous, dielectric. Prove that the potential difference between them is proportional to the charge Q. — *Hint: note that, since the dielectric is linear, the electric field vector at any point of the dielectric is proportional to the charge Q, although it is due both to the free charges Q and $-Q$ on the electrodes, and to the polarization charges over the surfaces and inside the inhomogeneous dielectric.*

P8.7. A parallel-plate capacitor has plates of area S and a dielectric consisting of n layers as in Fig. P8.3, with permittivities $\epsilon_1, \ldots, \epsilon_n$, and thicknesses $d_1, \ldots, d_n$. Evaluate the capacitance of the capacitor. — *(a)* $1/C = 1/C_1 + 1/C_2 + \ldots + 1/C_k + \ldots + 1/C_n$, $C_k = \epsilon_k S/d_k$. *(a)* $C = C_1 + C_2 + \ldots + C_k + \ldots + C_n$, $C_k = \epsilon_k S/d_k$. *(a)* $1/C = 1/C_1 + 1/C_2 + \ldots + 1/C_k + \ldots + 1/C_n$, $C_k = \epsilon_k d_k/S$.

P8.8. Repeat problem P8.7 assuming that the layers are as in Fig. P8.8, and if each layer takes the same amount of the capacitor plate area. — (a) $C = C_1 + \ldots + C_k + \ldots + C_n$, $C_k = \epsilon_k(S/n)/d$. (b) $1/C = 1/C_1 + \ldots + 1/C_k + \ldots + 1/C_n$, $C_k = \epsilon_k(S/n)/d$. (c) $C = C_1 + \ldots + C_k + \ldots + C_n$, $C_k = \epsilon_k(nS)/d$.

P8.9. Evaluate the maximal capacitance of the variable capacitor with n stationary plates, sketched in Fig. P8.9 (where $n = 3$), if the plates are semicircular, of radius R, and the distance between adjacent plates is d. The dielectric is air. — (a) $C_{\max} = n\epsilon_0 R^2 \pi/(2d)$. (b) $C_{\max} = 2n\epsilon_0 R^2 \pi/(2d)$. (c) $C_{\max} = 2(n-1)\epsilon_0 R^2 \pi/(2d)$.

P8.10. Evaluate the capacitance of the capacitor in Fig. P8.10, if the dielectric and aluminum ribbons are $a = 5\,\text{cm}$ wide, $b = 2\,\text{m}$ long, and $d = 0.1\,\text{mm}$ thick. Assume the dielectric has a relative permittivity $\epsilon_r = 2.7$. — *Hint: note that the capacitance of the rolled capacitor is twice that of the unrolled capacitor.* (a) $C = 57.8$ nF. (b) $C = 47.8$ nF. (c) $C = 67.8$ nF.

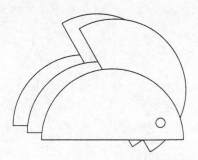

Fig. P8.9. A variable parallel-plate capacitor.

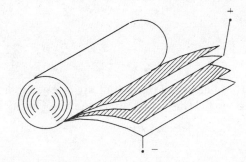

Fig. P8.10. A paper insulated capacitor.

P8.11. Determine the polarization charges on all surfaces in Fig. P8.3. — *Hint: the polarization charges exist on all three boundary surfaces.*

P8.12. Determine the polarization charges on all dielectric surfaces in Fig. P8.12. Are there volume polarization charges anywhere? If so, where? — *Hint: note that vector **D** is radial, and that there are no volume polarization charges (prove the last statement!).*

S **P8.13.** One of two long straight parallel wires is charged with a charge Q' per unit length, and the other with $-Q'$. The wires have radii a and are d ($d \gg a$) apart. (1) Find the expression for the voltage between the wires and the capacitance per unit length of the line. Plot the magnitude of the electric field in a cross section of this two-wire line along the straight line joining the two wires. (2) At which points is it likely that the surrounding air will break down and ionize, given that a high voltage generator is connected to the two wires? (3) If the wire radius is $a = 0.5\,\text{mm}$, and the wires are $d = 1\,\text{cm}$ apart, how large is the voltage of a voltage generator connected to the wires if the air at the wire surfaces breaks down? — (a) $C' = \pi\epsilon_0/[\ln(d/a)]$, $V_{\max} = 8987\,V$ *(for $E_0 = 30\,kV/cm$)*. (b) $C' = 2\pi\epsilon_0/[\ln(d/a)]$, $V_{\max} = 4493\,V$. (c) $C' = 4\pi\epsilon_0/[\ln(d/a)]$, $V_{\max} = 8987\,V$.

Solution. (1) Since the wires are far apart, charge is distributed practically uniformly around their circumference. The electric field of charges on the two wires at a point along the line joining them is in the same direction (from the positively charged wire, and towards the negatively charged wire). We need to integrate these two fields multiplied by dr from a to $(d - a)$. The two integrals are evidently the same, so that

$$V = 2 \int_a^{(d-a)} \frac{Q'}{2\pi\epsilon_0 r} dr \simeq \frac{Q'}{\pi\epsilon_0} \ln \frac{d}{a}.$$

So the capacitance per unit length of the thin two-wire line is

$$C' = \frac{\pi\epsilon_0}{\ln(d/a)}.$$

(2) The surrounding air will break down first at the conductor surfaces, since there the electric field strength is maximal.

(3) If the line is connected to voltage V, the electric field strength at the conductor surfaces is

$$E(a) = \frac{Q'}{2\pi\epsilon_0 a} = \frac{C'V}{2\pi\epsilon_0 a} = \frac{V}{2a \ln(d/a)}.$$

If the dielectric strength of air is E_0, for $E(a) = E_0$ the air around the conductors breaks down. This happens if the voltage between the conductors is

$$V_{\max} = 2aE_0 \ln \frac{d}{a}.$$

For the numerical values given in the problem, $V_{\max} = 8987\,\mathrm{V}$ (for $E_0 = 30\,\mathrm{kV/cm}$).

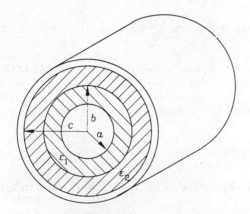

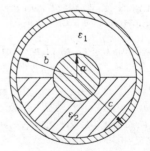

Fig. P8.12. A two-dielectric coaxial cable.   **Fig. P8.14.** A two-dielectric spherical capacitor.

S **P8.14.** A spherical capacitor with two dielectrics is shown in Fig. P8.14. The inner radius is a, the outer radius is b, and the outer radius of the shell is c. The inner sphere is charged with Q ($Q > 0$), and the outer shell with $-Q$. (1) Find the expression for the electric field everywhere and present your result graphically. (2) Find the expression for the capacitance of the capacitor. (3) If the outer shell is made to be much larger than the inner shell, what does the capacitance become and what does this mean physically? — *Hint: note that the field lines are radial and use the generalized Gauss' law.*

Solution. (1) The lines of vectors $\boldsymbol{E}$ and $\boldsymbol{D}$ are radial. Since $\boldsymbol{E}$ is tangential to the boundary surface, it is the same in both dielectrics, and depends on the distance r from the center of the capacitor only. From the generalized Gauss' law in integral form, applied to a sphere of radius r, $(a < r < b)$, we therefore have

$$\oint_S D(r)\mathrm{d}S = 2\pi r^2 \epsilon_1 E(r) + 2\pi r^2 \epsilon_2 E(r) = Q,$$

from which

$$E(r) = \frac{Q}{2\pi(\epsilon_1 + \epsilon_2)r^2} \quad (a < r < b).$$

So, inside the capacitor, $E(r)$ varies as $1/r^2$ with the distance r from the capacitor center. Inside the inner conductor (for $r < a$), inside the shell wall (for $b < r < c$), and outside the capacitor (for $r > c$), $E = 0$.

(2) To determine the capacitance, we need the voltage between the capacitor electrodes. From the last equation,

$$V_a - V_b = \int_a^b E(r)\mathrm{d}r = \frac{Q}{2\pi(\epsilon_1 + \epsilon_2)} \left(\frac{1}{a} - \frac{1}{b} \right),$$

so that the capacitor capacitance is

$$C = C_1 + C_2, \quad C_1 = \frac{2\pi\epsilon_1 ab}{b - a}, \quad C_2 = \frac{2\pi\epsilon_2 ab}{b - a}.$$

The capacitance is as that of two capacitors made by the two capacitor halves, connected in parallel. (Was this expected?)

(3) If the outer shell is of a radius $b \gg a$, the above expressions for C_1 and C_2 reduce to

$$C_1 = 2\pi\epsilon_1 a, \quad C_2 = 2\pi\epsilon_2 a.$$

The capacitance reduces to that of a sphere of radius a embedded half-way in each of the two infinite dielectrics.

P8.15. Two flat parallel conductive plates of surface $S = 0.05\,\mathrm{m}^2$ are charged with $Q_1 = 5 \cdot 10^{-8}\,\mathrm{C}$ and $Q_2 = -Q_1$. The distance between the plates is $D = 1\,\mathrm{cm}$. Find the electric field strength vector at all points if a third, uncharged metal plate, $d = 5\,\mathrm{mm}$ thick, is placed between the two plates $a = 2\,\mathrm{mm}$ away from one of the charged plates and parallel to it. Plot the electric field strength before and after the third plate is inserted. Compare and explain. Find the capacitance between the charged plates without and with the third plate between them. — *(a)* $C_{\text{without plate}} = 44.3\,pF$, $C_{\text{with plate}} = 88.6\,pF$. *(b)* $C_{\text{without plate}} = 22.1\,pF$, $C_{\text{with plate}} = 44.3\,pF$. *(c)* $C_{\text{without plate}} = 11.1\,pF$, $C_{\text{with plate}} = 22.2\,pF$.

P8.16. The dielectric in a parallel-plate capacitor of plate area $S = 100 \, \text{cm}^2$ consists of three parallel layers of relative permittivities $\epsilon_{1r} = 2, \epsilon_{2r} = 3$ and $\epsilon_{3r} = 4$. All three layers are $d = 1 \, \text{mm}$ thick. The capacitor is connected to a voltage $V = 100 \, \text{V}$. (1) Find the capacitance of the capacitor. (2) Find the magnitude of the vectors $\boldsymbol{D}$, $\boldsymbol{E}$, and $\boldsymbol{P}$ in all dielectrics. (3) Find the free and polarization charge densities on all boundary surfaces. — *The capacitance is given by one of the answers to problem P8.7, with $n = 3$.*

P8.17. The surface area of each plate of a parallel-plate capacitor is $S = 100 \, \text{cm}^2$, the distance between the plates is $d = 1 \, \text{mm}$, and it is filled with a liquid dielectric of unknown permittivity. In order to measure the permittivity, we connect the capacitor to a source of voltage $V = 200 \, \text{V}$. When the capacitor is connected to the source, it charges up, and the amount of charge is measured as $Q = 5.23 \cdot 10^{-8} \, \text{C}$ (the instrument that can measure this is called a ballistic galvanometer). Find the relative permittivity of the liquid dielectric. — *(a) $\epsilon_r = Qd/(\epsilon_0 SV)$. (b) $\epsilon_r = VQd/(\epsilon_0 S)$. (c) $\epsilon_r = Q/(\epsilon_0 SVd)$.*

P8.18. We wish to make a coaxial cable which has an electric field of constant magnitude. How does the relative permittivity of the dielectric inside the coaxial cable need to change as a function of radial distance in order to achieve this? The radius of the inner conductor is a and the value of the relative permittivity right next to the inner conductor is $\epsilon_r(a)$. Find the capacitance per unit length of this cable. — *(a) $\epsilon(r) = a^2 \epsilon_r(a)\epsilon_0/r^2$, $C' = 4\pi a \epsilon_r(a)\epsilon_0$. (b) $\epsilon(r) = a\epsilon_r(a)\epsilon_0/r$, $C' = 2\pi a \epsilon_r(a)\epsilon_0$. (c) $\epsilon(r) = (r-a)\epsilon_r(a)\epsilon_0/r$, $C' = 2\pi(r-a)\epsilon_r(a)\epsilon_0$.*

P8.19. A capacitor in the form of rolled metal and insulator foils, Fig. P8.10, needs to have a capacitance of $C = 10 \, \text{nF}$. Aluminum and oily paper foils $a = 3 \, \text{cm}$ wide are available. The thickness of the paper is $d = 0.05 \, \text{mm}$, and its relative permittivity is $\epsilon_r = 3.5$. The thickness of the aluminum foil is also $0.05 \, \text{mm}$. Find the needed length of the foil strips, as well as the maximum voltage to which such a capacitor can be connected. (Note that when rolled, the capacitance of the capacitor is twice that when the strips are not rolled.) — *The length of the strips is: (a) $b = 26.9 \, cm$, (b) $b = 56.9 \, cm$, (c) $b = 36.9 \, cm$.*

S **P8.20.** A coaxial cable has two dielectric layers with relative permittivities $\epsilon_{1r} = 2.5$ and $\epsilon_{2r} = 4$. The inner conductor radius is $a = 5 \, \text{mm}$, and the inner radius of the outer conductor is $b = 25 \, \text{mm}$. (1) Find how the dielectrics need to placed and how thick they need to be in order that the maximum electric field strength be the same in both layers. (2) What is the capacitance per unit length of the cable in this case? (3) What is the largest voltage that the cable can be connected to if the dielectrics have a breakdown field of $200 \, \text{kV/cm}$? — *Hint: note that the inner layer needs to have higher permittivity, and that the maximum electric field strength in the inner layer is for $r = a$, and in the outer layer for $r = c$ (the radius of the boundary between the two dielectrics).*

Solution. (1) Assume a charge Q' per unit length of the inner cable conductor, and a charge $-Q'$ per unit length of the outer conductor. The electric field strength in the two layers is

$$E_1(r) = \frac{Q'}{2\pi\epsilon_1 r}, \qquad E_2(r) = \frac{Q'}{2\pi\epsilon_2 r}.$$

The maximum electric field strength in the inner layer is for $r = a$, and in the outer layer for $r = c$, where c, the radius of the boundary between the two dielectrics, needs to be determined. Evidently, the inner layer needs to have higher permittivity, so that c is determined from the equation

$$\frac{Q'}{2\pi\epsilon_2 a} = \frac{Q'}{2\pi\epsilon_1 c},$$

from which $c = \epsilon_2 a/\epsilon_1$.

(2) The capacitance of the cable per unit length is

$$C = \frac{C_1' C_2'}{C_1' + C_2'}, \qquad C_1' = \frac{2\pi\epsilon_2}{\ln(c/a)}, \qquad C_2' = \frac{2\pi\epsilon_1}{\ln(b/c)}.$$

(3) The greatest field is at $r = a$ in the inner layer, and the same field is at $r = c$ in the outer layer. So the largest voltage is determined by the equation

$$E(a) = \frac{C' V_{\max}}{2\pi\epsilon_2 a} \leq E_0,$$

where E_0 is the dielectric strength of the dielectrics ($200\,\text{kV/cm} = 2 \cdot 10^7 \text{ V/m}$).

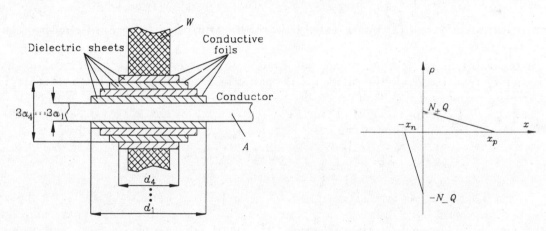

Fig. P8.21. A capacitor bushing.   Fig. P8.22. Linear charge profile in pn diode.

P8.21. Shown in Fig. P8.21 is what is known as *capacitor bushing*, which is used to insulate a high-potential conductor A at its passage through the grounded wall W. The hatched surfaces represent thin dielectric sheets of permittivity ϵ, and the thicker lines represent conducting foils placed between these sheets. Referring to Fig. P8.21, prove that the electric field intensity throughout the bushing is approximately the same, provided that $a_1 d_1 = a_2 d_2 = \ldots = a_4 d_4$. — *Hint: note that the bushing is a series connection of capacitors. The voltage is divided equally between the bushing layers if their capacitances are the same.*

*P8.22. Find the capacitance of a pn diode with a linear charge gradient, i.e., when the charge distribution on the p and n sides is as shown in Fig. P8.22. Assume that the charge on one side is much denser than that of the other side, and can therefore be considered to be a charge sheet. — *Note: this is an example of a space charge. In order that the capacitance can be defined, it is necessary that the total charge of the system is zero, which implies that $N_+ x_p = N_- x_n$. The charge in the positive layer, per unit area, is $Q_s = N_+ Q x_p/2$. The same* *charge per unit area, but of opposite sign, is in the other diode part. To find the capacitance, it is necessary to find the potential difference between the end planes.*

P8.23. Plot the capacitance of a varactor diode as a function of the voltage across the diode. The capacitance of this diode is nonlinear and can be approximated with the following function of the voltage across the diode:

$$C_d(V) = \frac{C_0}{\sqrt{1 + (V/V_d)}},$$

where C_0 is the built-in capacitance (given) and V_d is the built-in voltage of the diode (given). (This diode is used as an electrically-variable capacitor, because its capacitance can change significantly with applied voltage.) — *Hint: note that $C_d(0) = C_0$, $C_d(V_d) = C_0/\sqrt{2}$, and that for $V \gg V_0$ we have that $C_d(V) \simeq C_0 V_d/V$.*

S **P8.24.** Find the expression for the capacitance C_{12} between two bodies in terms of the coefficients of potential a_{ij} defined by Eqs. (8.7). The two bodies have potentials V_1 and V_2, and the reference potential is the ground potential, as in Fig. P8.24. — (a) $C = 1/(a_{11} + a_{22} - 2a_{12})$. (b) $C = a_{11} + a_{22} - 2a_{12}$. (c) $C = 2/(a_{11} + a_{22} + 2a_{12})$.

Solution. The equations (8.13) in the case of two bodies with equal charges of opposite sign (a capacitor) reduce to

$$V_1 = a_{11}Q - a_{12}Q = (a_{11} - a_{12})Q, \qquad V_2 = a_{21}Q - a_{22}Q = (a_{21} - a_{22})Q.$$

Thus, since $a_{21} = a_{12}$,

$$C = \frac{Q}{V_1 - V_2} = \frac{1}{a_{11} + a_{22} - 2a_{12}}.$$

S **P8.25.** A two-wire line with charges Q' and $-Q'$ runs parallel to the ground, with the two wires at different heights. The positively charged wire is h_1 above ground, and the negatively charged wire is h_2 above ground. The radii of both wires are a. Find the capacitance per unit length of such a line directly (from the definition of capacitance, and making use of images), and via the coefficients of potential defined by Eq. (8.7), as follows:

(a) Assuming the earth is at zero potential, that the left wire is charged with Q', and that the other is uncharged, find the potential of both wires.

(b) Repeat part (a) if the right wire is charged with $-Q'$ and the left wire is uncharged.

(c) From the preceding and Eqs. (8.7), write down the expressions for the coefficients of potential a_{ij} of the system.

(d) Find the capacitance per unit length of the line.

Solution. To determine the capacitance directly, we need to determine the potential of one and the other conductor due to charges Q' and $-Q'$ on them, taking into account their images. Adopting the ground surface for the reference of potential, and recalling that the potential of a line charge at a point a distance r from the charge is then

$$V(r) = \frac{Q'}{2\pi\epsilon_0} \ln \frac{h}{r},$$

where h is the height of the charge above ground, we find that

$$C' = \frac{2\pi\epsilon_0}{\ln \frac{4h_1 h_2 r_{12}^2}{a^2 r_{1'2}^2}},$$

where r_{12} is the distance between the wires, and $r_{1'2}$ is the distance between the image of wire 1 and wire 2.

To find the capacitance in terms of the coefficients of potential, from the solution to problem P8.24 the capacitance per unit length is

$$C' = \frac{1}{a'_{11} + a'_{22} - 2a'_{12}}.$$

The coefficients of potential per unit length are obtained from their definition:

$$a'_{jj} = \frac{V_j}{Q'_j}, \quad \text{for} \quad Q'_k = 0, \ k \neq j,$$

and

$$a'_{jk} = a'_{kj} = \frac{V_k}{Q'_j}, \quad \text{for} \quad Q'_k = 0, \ k \neq j.$$

From the above formula for the potential, we thus obtain

$$a'_{jj} = \frac{1}{2\pi\epsilon_0} \ln \frac{2h_j}{a}, \quad a'_{jk} = \frac{1}{2\pi\epsilon_0} \ln \frac{r_{jk}}{r_{j'k}},$$

r_{jk} being the distance between conductors j and k, and $r_{j'k}$ the distance between the image of conductor j and conductor k. Substituting the coefficients of potential into the expression for capacitance per unit length, the same expression is obtained as before.

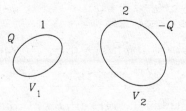

Fig. P8.24. A capacitor above ground.

*P8.26. Prove that Eqs.(8.8) follow from Eqs. (8.9). — *Hint: substitute Eqs. (8.10) in Eqs. (8.9) and group the terms.*

*P8.27. Prove that from Eqs. (8.10) it follows that $c_{ij} = -C_{ij}$, and $c_{ii} = C_{i1} + \ldots + C_{in}$. — *Hint: note that $c_{ij} = -C_{ij}$, $i \neq j$, and substitute these values into the expression for C_{ii}.*

9. Energy, Forces, and Pressure in the Electrostatic Field

• By average human standards, electric energy, forces and pressures are small. Nevertheless, they are of high practical importance. For example, electrical forces are used in extracting solid particles from smoke or dusty air, or spreading of the toner in xerographic copying machines. Sufficient electric energy to destroy virtually any semiconductor device can easily be created if a person is charged by walking on a carpet.

• A capacitor of capacitance C charged with charges Q and $-Q$ has an energy

$$W_e = \frac{Q^2}{2C} = \frac{1}{2}QV = \frac{1}{2}CV^2 \qquad \text{(joule } - \text{ J)}. \qquad (9.1)$$

• Energy is distributed throughout the electric field with a density

$$w_e = \frac{dW_e}{dv} = \frac{1}{2}\epsilon E^2 \qquad \text{(J/m}^3), \qquad (9.2)$$

so that the total energy contained in the field can be obtained as

$$W_e = \int_v \frac{1}{2}\epsilon E^2 dv \qquad \text{(J)}, \qquad (9.3)$$

where the integral refers to the domain of the entire field.

• Energy contained in the field can also be expressed in terms of charges and potential. For a volume distribution of charge,

$$W_e = \frac{1}{2}\int_{v_{\text{charges}}} \rho V \, dv \qquad \text{(J)}, \qquad (9.4)$$

and for charges $Q_1, Q_2, \ldots, Q_n$ distributed over surfaces of conducting bodies at potentials $V_1, V_2, \ldots, V_n$ it is given by

$$W_e = \frac{1}{2}\sum_{k=1}^{n} Q_k V_k \qquad \text{(J)}. \qquad (9.5)$$

• If the distribution of charges is known, the electric forces and moments can be determined from Coulomb's law. In some instances where this distribution is not known, they can be determined from formulas derived from the law of conservation of energy. If we know the electric energy of a system as a function of the position x of a body, and assume that charges on all bodies are constant, the x-component of the electric force on the body is obtained as

$$F_x = -\frac{dW_e(x)}{dx} \qquad \text{(charges kept constant)} \qquad \text{(N)}, \qquad (9.6)$$

and if the potentials of the bodies are kept constant instead,

$$F_x = +\frac{dW_e(x)}{dx} \qquad \text{(potentials kept constant)} \qquad \text{(N)}. \qquad (9.7)$$

• In an electrostatic field, there is pressure of electric forces on all boundary surfaces. It can be evaluated from the formula

$$p = \frac{1}{2}(\epsilon_2 - \epsilon_1)\left(E_{\text{tang}}^2 + \frac{D_{\text{norm}}^2}{\epsilon_1 \epsilon_2}\right) \qquad \text{(directed into medium 1)} \qquad \text{(N/m}^2\text{)}. \qquad (9.8)$$

The pressure on a charged conductor is obtained as

$$p_{\text{on conductor surface}} = \frac{1}{2}\frac{D_{\text{norm}}^2}{\epsilon} = \frac{1}{2}\boldsymbol{E} \cdot \boldsymbol{D} \qquad \text{(N/m}^2\text{)}, \qquad (9.9)$$

and is directed towards the dielectric.

QUESTIONS

S **Q9.1.** What force drives electric charges which form electric current through circuit wires? — (a) *Diffusion force.* (b) *Electric force.* (c) *Gravitational force.*

Answer. The electric force, due to the field of charges distributed along the wires, over their surfaces.

Q9.2. Capacitors of capacitances $C_1, C_2, \ldots, C_n$ are connected (1) in parallel, or (2) in series with a source of voltage V. Determine the energy in the capacitors in both cases. — *Hint: find the equivalent capacitance in both cases and use the formula for the energy of a charged capacitor.*

S **Q9.3.** A parallel-plate capacitor, with an air dielectric, of plate area S, and distance between them d is charged with a fixed charge Q. If the distance between the plates is increased by dx ($dx > 0$), what is the change in electric energy stored in the capacitor? Explain the result. — (a) *It decreased.* (b) *It increased.* (c) *It remained unchanged.*

Answer. Energy contained in the capacitor is $W_e = Q^2/(2C)$. Since $C = \epsilon_0 S/d$, for fixed charge on the plates energy is proportional to the distance between the plates. (Can you explain why?) So, energy will increase if the distance between the plates is increased. Physically, we had to do some work to increase the distance between oppositely charged plates.

Q9.4. Repeat question Q9.3 assuming that $dx < 0$. — *(a) It decreased. (b) It increased. (c) It remained unchanged.*

Q9.5. A parallel-plate capacitor with an air dielectric and capacitance C_0 is charged with a charge Q. The space between the electrodes is then filled with a liquid dielectric of permittivity ϵ. Determine the change in the electrostatic energy stored in the capacitor. Explain the result. — *(a) The energy decreased. (b) The energy increased. (c) The energy remained the same.*

Q9.6. Can the density of electric energy be negative? Explain. — *(c) Only in deep cavities in conductors. (b) No, because energy is always positive. (c) No, because it is given by a square of a quantity.*

S **Q9.7.** If you charge a 1 pF capacitor by connecting it to a source of 100 V, do you think the energy contained in the capacitor can damage a semiconductor device if discharged through it? Explain. — $W_e = CV^2/2 = 0.5 \cdot 10^{-8}$ *J. (a) No, it is too small. (b) Yes, it will damage any semiconductor device. (c) It can damage more sensitive semiconductor devices.*

Answer. The energy contained in the capacitor is $W_e = CV^2/2 = 0.5 \cdot 10^{-8}$ J. This energy looks tiny, but it *can* damage a semiconductor device if dissipated totally inside its very small volume.

Q9.8. If you touch your two hands to the electrodes of a charged high-voltage capacitor, what do you think are the principal dangers to your body? — *(a) It will burn your body. (b) It will heat up your blood. (c) The large electric pulse may damage the mechanism which produces electric pulses that drive the heart.*

Q9.9. Explain in your own words why a polarized dielectric contains energy distributed throughout the dielectric. — *(a) Polarized molecules are deformed and contain energy. (b) Polarized molecules are surrounded by other polarized molecules and therefore contain energy. (c) Because there is energy in a vacuum, between the polarized molecules, which do not contain energy.*

Q9.10. Discuss whether a system of charged bodies can have zero total electric energy. — *(a) Depends on the charge distribution. (b) In no circumstances it can be zero. (c) It can be zero for charges of equal amount and opposite sign.*

Q9.11. Can the electric energy of a system of charges be negative? — *(a) Yes, if the charges are negative. (b) Yes, if there are more negative than positive charges. (c) It cannot be negative.*

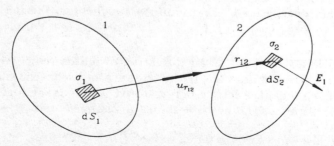

Fig. Q9.12. Two charged bodies with known charge distribution.

Q9.12. Explain in detail how you would calculate approximately the force F_{12} on the two bodies in Fig. Q9.12, assuming that you know the charge distribution on the two bodies. — *(a) Subdivide the two bodies into small surfaces, and perform a double sum. (b) The integral cannot be evaluated approximately. (c) Find the electrical centers of the two bodies and use Coulomb's law.*

Q9.13. If the field induces a dipole moment in a small body, it will also tend to move the body towards the region of stronger field. Sketch an inhomogeneous field and the dipole, and explain. — *Hint: assume that the lines of electric field are directed away from a small region. A body introduced in this field will acquire an induced charge with the negative charge in the higher field.*

S **Q9.14.** Under the influence of electric forces in a system, a body is rotated by a small angle. The system consists of charged insulated conducting bodies. Is the energy of the system after the rotation the same as before, larger than before, or smaller than before? Explain. — *(a) Larger than before. (b) Smaller than before. (c) The same as before.*

Answer. It is smaller than before, because conducting bodies are not connected to sources, and some energy had to be used for rotating the body.

S **Q9.15.** If we say that dW_e is negative, what does this mean? — *(a) That the charges in the system are negative. (b) This small work was done against the electric forces. (c) It can never be negative.*

Answer. This small amount of work was done *against* electric forces, by some external, nonelectric forces.

Q9.16. Is weight a force? If it is, what kind of a force? If it is not, what else might it be? — *(a) It is a force, due to gravitation. (b) It is a force, due to the earth being charged. (c) It is the mass of a body.*

Q9.17. Is it possible to have a system of three point charges that are in equilibrium under the influence of their own mutual electric forces? If you can find such a system, is the equilibrium stable, or unstable? — *(a) It is not possible. (b) It is possible, stable equilibrium (give an example). (c) It is possible, unstable equilibrium (give an example).*

Q9.18. A soap bubble can be viewed as a small stretchable conducting ball. If charged, will it stretch or shrink? Do you think the change in size can be observed? — *(a) It will stretch considerably. (b) It will shrink, observable possibly only by optical means. (c) It will stretch, observable possibly only by optical means.*

Q9.19. Explain why a charged body attracts *uncharged* small bodies of any kind. — *(a) Due to gravitational attraction. (b) Induces dipole moments in the small bodies, with the charge of opposite sign closer to the large body. (c) The small body induces a charge on the large body.*

Q9.20. Assume that of two dielectrics which are in contact the dielectric of permittivity ϵ_1 expanded a distance dx over a surface ΔS at the expense of the other, of permittivity ϵ_2. Assuming that vector E is tangential to the surface, what is the increase of the energy in the system? — *Hint: the energy in the small domain $\Delta S dx$ is obtained by multiplying this volume with the energy density. After the expansion, the energy is therefore different. Note that the increment of a quantity is always defined as its value at the final state, minus its value at the initial state.*

Q9.21. A glass of water is introduced into an arbitrary inhomogeneous electric field. What is the direction of the pressure on the water surface? — *(a) Towards the region of stronger field. (b) Towards the region of weaker field. (c) If the water is not charged, there will be no pressure.*

Q9.22. Derive Eq. (9.9) from Eq. (9.8). — *Hint: formally, a conducting body can be considered as a dielectric with infinite permittivity (why?).*

PROBLEMS

S **P9.1.** A bullet of mass $10\,\mathrm{g}$ is fired with a velocity of $800\,\mathrm{m/s}$. How many high-voltage capacitors of capacitance $1\,\mu\mathrm{F}$ can you charge to a voltage of $10\,\mathrm{kV}$ with the energy of the bullet? — *(a) 28. (b) 64. (c) 142.*

Solution. The kinetic energy of the bullet is $W_\mathrm{k} = mv^2/2 = 10 \cdot 10^{-3} \cdot 800^2/2 = 3200\,\mathrm{J}$. The electric energy in the specified high-voltage capacitor is $W_\mathrm{e} = CV^2/2 = 50\,\mathrm{J}$. So the energy of the bullet is 64 times that of the capacitor.

P9.2. A coaxial cable h long, of inner radius a and outer radius b, is first filled with a liquid dielectric of permittivity ϵ. Then it is connected for a short time to a battery of voltage V. After the battery is disconnected, the dielectric is drained out of the cable. (1) Find the voltage between the cable conductors after the dielectric is drained out of the cable. (2) Find the energy in the cable before and after the dielectric is drained. — *(a) $V_\mathrm{air} = V$, $W_\mathrm{dielectric} = C_\mathrm{dielectric} V^2/2$, $W_\mathrm{air} = \epsilon_r W_\mathrm{dielectric}$. (b) $V_\mathrm{air} = (\epsilon/\epsilon_0)V$, $W_\mathrm{dielectric} = C_\mathrm{dielectric} V^2/2$, $W_\mathrm{air} = \epsilon_r W_\mathrm{dielectric}$. (c) $V_\mathrm{air} = (\epsilon_0/\epsilon)V$, $W_\mathrm{dielectric} = C_\mathrm{dielectric} V^2/2$, $W_\mathrm{air} = W_\mathrm{dielectric}/\epsilon_r$.*

S **P9.3.** A spherical capacitor with an air dielectric, of electrode radii $a = 10\,\mathrm{cm}$ and $b = 20\,\mathrm{cm}$, is charged with a maximum charge for which there is still no air breakdown around the inner electrode of the capacitor. Determine the electric energy of the system. — *(a) $W_\mathrm{e\ max} = 0.25\,J$. (b) $W_\mathrm{e\ max} = 0.35\,J$. (c) $W_\mathrm{e\ max} = 0.45\,J$.*

Solution. The maximum charge Q_max is determined from

$$E(a) = \frac{Q_\mathrm{max}}{4\pi\epsilon_0 a^2} = E_0,$$

where E_0 is the dielectric strength of air. The energy in the capacitor for this charge is

$$W_\mathrm{e\ max} = \frac{Q_\mathrm{max}^2}{2C},$$

where C is the capacitor capacitance,

$$C = \frac{4\pi\epsilon_0 ab}{b - a}.$$

So we find $W_\mathrm{e\ max} = Q_\mathrm{max}^2/(2C) = 0.25\,\mathrm{J}$.

P9.4. Repeat the preceding problem for a coaxial cable of length $d = 10\,\mathrm{km}$, of conductor radii $a = 0.5\,\mathrm{cm}$ and $b = 1.2\,\mathrm{cm}$. — *(a) $W_\mathrm{e\ max} = 34.8\,J$. (b) $W_\mathrm{e\ max} = 44.8\,J$. (c) $W_\mathrm{e\ max} = 54.8\,J$.*

P9.5. Calculate the largest possible electric energy density in air. How does this energy density compare with a $0.5\,J/cm^3$ chemical energy density of a mixture of some fuel and compressed air? — (a) $w_e = 0.398\,J/cm^3$. (b) $w_e = 0.00398\,J/cm^3$. (c) $w_e = 0.0000398\,J/cm^3$.

P9.6. Show that half of the energy inside a coaxial cable with a homogeneous dielectric, of inner conductor radius a and outer conductor radius b, is contained inside a cylinder of radius $a < r < \sqrt{ab}$. — *Hint: assume charges Q' and $-Q'$ on the cable conductors, evaluate the electric field in the cable, and hence the energy per unit length contained between a and $\sqrt{ab}$.*

P9.7. A metal ball of radius $a = 10\,cm$ is placed in distilled water ($\epsilon_r = 81$) and charged with $Q = 10^{-9}\,C$. Find the energy which was used up to charge the ball. — (a) $455\,pJ$. (b) $555\,pJ$. (c) $755\,pJ$.

S   **P9.8.** A dielectric sphere of radius a and permittivity ϵ is situated in a vacuum and is charged throughout its volume with volume density of free charges $\rho(r) = \rho_0 a/r$, where r is the distance from the sphere center. Determine the electric energy of the sphere. — *Hint: the energy can be obtained from either Eq. (9.3), or Eq. (9.4). In both cases it is necessary to find first the electric field strength at all points, which can be done by using the generalized Gauss' law.*

Solution. The electric energy of the sphere is given by

$$W_e = \frac{1}{2}\int_v \rho V dv.$$

We know ρ inside the sphere, but not the potential. To determine the potential, we need to determine the electric field strength.

Due to symmetry, the lines of vector $\boldsymbol{E}$ are radial. Using the generalized integral form of Gauss' law we obtain

$$E(r) = \frac{\rho_0 a}{2\epsilon}\ (r \le a),\qquad E(r) = \frac{\rho_0 a^3}{2\epsilon_0 r^2}\ (r \ge a).$$

The potential at a point a distance r ($r \le a$) from the center of the sphere is

$$V(r) = \int_r^a E(r)dr + \int_a^\infty E(r)dr = \frac{\rho_0 a}{2\epsilon}(a - r) + \frac{\rho_0 a^2}{2\epsilon_0},$$

from which the energy of the system is obtained as

$$W_e = \int_0^a \frac{\rho_0 a}{r}\left[\frac{\rho_0 a}{2\epsilon}(a - r) + \frac{\rho_0 a^2}{2\epsilon_0}\right] 4\pi r^2 dr = \frac{\rho_0^2 a^5 \pi}{6\epsilon} + \frac{\rho_0^2 a^5 \pi}{2\epsilon_0}.$$

Since we know the electric field strength at all points, the electric energy can also be calculated as the integral over all field of the electric energy density.

P9.9. Repeat the preceding problem if the volume density of free charges is constant, equal to ρ. — *Hint: use the procedure suggested in the preceding problem.*

P9.10. Inside a hollow metal sphere, of inner radius b and outer radius c, is a metal sphere of radius a. The centers of the two spheres coincide (concentric spheres), and the dielectric is air. If the inner sphere carries a charge Q_1 and the outer sphere a charge Q_2, what is the energy stored in the system? — *Hint: use Eq. (9.5).*

*P9.11. Prove *Thomson's theorem*: the distribution of static charges on conductors is such that the energy of the system of charged conductors is minimal. — *Hint: there is no tangential electric force on charges spread over surfaces of conductors. Consequently, no energy is required in the first approximation to move any small patch of charges an infinitesimal distance along the conductor surface. This means that the first variation of energy corresponding to a slightly disturbed charge distribution is zero, so it is either maximal or minimal.*

*P9.12. Prove that, if an uncharged conductor, or a conductor at potential zero, is introduced in an electrostatic field produced by charges distributed on conducting bodies, the energy of the system decreases. — *Hint: use Eq. (9.5), and assume that the $(n+1)$th body, uncharged or at potential zero, is pulled into the field by electric forces.*

P9.13. An electric dipole of moment p is situated in a uniform electric field E. If the angle between the vectors p and E is α, find the torque of the electric forces acting on the dipole. What do the electric forces tend to do? — *(a) $M = p \times E$, tends to align vectors p and E. (b) $M = p \times E$, tends to orient p to be normal to E. (c) $M = -p \times E$, tends to align p and E.*

P9.14. An electric dipole of moment $p = Qd$ is situated in an electric field of a negative point charge Q_0, at a distance $r \gg d$ from the point charge. If the vector p is oriented towards the point charge, find the total electric force acting on the dipole. — *(a) Attractive, of magnitude $F = |Q_0|p/(2\pi\epsilon_0 r^3)$. (b) Repulsive, of the same magnitude. (c) Attractive, of magnitude $F = |Q_0|p/(4\pi\epsilon_0 r^2)$.*

S **P9.15.** A two-wire line has conductors with radii $a = 3\,\text{mm}$ and the wires are $d = 30\,\text{cm}$ apart. The wires are connected to a voltage generator such that the voltage between them is on the verge of onsetting air ionization. (1) Find the electric energy per unit length of this line. (2) Find the force per unit length acting on each of the line wires. — *(a) $W_e' = C'V_{max}^2/4$, the force is attractive, of magnitude $F' = C'^2 V_{max}^2/(4\pi\epsilon_0 d)$. (b) $W_e' = C'V_{max}^2/2$, the force is attractive, of magnitude $F' = C'^2 V_{max}^2/(2\pi\epsilon_0 d)$. (c) $W_e' = C'V_{max}^2/2$, the force is repulsive, of magnitude $F' = C'^2 V_{max}^2/(2\pi\epsilon_0 d)$. The expression for V_{max} was obtained in problem P8.13.*

Solution. (1) From the result of problem P8.13, the voltage on the verge of onsetting air ionization is

$$V_{max} = 2aE_0 \ln \frac{d}{a},$$

where E_0 is the dielectric strength of air (about $30\,\text{kV/cm}$). The two-wire line capacitance per unit length is (see solution to problem P8.13)

$$C' = \frac{\pi\epsilon_0}{\ln(d/a)}.$$

The electric energy per unit length of the line for maximum voltage between its conductors for no breakdown of air around the wires is

$$W'_e = \frac{1}{2}C'V^2_{\text{max}}.$$

(2) The force on the wires due to their charge is attractive, of magnitude

$$F' = \frac{Q'^2}{2\pi\epsilon_0 d} = \frac{C'^2 V^2_{\text{max}}}{2\pi\epsilon_0 d}.$$

P9.16. A conducting sphere of radius a is cut into two halves, which are pressed together by a spring inside the sphere. The sphere is situated in air and is charged with a charge Q.. Determine the force on the spring due to the charge on the sphere. In particular, if $a = 10\,\text{cm}$, determine the force corresponding to the maximal charge of the sphere in air for which there is no air breakdown on the sphere surface. — Hint: use Eq. (9.9). (a) $F_{x\ \text{max}} = 0.313\,N$. (b) $F_{x\ \text{max}} = 0.213\,N$. (c) $F_{x\ \text{max}} = 0.113\,N$.

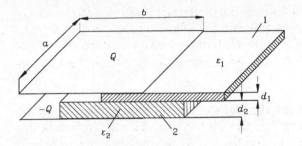

Fig. P9.17. Three-dielectric capacitor.   **Fig. P9.18.** Coaxial cable with sliding conductor.

S **P9.17.** Find the electric force acting on the dielectrics labeled 1 and 2 in a parallel-plate capacitor in Fig. P9.17. The capacitor plates are charged with Q and $-Q$. Neglect edge effects. — Hint: let x be the distance of the slab of permittivity ϵ_1 from the left edge of the capacitor in Fig. P9.17, and y be the distance of the slab of permittivity ϵ_2 from that edge. Find the capacitance, and hence the energy, of the capacitor as a function of x and y, and use Eq. (9.6).

Solution. Let x be the distance of the slab of permittivity ϵ_2 from the left edge of the capacitor in Fig. P9.8, and y be the distance of the slab of permittivity ϵ_1 from that edge. The capacitance of the capacitor is then found to be

$$C(x,y) = \epsilon_0 \frac{xa}{d_1 + d_2} + \frac{\epsilon_0 \epsilon_2 a(y-x)}{\epsilon_0 d_2 + \epsilon_2 d_1} + \frac{\epsilon_1 \epsilon_2 a(b-y)}{\epsilon_1 d_2 + \epsilon_2 d_1}.$$

Assuming the voltage V between the two plates is kept constant, the force on dielectric 1, directed to the right, is obtained as

$$F_1 = \frac{\partial W_e(x,y)}{\partial y} = \frac{1}{2}V^2 \frac{\partial}{\partial y} C(x,y).$$

The force on dielectric 2 is

$$F_2 = \frac{\partial W_e(x,y)}{\partial x} = \frac{1}{2}V^2 \frac{\partial}{\partial x}C(x,y).$$

The evaluation of the two derivatives is straightforward, $C(x,y)$ being a linear function in both x and y.

S **P9.18.** The inner conductor of a coaxial cable shown in Fig. P9.18 can slide along the cylindrical hole inside the dielectric filling. If the cable is connected to a voltage V, find the electric force acting on the inner conductor. — *Hint: use Eq. (9.7).*

Solution. The capacitance of a length L ($L > x$) of the cable is

$$C = \frac{2\pi\epsilon x}{\ln(b/a)} + \frac{2\pi\epsilon_0(L-x)}{\ln(b/a)},$$

so that the energy in the cable for a voltage V between its conductors is

$$W_e(x) = \frac{1}{2}CV^2 = \left[\frac{\pi\epsilon x}{\ln(b/a)} + \frac{\pi\epsilon_0(L-x)}{\ln(b/a)}\right]V^2.$$

The force on the inner cable conductor is hence

$$F_x = \frac{dW_e(x)}{dx} = \frac{\pi(\epsilon - \epsilon_0)V^2}{\ln(b/a)}.$$

P9.19. One of the ends of an air-filled coaxial cable with inner radius $a = 1.2\,\text{mm}$ and an outer radius of $b = 1.5\,\text{mm}$ is dipped into a liquid dielectric. The dielectric has a density of mass equal to $\rho_m = 0.8\,\text{g/cm}^3$, and an unknown permittivity. The cable is connected to a voltage $V = 1000\,\text{V}$. Due to electric forces, the level of liquid dielectric in the cable is $h = 3.29\,\text{cm}$ higher than the level outside of the cable. Find the approximate relative permittivity of the liquid dielectric, assuming the surface of the liquid in the cable is flat. — *Hint: use Eq. (9.7), and note that equilibrium is attained when the gravitational force on the liquid above the surrounding level equals the electric force.* (a) $\epsilon_r = 2.27$. (b) $\epsilon_r = 4.27$. (c) $\epsilon_r = 6.27$.

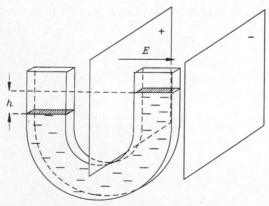

Fig. P9.21. Dielectric tube which is partially between the plates of a parallel-plate capacitor.

P9.20. The end of a coaxial cable is closed by a dielectric piston of permittivity ϵ and length x. The radii of the cable conductors are a and b, and the dielectric in the other part of the cable is air. What is the magnitude and direction of the axial force acting on the dielectric piston, if the potential difference between the conductors is V? — *Hint: this is essentially the same problem as problem P9.19.*

P9.21. One branch of a U-shaped dielectric tube filled with a liquid dielectric of unknown permittivity is situated between the plates of a parallel-plate capacitor (Fig. P9.21). The voltage between the capacitor plates is V, and the distance between them d. The cross section of the U-tube is a very thin rectangle, with the larger side parallel to the electric field intensity vector in the charged capacitor. The dielectric in the tube above the liquid dielectric is air, and the mass density of the liquid dielectric is ρ_m. Assume that h is the measured difference between the levels of the liquid dielectric in the two branches of the U-tube. Determine the permittivity of the dielectric. — *Hint: use Eq. (9.8).* (a) $\epsilon = \epsilon_0 + \rho_m ghd^2/V^2$. (b) $\epsilon = \epsilon_0 + 4\rho_m ghd^2/V^2$. (c) $\epsilon = \epsilon_0 + 2\rho_m ghd^2/V^2$.

P9.22. A soap bubble of radius $R = 2\,cm$ is charged with a maximal charge for which breakdown of air on its surface does not occur. Calculate the electrostatic pressure on the bubble. — (a) $p_{max} = 39.8\,N/m^2$. (b) $p_{max} = 59.8\,N/m^2$. (c) $p_{max} = 69.8\,N/m^2$.

10. Time-Invariant Electric Current in Solid and Liquid Conductors

• "Time-invariant electric current", or "direct current (dc)", implies a steady, time-constant motion of a very large number of small charged particles. Inside conductors, electric current is produced by an electric field, which makes free electric charges acquire an average "drift" velocity along the field lines. A domain in which currents exist is known as the *current field*. When they collide with neutral atoms drifting charges loose the acquired kinetic energy, which is transformed into heat (resistive, or "Joule's losses"). The electric field that produces a dc current is identical to the electrostatic field of charges distributed in the same manner, so that the concepts of scalar potential, voltage, etc., are valid for time-invariant currents.

• In liquid conductors there are pairs of positive and negative *ions*, which move in opposite directions under the influence of the electric field, with an additional effect of *electrolysis*. In gases, the average path length between two collisions is much longer than for solid and liquid conductors, and the mechanism of current flow is quite different.

• Electric current in conductors is described by two quantities. For a conductor with N free charges per unit volume, each carrying a charge Q and having an average (drift) velocity v at a given point, the *current density vector* at that point is defined as

$$\boldsymbol{J} = NQ\boldsymbol{v} \qquad \text{(amperes/m}^2 - \text{A/m}^2\text{)}. \qquad (10.1)$$

The *current intensity*, I, through a surface S is defined as

$$I = \frac{dQ_{\text{through } S \text{ in } dt}}{dt} \qquad \text{(C/s = A)}, \qquad (10.2)$$

which can also be expressed in terms of the current density vector,

$$I = \int_S \mathbf{J} \cdot \mathrm{d}\mathbf{S} \qquad \text{(A)}. \qquad\qquad (10.3)$$

- The vectors $\mathbf{J}$ and $\mathbf{E}$ in solid and liquid conductors in most cases are connected by the relationship

$$\mathbf{J} = \sigma\mathbf{E} \qquad (\sigma : \text{ siemens per meter } - \text{S/m}), \qquad\qquad (10.4)$$

known as the *point (local) form of Ohm's law.* Conductors for which Eq. (10.4) is valid are called *linear conductors.* The constant σ is known as the *conductivity* of the conductor. Its reciprocal is the conductor *resistivity,* ρ (ohm-meter — $\Omega{\cdot}$m), so that $\mathbf{E} = \rho\mathbf{J}$.

- Joule's law in point form reads

$$p_J = \frac{\mathrm{d}P_J}{\mathrm{d}v} = \mathbf{J} \cdot \mathbf{E} = \frac{J^2}{\sigma} = \sigma E^2 \qquad (\text{watt/m}^3 - \text{W/m}^3). \qquad\qquad (10.5)$$

If we wish to determine the power of Joule's losses in a domain of space, we integrate the above power density over that domain,

$$P_J = \int_v \mathbf{J} \cdot \mathbf{E}\,\mathrm{d}v \qquad (\text{watt} - \text{W}). \qquad\qquad (10.6)$$

- Experiments tell us that electric charge cannot be created or destroyed. This is the *law of conservation of electric charge.* The *continuity equation* is the mathematical expression of this law, and relates to a closed surface in the current field:

$$\oint_S \mathbf{J} \cdot \mathrm{d}\mathbf{S} = -\frac{\mathrm{d}}{\mathrm{d}t} \int_v \rho(t)\,\mathrm{d}v. \qquad\qquad (10.7)$$

If the surface is time-invariant (the most frequent case), this becomes

$$\oint_S \mathbf{J} \cdot \mathrm{d}\mathbf{S} = -\int_v \frac{\partial \rho(t)}{\partial t}\,\mathrm{d}v. \qquad\qquad (10.8)$$

If the current field is constant in time, this simplifies to

$$\oint_S \mathbf{J} \cdot \mathrm{d}\mathbf{S} = 0, \qquad\qquad (10.9)$$

which is known as the generalized *Kirchhoff Current Law* from circuit theory,

$$\sum_{k=1}^{n} I_k = 0. \tag{10.10}$$

- A resistor is a resistive body with two equipotential contacts. For a linear resistor, the potential difference between its terminals is proportional to the current intensity,

$$V_+ - V_- = RI \qquad (R : \text{ohm} - \Omega), \tag{10.11}$$

where R is a constant. This is *Ohm's law*, and R is called the *resistance* of the resistor. The reciprocal of resistance is the *conductance*, G (siemens — S). Joule's law for resistors has the form

$$P = VI = RI^2 = \frac{V^2}{R} \qquad (\text{W}). \tag{10.12}$$

- A frequent resistor is in the form of a wire of length l, cross-sectional area S and resistivity ρ. Its resistance is given by

$$R = \rho \frac{l}{S}. \tag{10.13}$$

- Devices able to separate electric charges and to charge two electrodes are known as *electric generators*. Inside them, there are "impressed" (nonelectric) forces on electric charges. Since they act on electric charges, these forces can also be represented as an *impressed electric field*, E_i, defined by

$$F_{\text{impressed}} = QE_i. \tag{10.14}$$

- The *electromotive force*, or briefly *emf*, of a generator is defined as the work done by the impressed forces in taking a unit charge through the generator, from its negative to its positive terminal. It is also equal to the *open-circuit voltage of the generator*,

$$\mathcal{E} = \left\{ \int_+^- E \cdot dl \right\}_{\text{any path}} = V_+ - V_- \qquad (\text{V}). \tag{10.15}$$

Since the generator is always made of a material with nonzero resistivity, it has an *internal resistance*.

- At the boundary surface between two media in a current field the following boundary conditions hold:

$$J_{1n} = J_{2n}, \quad \text{or} \quad \sigma_1 E_{1n} = \sigma_2 E_{2n}, \tag{10.16}$$

$$E_{1t} = E_{2t}, \text{ or } \frac{J_{1t}}{\sigma_1} = \frac{J_{2t}}{\sigma_2}. \tag{10.17}$$

• A *grounding electrode* is a metal body buried in the ground, connected by a wire to a generator. The other terminal of the generator is connected to another, large grounding electrode. A grounding electrode is characterized by the *grounding resistance*, defined as the ratio of the electrode potential and current intensity in it. The image method enables the air above ground to be replaced by the image of the original electrode, with current *of the same intensity and direction*, and with the entire space filled with the ground.

QUESTIONS

Q10.1. What do you think is the main difference between the motion of a fluid and the motion of charges constituting an electric current in conductors? — *(a) There is no difference. (b) All fluid particles move, while electric current is a motion of free charges inside stationary substance. (c) Due to friction with the tube wall, the velocity of the fluid is not constant over the tube cross section, while the velocity of free charges is constant over conductor cross section.*

S **Q10.2.** Describe in your own words the mechanism of transformation of electric energy into heat in current-carrying conductors. — *(a) It is due to friction between moving free charges. (b) Energy must be spent to detach the charges from the rest of the atom, and is transformed into heat. (c) When a charge accelerated by the electric field is stopped by a neutral atom, its kinetic energy is transferred to the atom, which then vibrates more vigorously.*

Answer. The electric field does work when accelerating free charges in the conductor. When an accelerated charge is stopped by a neutral atom, the kinetic energy of the moving charge is transferred to the atom. The atom starts to vibrate more vigorously, i.e., it becomes warmer.

Q10.3. Is Eq. (10.2) valid also for a closed surface, or must the surface be open? Explain. — *(a) It is valid for any surface. (b) The surface must be open. (c) It is valid for a closed surface also, provided it has no holes in it.*

Q10.4. A closed surface S is situated in the field of time-invariant currents. What is the charge that passes through S in a time interval dt? — *(a) Depends on the surface shape and size. (b) $dQ_{ind} t I_{through\ S} dt$. (c) Zero.*

S **Q10.5.** Is a current intensity on the order of $1\,A$ frequent in engineering applications? Are current densities on the order of $1\,mA/m^2$ or $1\,kA/m^2$ frequent in engineering applications? Explain. — *(a) $1\,A$ is infrequent, $1\,mA/m^2$ is frequent, $1\,kA/m^2$ infrequent. (b) $1\,A$ is frequent, $1\,mA/m^2$ is infrequent, $1\,kA/m^2$ is infrequent. (c) $1\,A$ is frequent, $1\,mA/m^2$ is infrequent, $1\,kA/m^2$ is frequent.*

Answer. A current intensity on the order of $1\,A$ is quite frequent. For example, current intensity in a 100 W electric bulb connected to 110 V is about $1\,A$.

Current density of $1\,mA/m^2$ is quite small. For example, if it exists in a wire of $1\,mm$ cross-sectional area, the current intensity in the wire is only $1\,nA$ ($10^{-9}\,A$). The other current density is much more frequent. In a wire of $1\,mm$ cross-sectional area, the current intensity is then $1\,mA$, which is the order of magnitude of current intensity in electronic circuits.

Q10.6. What is the difference between linear and nonlinear resistors? Can you think of an example of a nonlinear resistor? — *(a) There is no significant difference. (b) For nonlinear resistors the Ohm law, V = RI, cannot be written. (c) For nonlinear resistors, R is a function of the current intensity in the resistor, or of the voltage between its terminals.*

Q10.7. A wire of length l, cross-sectional area S, and resistivity ρ is made to meander very densely. The lengths of the successive parts of the meander are on the order of the wire radius. Is it possible to evaluate the resistance of such a wire accurately using Eq. (10.13)? Explain. — *(a) Yes, if the wire is thin. (b) Yes, if the wire resistance is high. (c) No, because the field in the wire is not uniform.*

Q10.8. Explain in your own words the statement in the equation $A_{\text{electric forces}} = QV = VIt$. — *Hint: recall the definition of the potential.*

S   **Q10.9.** Assume that you made a resistor in the form of an uninsulated metal container (one resistor contact) with a conducting liquid (e.g., tap water with a small amount of salt), and a thin wire dipped into the liquid (the other resistor contact). If you change the level of water, but keep the length of the wire in the liquid constant, will this produce a substantial variation of the resistor resistance? If you change the length of the wire in the liquid, will this produce a substantial variation of the resistor resistance? Explain. — *(a) The principal part of the resistance comes from the size of the container. (b) The principal part of the resistance is due to the small volume of the conducting liquid around the wire. (c) The size of the container and the length of the wire have approximately the same influence on the resistance.*

Answer. The principal part of the resistance is due to the small volume of the conducting liquid around the wire. If we keep the length of the wire dipped in the liquid conductor fixed, changing the container size will change the resistance of the resistor very little. If, however, we dip the wire deeper into the liquid conductor, the resistance will be substantially reduced.

Q10.10. What is meant by "nonelectric forces" acting on electric charges inside electric generators? — *(a) The forces not acting on electric charges. (b) The forces on electric charges not due to the electric field. (c) The forces acting on all particles, charged or uncharged, not due to the electric field.*

Q10.11. List a few types of electric generators, and explain the nonelectric (impressed) forces acting in them. — *Hint: recall various chemical generators, light cells, rotating generators. . . .*

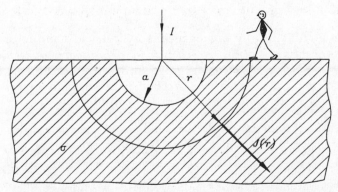

Fig. Q10.14. A hemispherical grounding electrode.

Q10.12. Does the impressed electric field strength describe an *electric* field? What is the unit of the impressed electric field strength? — *(a) It does. Its unit is V/m. (b) It does not. Its unit is V/m. (c) Only in some instances. Its unit is V/m.*

S **Q10.13.** Prove that on a boundary surface in a time-invariant current field, $J_{1n} = J_{2n}$. — *Hint: imagine a small flat cylinder with bases in the two media, and apply to it the current-continuity equation.*

Answer. If the current is time invariant, the surface charge density on the boundary surface is constant in time. This can stay so only if the normal components of the current density vector on the two sides of the boundary are equal.

Q10.14. Where do you think the charges producing the electric field in the ground in Fig. Q10.14 are located? (These charges cause the current flow in the ground.) — *(a) Along the wire feeding the grounding electrode. (b) At the generator terminals. (c) On the boundary hemisphere/ground, and on the boundary ground/air.*

S **Q10.15.** Explain in your own words what is meant by the grounding resistance, and what this resistor is physically. — *(a) Resistance of the grounding electrode. (b) Resistance between the grounding electrode and a distant electrode. (c) Resistance of a distant, collecting electrode.*

Answer. The grounding resistance is the resistance between the grounding electrode and a distant electrode. The resistor having this resistance is the entire earth with the two contact electrodes buried in it. Of course, what matters most is the ground between and in the vicinity of the two electrodes.

Q10.16. Is it possible to define the grounding resistance if the generator is not ground- ed? Explain. — *(a) Yes, it does not depend on the generator grounding state. (b) Yes, if the generator is very far. (c) No, because we need two electrodes for a resistor.*

Q10.17. Assume that there is a large current flowing through the grounding electrode. Propose at least three different ways how to approach the electrode with a minimum danger of electric shock. — *(a) 1. Running (never touching the ground simultaneously by both feet); 2. jumping towards the grounding electrode with the two feed close together; 3. approaching the grounding electrode in a dense spiral, thus avoiding large radial electric field on the surface of the ground. (b) Only the first and the second answer in (a) are acceptable. (c) Only the last answer in (a) is acceptable.*

PROBLEMS

S **P10.1.** Prove that the current in any homogeneous cylindrical conductor is distributed uniformly over the conductor cross section. — *Hint: in a homogeneous cylindrical conductor, the lines of the current density vector inside the cylinder are certainly parallel to the cylinder. Since $E = J/\sigma$, so are the lines of the electric field strength vector. Apply the law of conservation of energy to a rectangular contour in the conductor.*

Solution. In a homogeneous cylindrical conductor, the lines of the current density vector inside the cylinder are certainly parallel to the cylinder. Since $E = J/\sigma$, so are the lines of the electric field strength vector.

Imagine a rectangular contour inside the conductor, with two sides parallel to the cylinder. Since the line integral of E around the contour must be zero, it follows that E is the same along the two sides. The contour being otherwise arbitrary, E (and therefore also J) is the same at all points of the cylindrical conductor.

P10.2. Uniformly distributed charged particles are placed in a liquid dielectric. The number of particles per unit volume is $N = 10^9$ m^{-3}, and each is charged with $Q = 10^{-16}$ C. Calculate the current density and the current magnitude obtained when such a liquid moves with a velocity of $v = 1.2$ m/s through a pipe of cross-sectional area $S = 1$ cm^2. Is this current produced by an electric field? — *(a) $J = 1.2 \cdot 10^{-7}$ A/m^2, $I = 1.2 \cdot 10^{-11}$ A. This current is not produced by an electric field. (b) $J = 1.2 \cdot 10^{-10}$ A/m^2, $I = 1.2 \cdot 10^{-14}$ A. This current is produced by an electric field. (c) $J = 1.8 \cdot 10^{-10}$ A/m^2, $I = 1.8 \cdot 10^{-14}$ A. This current is not produced by an electric field.*

P10.3. A conductive wire has the shape of a hollow cylinder with inner radius a and outer radius b. A current I flows through the wire. Plot the current density as a function of radius, $J(r)$. If the conductivity of the wire is σ, what is the resistance of the wire per unit length? — *Hint: note that inside the cavity and outside the tube $J = 0$. Use Eq. (10.13) for calculating resistance.*

P10.4. A conductor of radius a is connected to one with radius b. If a current I is flowing through the conductor, find the ratio of the current densities and of the densities of Joule's losses in both parts of the conductor if the conductivity for both parts is σ. — *(a) $J_a/J_b = b/a$, $(dP_{Ja}/dv)/(dP_{Jb}/dv) = b^2/a^2$. (b) $J_a/J_b = b^2/a^2$, $(dP_{Ja}/dv)/(dP_{Jb}/dv) = b^4/a^4$. (c) $J_a/J_b = a^2/b^2$, $(dP_{Ja}/dv)/(dP_{Jb}/dv) = a^4/b^4$.*

P10.5. The homogeneous dielectric inside a coaxial cable is not perfect. Therefore, there is some current, $I = 50\,\mu$A, flowing through the dielectric from the inner toward the outer conductor. Plot the current density inside the cable dielectric, if the inner conductor radius is $a = 1$ mm, the outer radius $b = 7$ mm, and the cable length $l = 10$ m. — *(a) Current is along the cable axis, of density $J(r) = I/(2\pi r l)$. (b) Current is radial, of density $J(r) = I/(2\pi r l)$. (c) Current is radial, of density $J(r) = I/(4\pi r^2 l)$.*

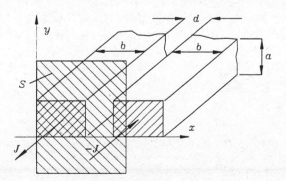

Fig. P10.6. Calculating current intensity.

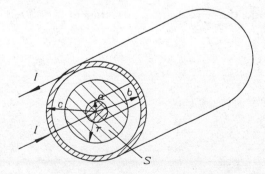

Fig. P10.7. A coaxial cable.

S **P10.6.** Find the expression for the current through the rectangular surface S in Fig. P10.6 as a function of the surface width x. — *Hint: consider the rectangle of width x, with x ranging from zero to $(2b + d)$, and use the definition of current intensity in Eq. (10.3).*

Solution. If $0 \leq x \leq b$, the current through S is

$$I(x) = \int_0^x Ja\,dx = Jax \quad (0 \le x \le b).$$

The value of current remains the same for $b \le x \le (b+d)$, equal to Jab. For $x \ge (b+d)$,

$$I(x) = \int_0^x Ja\,dx = \int_0^b Ja\,dx - \int_{b+d}^x Ja\,dx = Jab - Ja[x.-(b+d)] \quad (b+d \le x \le 2b+d).$$

For $x \ge (2b+d)$, the current intensity is zero. Are the $\le$ and $\ge$ signs appropriate? Explain.

P10.7. Find the expression for the current intensity through a circular surface S shown in Fig. P10.7 for $0 < r < \infty$. — *Hint: consider a circular surface of radius $0 \le r \le \infty$ and use the definition of current intensity in Eq. (10.3).*

S **P10.8.** The resistivity of a wire segment of length l and cross-sectional area S varies along its length as $\rho(x) = \rho_0(1 + x/l)$. Determine the wire segment resistance. — *(a)* $R = 3\rho_0 l/S$. *(b)* $R = 3\rho_0 l/(2S)$. *(c)* $R = \rho_0 l/(2S)$.

Solution. Assume a current I in the wire. The current density vector is parallel to the wire axis and constant along the wire and over its cross section, $J = I/S$. The electric field strength is thus $E(x) = \rho(x)J = \rho_0(1 + x/l)I/S$. The potential difference between the ends of the wire segment is

$$V(l) - V(0) = \int_0^l E(x)\,dx = \frac{3\rho_0 Il}{2S},$$

and the wire segment resistance

$$R = \frac{V(l) - V(0)}{I} = \frac{3\rho_0 l}{2S}.$$

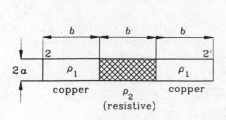

Fig. P10.9. An idealized resistor.

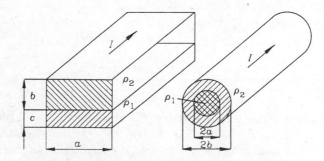

Fig. P10.10. Two inhomogeneous conductors.

P10.9. Find the resistance between points 2 and 2' of the resistor shown in Fig. P10.9. — (a) $R = b(2\rho_1 + \rho_2)/(a^2\pi)$. (b) $R = b(\rho_1 + \rho_2)/(a^2\pi)$. (c) $R = b(\rho_1 + \rho_2)/(2a^2\pi)$.

P10.10. Show that the electric field is uniform in the case of both inhomogeneous conductors in Fig. P10.10. Find the resistance per unit length of these conductors, the ratio of currents in the two layers, and the ratio of Joule's losses in the two layers. — *Hint: note that the tangential electric field on the interface between the two conductors in both cases is the same (boundary condition), and that the lines of vectors **J** and **E** are parallel to the conductors. Therefore the electric field is uniform in both inhomogeneous conductors.*

P10.11. The dielectric in a coaxial cable with inner radius a and outer radius b has a very large, but finite, resistivity ρ. Find the conductance per unit length between the cable conductors. Specifically, find the conductance between the conductors of a cable $L = 1\,\text{km}$ long with $a = 1\,\text{cm}$, $b = 3\,\text{cm}$, and $\rho = 10^{11}\,\Omega\cdot\text{m}$. — (a) $G' = 2.72\cdot10^{-11}\,S/m$. (b) $G' = 4.72\cdot10^{-11}\,S/m$. (c) $G' = 5.72\cdot10^{-11}\,S/m$.

P10.12. A lead battery is shown schematically in Fig. P10.12. The total surface area of the lead plates is $S = 3.2\,\text{dm}^2$, and the distance between the plates is $d = 5\,\text{mm}$. Find the approximate internal resistance of the battery, if the resistivity of the electrolyte is $\rho = 0.016\,\Omega\cdot\text{m}$. — (a) $R = 5\cdot10^{-4}\,\Omega$. (b) $R = 4\cdot10^{-4}\,\Omega$. (c) $R = 6\cdot10^{-4}\,\Omega$.

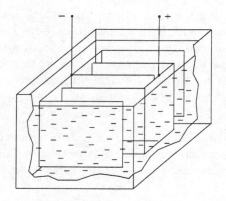

Fig. P10.12. A lead battery.

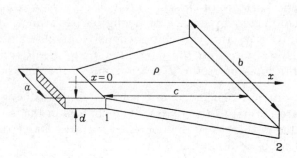

Fig. P10.13. A nonuniform strip conductor.

S **P10.13.** Calculate approximately the resistance between cross sections 1 and 2 of the nonuniform strip conductor sketched in Fig. P10.13. The resistivity of the conductor is ρ. Why can the resistance be calculated only approximately? — (a) $R = [\rho c \ln(b/a)]/[(d-a)d]$. (b) $R = [\rho c \ln(d/a)]/[(b-a)d]$. (c) $R = [\rho c \ln(b/a)]/[(b-a)d]$.

Solution. The resistance can be calculated only approximately because the cross sections 1 and 2 are not strictly equipotential. (Sketch the lines of vector **E** if you do not understand this statement.)

Let a current of intensity I flow through the conductor. The current density in the conical section is approximately given by

$$J(x) = \frac{I}{S(x)}, \quad \text{where} \quad S(x) = ad + \frac{bd - ad}{c}x.$$

The electric field strength

$$E(x) = \rho J(x),$$

whence the potential difference between cross sections 1 and 2 is

$$V_1 - V_2 = \int_0^c E(x)\mathrm{d}x = \frac{\rho c I}{(b-a)d}\ln\frac{b}{a}.$$

Finally, the resistance between cross sections 1 and 2

$$R = \frac{V_1 - V_2}{I} = \frac{\rho c}{(b-a)d}\ln\frac{b}{a}.$$

What does this expression reduce to when $b \to a$?

P10.14. Calculate approximately the resistance between cross sections 1 and 2 of the conical part of the conductor sketched in Fig. P10.14. The resistivity of the conductor is ρ. — *Hint: the solution parallels that of the preceding problem.*

S **P10.15.** In the crosshatched region of the very large conducting slab of conductivity σ and permittivity ϵ_0 shown in Fig. P10.15, a uniform impressed field E_i acts as indicated. The end surfaces of the slab are coated with a conductor of conductivity much greater than σ, and connected by a wire of negligible resistance. Determine the current density, the electric field intensity, and the charge density at all points of the system. Ignore the fringing effect. — *Hint: note that the current density at all points between the end plates is constant, and that the lines of vector J are normal to the plates.*

Solution. The current density at all points between the end plates is constant, and lines of vector J are normal to the plates. The charges in the two hatched regions are propelled by the electric field due to charges on the boundary surfaces, and in the crosshatched region there is the impressed electric field in addition. So we have

$$J = \sigma E_2 \quad \text{(regions with no impressed field), and}$$

$$J = \sigma(E_i + E_1) \quad \text{(region with impressed field).}$$

From these two equations,

$$E_2 = E_i + E_1.$$

The electric field due to charges (E_1 and E_2) satisfies the equation

$$\oint_C \boldsymbol{E} \cdot \mathrm{d}\boldsymbol{l} = E_2(d-a) + E_1 a = 0, \quad \text{from which} \quad E_1 = -\frac{d-a}{a}E_2.$$

Combining the equations, we have

$$E_2 = \frac{a}{d} E_i, \qquad E_1 = -\frac{d-a}{d} E_i.$$

The current density between the end surfaces is therefore

$$J = \sigma \frac{a}{d} E_i = \frac{\sigma}{d}(a E_i).$$

This was to be expected, since $a E_i$ is the electromotive force of the slab generator, and σ/d is the conductance of the slab between the end coatings per unit area.

The surface charges on the four boundary surfaces are as follows. On the two end surfaces they are of opposite sign, and of magnitude (we use the symbol ρ_s instead of σ, the latter being used for conductivity)

$$\rho_s = \epsilon_0 E_2 = \epsilon_0 \frac{a}{d} E_i.$$

On the two boundary surfaces of the generating region they are also of opposite sign, and of magnitude

$$\rho_{s \text{ gen}} = D_2 - D_1 = \epsilon_0(E_2 - E_1) = \epsilon_0 E_i.$$

It is left to the reader to prove that E_1 and E_2 are indeed due to the charges ρ_s and $\rho_{s \text{ gen}}$.

P10.16. Repeat problem P10.15 assuming that the permittivity of the slab is ϵ (different from ϵ_0). Note that in that case polarization charges are also present. *Hint: similar to that for the preceding problem.*

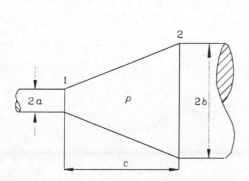

Fig. P10.14. A conical conductor.

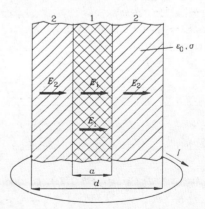

Fig. P10.15. A resistor with impressed field.

S **P10.17.** Determine the resistance of a hemispherical grounding electrode of radius a if the ground is not homogeneous, but has a conductivity σ_1 for $a < r < b$, and σ_2 for $r > b$, where $b > a$, and r is the distance from the grounding electrode center. — (a) $R = [1/(\pi b)]/[(b -$

$a)/(\sigma_1 a)+1/\sigma_2]$. *(b)* $R = [1/(4\pi b)]/[(b-a)/(\sigma_1 a)+1/\sigma_2]$. *(c)* $R = [1/(2\pi b)]/[(b-a)/(\sigma_1 a)+1/\sigma_2]$.

Solution. Assume a current of intensity I to flow from the grounding electrode into the ground. The current density in the ground is radial, and as a function of the distance r from the electrode center, at all points in the ground, is given by the expression

$$J(r) = \frac{I}{2\pi r^2},$$

since $2\pi r^2$ is the area of a hemispherical surface of radius r.

The electric field strength is also radial, of intensity

$$E_1(r) = \frac{I}{2\pi\sigma_1 r^2} \quad (a < r < b), \qquad E_2(r) = \frac{I}{2\pi\sigma_2 r^2} \quad (b < r < \infty).$$

So the potential difference between the surface of the grounding electrode and a very distant electrode at the reference point is

$$V(a) = \int_a^b E_1(r)\mathrm{d}r + \int_b^\infty E_2(r)\mathrm{d}r = \frac{I}{2\pi}\left(\frac{b-a}{\sigma_1 ab} + \frac{1}{\sigma_2 b}\right),$$

from which the grounding resistance of the electrode is

$$R = \frac{1}{2\pi b}\left(\frac{b-a}{\sigma_1 a} + \frac{1}{\sigma_2}\right).$$

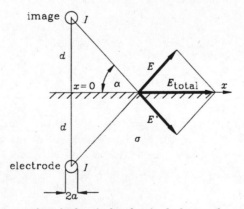

Fig. P10.19. A deeply buried spherical electrode and its image.

P10.18. Determine the resistance between two hemispherical grounding electrodes of radii R_1 and R_2, which are a distance d $(d \gg R_1, R_2)$ apart. The ground conductivity is σ. — *(a)* $R = 1/(\pi\sigma a)$. *(b)* $R = 1/(2\pi\sigma a)$. *(c)* $R = 1/(4\pi\sigma a)$.

P10.19. A grounding sphere of radius a is buried at a depth d $(d \gg a)$ below the surface of the ground of conductivity σ. Determine points at the surface of the ground at which the electric field intensity is the largest. Determine the electric field intensity at these points if the intensity of the current through the grounding sphere is I. — *Hint: make use of the method of images, Fig. P10.19.* (a) $x_{\max} = d/2$, $E_{\text{total max}} = I/(3\pi\sigma d^2)$. (b) $x_{\max} = d/\sqrt{2}$, $E_{\text{total max}} = I/(3\sqrt{3}\pi\sigma d^2)$. (c) $x_{\max} = d/\sqrt{3}$, $E_{\text{total max}} = I/(3\sqrt{2}\pi\sigma d^2)$.

11. Some Applications of Electrostatics

• Thunderstorms are the most obvious manifestations of electrical phenomena on our planet. A typical storm cloud carries about 10-20 coulombs of each type of charge, at an average height of 5 km above the earth's surface. The beginning of a cloud-to-ground lightning is an invisible discharge, which is called the "stepped leader." The stepped-leader air breakdown is initiated at the bottom of the cloud. It moves in discrete steps, each about 50 m long and lasting for about $1\,\mu$s. When the leader is about 100 m above ground, a spark (the visible "return stroke" of only about $100\,\mu$s duration) moves up from the ground to meet it and a conducting channel from the cloud to the ground lights up. The currents in the return stroke range from few kA to as much as 200 kA.

• Electric charges in a vacuum or in gases are propelled by the electric field of stationary charges. The point form of Ohm's law does not hold. In rarefied gases the paths of accelerated ions between two successive collisions are relatively long, so that they can acquire a considerable kinetic energy. As a consequence, various new effects can be produced. The best known is probably a chain production of new pairs of ions by collisions of high-velocity ions with neutral molecules, which may result either in a corona (ionized layer around charged bodies), or in breakdown of the gas (discharge), depending on the structure geometry and voltage.

• For a charge Q of mass m moving in an electrostatic field in a vacuum the equation of motion has the form

$$m\frac{\mathrm{d}^2 \boldsymbol{r}(t)}{\mathrm{d}t^2} = Q\boldsymbol{E}(\boldsymbol{r}),\qquad (11.1)$$

where $\boldsymbol{r}(t)$ is the position vector (variable in time) of the charge, and $\boldsymbol{E}$ is the electric field strength, a function of coordinates (i.e., of $\boldsymbol{r}$). In a uniform electric field along the x axis, for a charge starting to move at $t = 0$ from $x = 0$, this yields

$$x = \frac{QE}{2m}t^2.\qquad (11.2)$$

• Let a particle of charge Q and mass m leave a point 1 at a potential V_1 with a velocity of magnitude v_1. The magnitude of its velocity at a point 2 at potential V_2 is

$$v_2 = \sqrt{v_1^2 + \frac{2Q(V_1 - V_2)}{m}}.\qquad (11.3)$$

• Electrostatic filters are used for removing fine particles from exhaust gases. The particles are charged, separated from the rest of the gas by a strong electric field, and finally attracted to a pollutant-collecting electrode.

• The modern copier machines use the process known as *xerography* (from the Greek words *xeros* — dry and *graphos* — writing). Xerography uses a photosensitive material such as selenium. Selenium is normally a dielectric, but when illuminated, it becomes conductive. A photosensitive plate is first charged over its surface. It is then illuminated by an image of the document. A charge image of the document is thus obtained, with dark places (e.g., letters) charged, the rest of the plate being discharged by light. This is followed by a process of obtaining a copy of the image on a sheet of paper. In order that this be possible, the charge image, i.e., the surface charge of density σ, must remain on the plate for a sufficiently long time. This time is determined by the equation

$$\frac{d\sigma}{dt} + \frac{1}{\epsilon\rho}\sigma = 0, \tag{11.4}$$

where ρ and ϵ are resistivity and permittivity of dark selenium. Assuming that at $t = 0$ the surface charge density is σ_0, the solution of this equation is $\sigma = \sigma_0 e^{-t/(\epsilon\rho)}$. The quantity $(\epsilon\rho)$ is called the *charge transfer time constant*, or *dielectric relaxation constant* of dark selenium, and is similar to the RC time constant of a RC circuit.

• An important application of electrostatic fields is separation, used in industry for purification of food, purification of ores, sorting of reusable wastes and sizing (sorting according to size and weight). Basically, in these processes particles are charged, and then separated by an electric force, or by a combination of an electric force and some other force.

• Four-point probes are commonly used instruments in every semiconductor lab for measuring resistivity of a material, although they are used for other purposes as well. The basic idea is to create an electric field over the surface of the specimen by two current probes, and then to measure the voltage between two convenient points by two other (voltage) probes. In this way the contact between the current probes and the material, which is very difficult to control, practically does not influence the results. Two-point probes have only two probes, which are used both for injecting the current in the material and for measuring the voltage between these two points. As explained, the contact between the probes and the material is not well defined, and therefore two-point probes cannot yield accurate values of the material resistivity. (See also explanations in Lab 3.)

• There are many other applications of electrostatic fields: coating with paint or other material, highly sensitive charge-coupled device (CCD) cameras, non-impact printing (e.g., in ink-jet printers), electrostatic motors, electrostatic generators and electrophoresis (separation of charged colloidal particles by the electric field) used in biology, etc.

QUESTIONS

Q11.1. Describe the formation of a lightning stroke. — *(a) Thunderstorm clouds have large negative charge, and a spark is obtained when the clouds approach the ground. (b) The clouds have a large positive charge, and a spark is obtained when it is initiated somewhere on the ground. (c) A cloud is charged like a vertical dipole, its lower part being a large negative* *charge. The conducting channel starts there, and when it approaches the ground, there is a spark from the ground to that channel.*

Q11.2. How large are currents in a lightning stroke? — *(a) Between about 100 A and 1000 A. (b) Between about 5000 A and 200,000 A. (c) Between about 50,000 A and 1,000,000 A.*

S **Q11.3.** According to which physical law does thunder occur? — *(a) It is produced by the light of the lightning. (b) It is produced by the pressure of air heated up in the return stroke channel. (c) It is produced when the lightning hits the ground.*

Answer. Thunder is the result of rapid heating of the lightning channel, more precisely of rapid increase in air pressure in the channel resulting from this heating.

Q11.4. A spherical cloud of positive charges is allowed to disperse under the influence of its own repulsive forces. Will charges follow the lines of the electric field strength vector? — *(a) No (explain). (b) Yes (explain). (c) Depends on the initial conditions.*

S **Q11.5.** A cloud of identical, charged particles is situated in a vacuum in the gravitational field of the earth. Is there an impressed electric field in addition to the electric field of the charges themselves? Explain. — *(a) Yes, because there are nonelectric forces acting on the particles. (b) No, because there are no nonelectric forces acting on the particles. (c) Depends on the initial velocities of the particles.*

Answer. There is the gravitational force acting on the particles, which is not of electric origin. So, there is an impressed electric field acting on the particles. If the mass of the particles is m and their charge Q, the downward force acting on the particles is $F_i = mg$. The impressed electric field is $E_i = mg/Q$.

Q11.6. Is Eq. (11.1) valid if the charge Q from time to time collides with another particle? — *(a) It is valid at all times. (b) It is not valid at all. (c) It is valid for all time intervals between two collisions.*

Q11.7. Explain why the equation $mv_1^2/2 - mv_2^2/2 = Q(V_2 - V_1)$ is correct, and why $m(v_2 - v_1)^2/2 = Q(V_2 - V_1)$ is not. — *Hint: have in mind the definition of kinetic energy.*

Q11.8. Discuss the validity of Eqs. (11.3) if the charge Q is negative. — *(a) It is valid only for positive charges. (b) For negative charges, Q should be replaced by $|Q|$. (c) It is valid, because the potential difference $(V_1 - V_2)$ is then also negative.*

S **Q11.9.** An electron is emitted parallel to a large uncharged conducting flat plate. Describe qualitatively the motion of the electron. — *(a) It moves parallel to the plate, because the plate is uncharged. (b) It is repelled by the plate (why?). (c) The electron is attracted towards the plate, and eventually hits the plate (why?).*

Answer. Due to the induced charges of opposite sign on the plate, the electron is attracted towards the plate. The attraction increases as the distance between the electron and the plate decreases, and eventually the electron hits the plate.

Q11.10. If the voltage between the electrodes of an air-filled parallel-plate capacitor is increased so that corona starts on the plates, what will eventually happen without increasing the voltage further? — *(a) The corona will remain around the plates. (b) Breakdown of air in the entire capacitor will occur. (c) After a while the corona will disappear.*

Q11.11. Electric charge is continually brought on the inner surface of an isolated hollow metal sphere situated in air. Explain what will happen outside the sphere. — *(a) Nothing, since the sphere is a Faraday cage. (b) Depends on the thickness of the sphere wall (explain). (c) The electric field on the outer surface will increase, and corona will be formed for sufficiently large charge.*

S **Q11.12.** Give a few examples of desirable and undesirable (1) corona and (2) spark discharges. — *Hint: think of lightning rods, sharp spikes on airplanes, high-voltage transmission lines, and sparks between electrodes of spark plugs of internal combustion engines and electrostatic gas lighters, or between bodies in any environment containing explosive gas mixtures.*

Answer. (1) Corona is desirable on airplane pointed parts, to reduce the airplane charge, and at the tip of the lighting rod, to initiate the lightning. It is undesirable along high-voltage transmission lines, because it causes a loss of charge, i.e., of energy. (2) A spark discharge is desirable between the electrodes of spark plugs of internal combustion engines and between the electrodes of electrostatic gas lighters. It is undesirable between charged bodies in any environment containing explosive gas mixtures.

Q11.13. Describe how an electrostatic pollution-control filter works. — *(a) Dust and other particles are charged, so that they remain in filters like those of a vacuum cleaner. (b) Dust and other particles are charged, and are attracted towards the ground, which is always negatively charged. (c) Dust and other particles are charged by convenient means, and then attracted to an oppositely charged electrode.*

Q11.14. Sketch the field that results when an uncharged spherical conductive particle is brought into an originally uniform electric field. — *Hint: correct the sketch in Fig. Q11.14 if you think it needs to be corrected.*

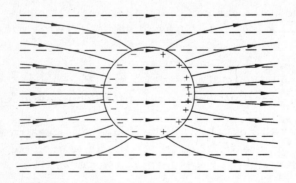

Fig. Q11.14. Lines of originally uniform electric field (dashed) and of the resultant field around an uncharged conducting sphere introduced in the uniform field (solid).

S **Q11.15.** Describe the process of making a xerographic copy. — *(a) Charge a selenium plate, and press the document over it to leave an image. (b) Charge a selenium plate, and expose the reverse image of the document onto the plate, to obtain dark places where the document is white, and conversely, which will produce a charge image on the plate. (c) Charge a selenium plate, expose the image of the document onto the plate, to discharge bright areas, and reproduce this charge image by convenient means.*

Answer. (1) Charge a selenium plate evenly over its surface. (2) Expose the image of the document onto the plate, to discharge bright areas. (3) Attract charged toner particles to the charged image of the document on the selenium plate. (4) Charge paper sheet with opposite charge, and place it over selenium plate to attract toner particles. (5) Bake toner on paper to fix it.

Q11.16. Explain the physical meaning of the charge transfer time constant, or dielectric relaxation constant. — *(a) The time it takes that the charge density at a point goes to zero. (b) The time it takes that the charge density at a point decreases to 1/e of its initial value. (c) The time it takes that the charge density at a point decreases to 0.1 of its initial value.*

Q11.17. Consider a thin film of resistivity ρ and permittivity ϵ residing over a good conductor. Assume there is a surface charge of density $\sigma(t)$ over the free surface of the film. Prove that $\sigma(t)$ satisfies the equation $d\sigma(t)/dt + \sigma(t)/(\epsilon\rho) = 0$. Find the solution of the equation. — *Hint: write the equation of continuity and the generalized Gauss' law to a coinlike surface with one base inside the film, and the other in air, and combine these two equations. Note that in the continuity equation you can write $J = E/\rho$. The solution of the equation is: (a) $\sigma(t) = \sigma_0 e^{-t/\epsilon\rho}$, (b) $\sigma(t) = \sigma_0$, (c) $\sigma(t) = \sigma_0 e^{t/\epsilon\rho}$.*

Q11.18. Describe the difference in the xerox image with and without the developer plate. — *(a) There is no difference. (b) With the developer plate, contours of images are better defined. (c) With the plate, the entire areas of images are covered with toner.*

Q11.19. Derive the equation of particle trajectory in a forming chute electrostatic separation process. — *Hint: the equation is that of the motion of a charged particle.*

S **Q11.20.** Why is a four-point probe measurement more precise than a two-point probe measurement? — *(a) It is not more precise. (b) It is more precise because it has four, instead of two, probes. (c) A two-point probe has inadequately defined radius of electrodes embedded in the substance. A four-point probe measures the potential difference between two points away from these undefined regions.*

Answer. A two-point probe has inadequately defined radius of electrodes touching the substance the resistivity of which we are measuring. A four-point probe measures the potential difference between two points away from these undefined regions, and is therefore much more accurate.

S **Q11.21.** Describe how a CCD camera works.

Answer. The receiving screen consists of a large number of tiny MOS capacitors. The capacitors consist of small metal patches on thin layer of oxide, and the oxide layer is over a p-type semiconductor (silicon). The metal patches are positively charged, and are not tightly packed.

The light photons incident on the interpatch regions pass through the thin oxide layer and create pairs of free charge carriers. The electrons are attracted to the positively charged capacitor electrodes, their number being proportional to the number of photons incident in the vicinity of the electrode. By measuring the number of electrons, it is possible to reconstruct the image.

PROBLEMS

S **P11.1.** Calculate the voltage between two feet of a person (0.5 m apart), standing $r = 20$ m away from a 10 kA lightning stroke, if the moderately wet homogeneous soil conductivity is 10^{-3} S/m. Do the calculations for the two cases when the person is standing in positions (A) and (B) as shown in Fig. P11.1. — *(a) In position A, $[V_1 - V_2]_A \simeq 1940$ V, and in position B it is close to zero. (b) 1940 V, 385 V. (c) 3750 V, approximately zero.*

Solution. The electric field strength on the surface of the ground due to the current of the lightning stroke is

$$E(r) = \frac{I}{2\pi\sigma r^2}.$$

The potential difference between a point a distance r from the point of the stroke, and a point a distance $r + d$ from it, is

$$V_1 - V_2 = \int_r^{r+d} E(r)\mathrm{d}r = \frac{I}{2\pi\sigma}\left(\frac{1}{r} - \frac{1}{r+d}\right).$$

For given data, we thus obtain that, for the person in position A, $[V_1 - V_2]_A \simeq 1940$ V! In position B, the person has feet close together and the voltage is approximately zero.

P11.2. Calculate the electric field strength above a tree that is $d = 1\,\text{km}$ away from the projection of the center of a cloud onto the earth (Fig. P11.2). Assume that because the tree is like a sharp point, the field above the tree is about 100 times that on the flat ground. As earlier, you can assume the cloud is an electric dipole above a perfectly conducting earth, with dimensions as shown in the figure, and with $Q = 4\,\text{C}$ of charge. (Note that the height of any tree is much smaller than the indicated height of the cloud.) — *Let the x axis be directed upward, with the origin at the position of the tree. (a) $E_{\text{tree}} \simeq 184,000\ V/m$. (b) $E_{\text{tree}} \simeq 368,000\ V/m$. (c) $E_{\text{tree}} \simeq 265,000\ V/m$.*

P11.3. Derive the equation $md^2x/dt^2 = QE$ for a x-directed motion of a charged particle. — *Hint: start from Eq. (11.1).*

P11.4. Assuming a nonzero, and x-directed initial velocity of a charged particle situated in a x-directed electric field, find the velocity and the position of the charge Q as a function of time. Plot your results. — *Hint: start from the differential equation of the preceding problem.*

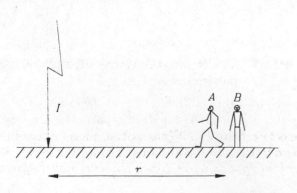

Fig. P11.1. A person near a lightning stroke.

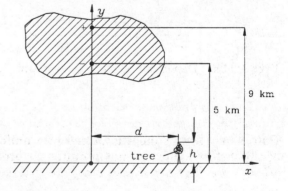

Fig. P11.2. Field above a tree in a storm.

P11.5. Assuming that the initial velocity in problem P11.4 is y-directed, solve for the velocity and the position of the charge Q as a function of time. Plot your results. — *Hint: in this case, Eq. (11.1) needs to be solved for two scalar unknowns, the x and y coordinates of the charge.*

P11.6. A thin electron beam is formed with some convenient electrode system. The electrons in the beam are accelerated by a voltage V_0. The beam passes between two parallel plates, which electrostatically deflect the beam, and later falls on the screen S (Fig. P11.6). Determine and plot the deflection y_0 of the beam as a function of the voltage V between the plates. (This method is used for electrostatic deflection of the electron beam in some cathode-ray tubes.) — (a) $y_0/V = l(L/2+l)/(2V_0d)$. (b) $y_0/V = l(L+l/2)/(V_0d)$. (c) $y_0/V = l(L+l/2)/(2V_0d)$.

S **P11.7.** A beam of charged particles of positive charge Q, mass m, and different velocities, enters between two closely spaced curved metal plates. The distance d between the plates is much smaller than the radius R of their curvature (Fig. P11.7). Determine the velocity v_0 of the particles which are deflected by the electric field between the plates so that they leave the plates without hitting any of them. Note that this is a kind of filter for charged particles, resulting in a beam of particles of the same velocity. — (a) $v_0 = \sqrt{QRV/(md)}$. (b) $v_0 = \sqrt{QRV/(md)}$. (c) $v_0 = \sqrt{QRV/(md)}$.

Solution. In the space between the plates, two forces act on the charged particles. The electric force, F_e, equal in magnitude to QV/d, acts towards the grounded plate. The centrifugal force, F_c, equal in magnitude to mv_0^2/R, is perpendicular to the trajectory of the charges. The forces are of equal magnitudes and in opposite directions. Only those charges will follow a circular arc of radius R which have a velocity v_0 determined by

$$\frac{mv_0^2}{R} = \frac{QV}{d}, \qquad \text{from which} \qquad v_0 = \sqrt{\frac{QRV}{md}}.$$

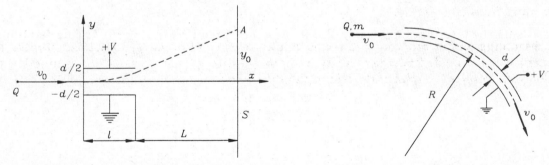

Fig. P11.6. Deflection of an electron beam. **Fig. P11.7.** An electrostatic velocity-filter of charged particles.

P11.8. A metal sphere is placed in a uniform electric field E_0. What is the maximum value of this field which does not produce air breakdown when the metal ball is brought into it? — *Hint: note that the maximum field on the sphere surface is three times the original uniform field.*

P11.9. Calculate the dielectric relaxation constants for selenium, n-doped silicon with carrier concentration $n = 10^{16}$ cm^{-3}, and n-doped galium arsenide with concentration $n = 10^{16}$ cm^{-3}. For semiconductors, such as silicon and galium arsenide, the conductivity is given by $\sigma = Q\mu n$, where Q is the electron charge. μ is a property of electrons inside a material, and it is called the mobility (defined as $v = \mu E$, where v is the velocity of charges that are moved by a field

E). For silicon, $\mu = 0.135\,\text{m}^2/\text{Vs}$ and $\epsilon_r = 12$, and for galium arsenide, $\mu = 0.86\,\text{m}^2/\text{Vs}$ and $\epsilon_r = 11$. For selenium, $\rho = 10^{12}\,\Omega\text{-m}$ and $\epsilon_r = 6.1$. — *Hint: use the definition of the relaxation constant following Eq. (11.4). The following three answers are correct, but you need to associate them with the material:* $\tau_1 = 54\,s$, $\tau_2 = 0.5\,ps$, *and* $\tau_3 = 0.073\,ps$.

P11.10. How far do 1-mm-diameter quartz particles charged with $Q = 1\,\text{pC}$ need to fall in a field $E = 2 \cdot 10^5\,\text{V/m}$ in order to be separated by $0.5\,\text{m}$ in a forming chute separation process? The mass density of quartz is $\rho_m = 2.2\,\text{g/cm}^3$. — *Hint: use the equations of motion for a charged particle in the electric field. (a)* $y = -18.34\,m$. *(b)* $y = -28.25\,m$. *(c)* $y = -7.28\,m$.

S **P11.11.** Find the expression for determining resistivity from a four point probe measurement, as in Fig. P11.11. — *Assume that the dimension of the contacts is very small when compared with a. (a)* $\rho = 2\pi a(V_1 - V_2)/I$. *(b)* $\rho = 4\pi a(V_1 - V_2)/I$. *(c)* $\rho = \pi a(V_1 - V_2)/I$.

Solution. Assume that the dimension of the contacts is very small when compared with a. The potential difference between the two middle contacts, as measured by the voltmeter, is

$$V_1 - V_2 = \int_a^{2a} \left[\frac{\rho I}{2\pi r^2} + \frac{\rho I}{2\pi(3a - r)^2} \right] dr = \frac{\rho I}{2\pi a},$$

from which

$$\rho = \frac{2\pi a(V_1 - V_2)}{I}.$$

P11.12. Using the information given in P11.9, for a measured resistivity of $10\,\Omega\text{-cm}$, determine the corresponding charge concentration of (1) silicon and (2) gallium arsenide. — *One of the following concentrations is that of silicon, and the other of gallium arsenide:* $n_1 = 0.46 \cdot 10^{15}\,cm^{-3}$, $n_2 = 0.0726 \cdot 10^{15}\,cm^{-3}$.

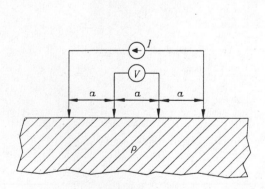

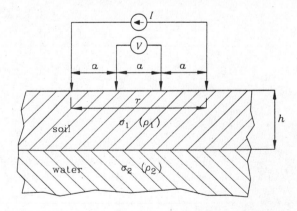

Fig. P11.11. A 4-point probe measurement.   **Fig. P11.13.** A Wenner array used in geology.

P11.13. A Wenner array used in geology is shown in Fig. P11.13. This instrument is used for determining approximately the depth of a water layer under ground. First the electrodes are placed close together, and the resistivity of soil is determined. Then the electrodes are moved further and further apart, until the resistivity measurement changes due to the effect of the water layer. Assuming that the top layer of soil has a very different conductivity than the water layer, what is the approximate spacing between the probes, r, that detects a water layer at depth h under the surface? The exact analysis is complicated, so think of an aproximate qualitative solution. — *Hint: try to conclude what happens with the measured voltage when r is increased from a value much smaller than h to approximately 2h.*

P11.14. A thin film of resistive material is deposited on a perfect insulator. Using a four-point probe measurement, determine the expression for surface resistivity ρ_s of the thin film. Assume the film is very thin. — (a) $\rho_s = 2\pi(V_1 - V_2)/(I \ln 2)$. (b) $\rho_s = 2\pi(V_1 - V_2)/(I \ln 4)$. (c) $\rho_s = \pi(V_1 - V_2)/(I \ln 2)$.

S **P11.15.** Consider an approximate circuit equivalent of a thin resistive film as in Fig. P11.15. The mesh is infinite, and all resistors are equal and have a value of $R = 1\,\Omega$. Using a two-point probe analogy, determine the resistance between any two adjacent nodes A and B in the mesh. — *Hint: assume two current generators of equal, but opposite, currents at the two adjacent nodes and use superposition.* (a) $R/2$. (b) $R/3$. (c) $R/4$.

Solution. Assume we inject a current of intensity I at point A, and collect it at infinity. Due to symmetry, the currents in the four resistors meeting at A are $I/4$ each (reference direction from point A). If a current $-I$ is inserted at B and collected at infinity, the currents in the four resistors meeting at B are $-I/4$ (reference direction from point B). Thus, if a generator of current I is connected between A and B, the current in the resistor joining the two points is $I/2$, and the voltage across the resistor is $RI/2$. The resistance between A and B is hence $R/2$.

P11.16. Find the resistance between nodes (1) A and C and (2) A and D in Fig. P11.15. — *Hint: follow the suggestion given in the preceding problem and use superposition.* (a) $R/2$ in case 1, $R/4$ in case 2. (b) $R/3$ in case 1, $R/2$ in case 2. (c) R in both cases.

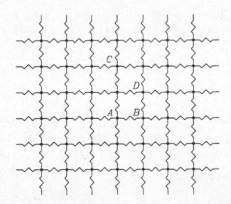

Fig. P11.15. An approximate equivalent circuit of a thin resistive film.

P11.17. Construct an approximate equivalent circuit for a block of homogeneous resistive material. Determine the resistance between two adjacent nodes of the equivalent circuit. — *Hint: assume two current generators of equal, but opposite, currents at the two adjacent nodes and use superposition.* (a) $R/2$. (b) $R/3$. (c) $R/4$.

Part 2: Time-Invariant Magnetic Field

12. Magnetic Field in a Vacuum

• If two charges are moving, there is a force between them in addition to the Coulomb force. This force is known as the *magnetic force*. For just a pair of charges, the magnetic force is extremely small when compared to the Coulomb force. However, the magnetic force between two current-carrying wire loops, where a magnetic interaction between very large number of moving charges occurs, can be much larger than the maximal obtainable electric force between them.

• Short segments of current loops are known as *current elements*. By convention, a current element $I\mathrm{d}l$ is oriented in the reference direction of the current in the loop. The magnetic force exerted by a current element $I_1\mathrm{d}l_1$ on a current element $I_2\mathrm{d}l_2$ a distance r away is given by

$$\mathrm{d}\boldsymbol{F}_{12} = I_2\mathrm{d}l_2 \times \left(\frac{\mu_0}{4\pi}\frac{I_1\mathrm{d}l_1 \times \boldsymbol{u}_r}{r^2}\right), \qquad (12.1)$$

where

$$\mu_0 = 4\pi \cdot 10^{-7} \qquad (\text{henry/m} - \text{H/m}) \qquad (12.2)$$

is known as the *permeability of a vacuum*, or of free space. The unit vector $\boldsymbol{u}_r$ is directed from the "source" element 1 towards element 2.

• If in a domain of space there is a force acting on a current element, or on a moving charge, we say that a *magnetic field* exists in the domain. The fundamental quantity for the description of the magnetic field is the *magnetic flux density vector*. For a single current element, the flux density vector is of the form

$$\mathrm{d}\boldsymbol{B} = \frac{\mu_0}{4\pi}\frac{I\mathrm{d}l \times \boldsymbol{u}_r}{r^2} \qquad (\text{tesla} - \text{T}). \qquad (12.3)$$

This formula is known as the *Biot-Savart law*. The unit vector $\boldsymbol{u}_r$ is directed *from the field source (the current element) towards the field point*. Note that the magnetic flux density vector is perpendicular to the plane of vectors $\boldsymbol{u}_r$ and $\mathrm{d}l$. Its orientation is determined by the right-hand rule when the vector $\mathrm{d}l$ is rotated by the shortest route towards the vector $\boldsymbol{u}_r$.

The magnetic flux density produced by the entire current loop C is given by

$$B = \frac{\mu_0}{4\pi} \oint_C \frac{I \mathrm{d}l \times \boldsymbol{u}_r}{r^2} \qquad \text{(T)}, \tag{12.4}$$

representing another formulation of the Biot-Savart law. For volume and surface currents this equation becomes

$$B = \frac{\mu_0}{4\pi} \int_v \frac{\boldsymbol{J} \times \boldsymbol{u}_r \, \mathrm{d}v}{r^2} \qquad \text{(T)}, \tag{12.5}$$

$$B = \frac{\mu_0}{4\pi} \int_S \frac{\boldsymbol{J}_s \times \boldsymbol{u}_r \, \mathrm{d}S}{r^2} \qquad \text{(T)}. \tag{12.6}$$

• From the definition of the magnetic flux density, it follows that the magnetic force on a current element $I \mathrm{d}l$ in a magnetic field of flux density $\boldsymbol{B}$ is given by

$$\mathrm{d}\boldsymbol{F} = I \mathrm{d}l \times \boldsymbol{B} \qquad \text{(N)}. \tag{12.7}$$

The *magnetic moment* of a flat loop with current I and of are S is defined as

$$\boldsymbol{m} = I\boldsymbol{S} \qquad (\mathrm{A} \cdot \mathrm{m}^2), \tag{12.8}$$

where $\boldsymbol{S} = S\boldsymbol{n}$, and $\boldsymbol{n}$ is the unit vector normal to the loop, its direction obtained by the right-hand rule with respect to the direction of current in the loop. If this loop is situated in a uniform magnetic field, the moment of magnetic forces on the loop is

$$\boldsymbol{M} = \boldsymbol{m} \times \boldsymbol{B} \qquad (\mathrm{N} \cdot \mathrm{m}). \tag{12.9}$$

• The flux of vector $\boldsymbol{B}$ through a surface,

$$\Phi = \int_S \boldsymbol{B} \cdot \mathrm{d}\boldsymbol{S} \qquad \text{(webers} - \text{Wb)} \tag{12.10}$$

is known as the "magnetic flux". The magnetic flux has a very simple and important property — it is zero through *any* closed surface:

$$\oint_S \boldsymbol{B} \cdot \mathrm{d}\boldsymbol{S} = 0. \tag{12.11}$$

This is known as the *law of conservation of magnetic flux*. As a result of this property, the magnetic flux through all surfaces spanned over the same contour is the same. This law also tells us that we cannot separate a S and N pole of a magnet.

• The magnetic force on a charge Q moving with a velocity $\boldsymbol{v}$ in the magnetic field of (local) flux density $\boldsymbol{B}$ is given by

$$\mathbf{F} = Q\mathbf{v} \times \mathbf{B} \qquad \text{(N)}. \qquad\qquad (12.12)$$

If a particle is moving both in an electric and a magnetic field, the total force (often called the *Lorentz force*) on the particle is

$$\mathbf{F} = Q\mathbf{E} + Q\mathbf{v} \times \mathbf{B} \qquad \text{(N)}. \qquad\qquad (12.13)$$

• The magnetic flux density vector $\mathbf{B}$ of time-invariant currents satisfies the *Ampère law* (for a vacuum):

$$\oint_C \mathbf{B} \cdot \mathrm{d}\mathbf{l} = \mu_0 \int_S \mathbf{J} \cdot \mathrm{d}\mathbf{S}. \qquad\qquad (12.14)$$

The reference direction of the vector surface element $\mathrm{d}\mathbf{S}$ of the surface S is adopted according to the right-hand rule with respect to the reference direction of the contour C (i.e., the direction of the contour element $\mathrm{d}\mathbf{l}$). The Ampère law follows from the Biot-Savart law.

QUESTIONS

Q12.1. If μ_0 were defined to have a different value, e.g., $\mu_0 = 1 \cdot 10^{-7}$, what would the expression for the force between two current elements be? — *(a) It would remain the same. (b) The factor 4π in the denominator would not exist. (c) There would be a factor $(4\pi)^2$ in the denominator.*

Q12.2. If we would like to have the term $I_2\mathrm{d}\mathbf{l}_2$ in Eq. (12.1) to be at the end on the right-hand side and not at the beginning, how would the expression read? — *(a) Nothing would change. (b) This cannot be done. (c) The right-hand side would have a minus sign.*

S **Q12.3.** Fig. Q12.3 shows four current elements (the contours they belong to are not shown). Determine the magnetic force between all possible pairs of the elements (a total of twelve expressions). — *Hint: use Eq. (12.1). Note that some are zero, and that in some cases the forces between two elements are not of the same magnitude and in opposite directions.*

Answer. We have first that $\mathrm{d}\mathbf{F}_{12} = \mathrm{d}\mathbf{F}_{23} = \mathrm{d}\mathbf{F}_{32} = \mathrm{d}\mathbf{F}_{43} = 0$. (Explain why in these cases the force is zero!) The rest are not zero, and are obtained from the general expression for the force between two current elements. For example,

$$\mathrm{d}\mathbf{F}_{21} = I_1\mathrm{d}\mathbf{l}_1 \times \left(\frac{\mu_0}{4\pi} \frac{I_2\mathrm{d}\mathbf{l}_2 \times \mathbf{u}_r}{a^2} \right),$$

where a is the distance between elements 2 and 1, and $\mathbf{u}_r$ is directed from element 2 to element 1.

Q12.4. What is the shape of the lines of vector $\mathbf{B}$ of a single current element? Is the magnitude of $\mathbf{B}$ constant along these lines? Is $\mathbf{B}$ constant along these lines? — *(a) Straight lines parallel to the element, the magnitude and vector $\mathbf{B}$ constant along the lines. (b) Circles centered on the line carrying the element, the magnitude of vector $\mathbf{B}$ constant along the line, the vector* *itself not (as in Fig. Q12.3). (c) As in (a), but with straight lines normal to the current element.*

S **Q12.5.** Prove that $I\mathrm{d}l$ for line currents is equivalent to $J\mathrm{d}v$ for volume currents, and to $J_s\mathrm{d}S$ for surface currents. — *Hint: recall the definitions of the three types of current.*

Answer. Assume that the current exists along a thin wire of cross-sectional area S. The product $J\mathrm{d}v$ then becomes $JS\mathrm{d}l$, where $\mathrm{d}l$ is the length of a wire element. We can transfer the vector sign from J to $\mathrm{d}l$, provided that we adopt the direction of $\mathrm{d}l$ to be in the direction of current, as we did. Since $JS = I$, $J\mathrm{d}v$ for a wire becomes $I\mathrm{d}l$.

If the current exists in a very thin layer of thickness d over a surface, for a surface element $\mathrm{d}S$ we have that $J\mathrm{d}v = Jd\,\mathrm{d}S$. By definition, if we let $d \to 0$, the product $Jd = J_s$, and so $J\mathrm{d}v$ for a surface current becomes $J_s\mathrm{d}S$.

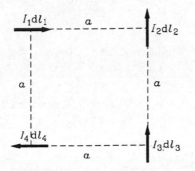

Fig. Q12.3. Four current elements.

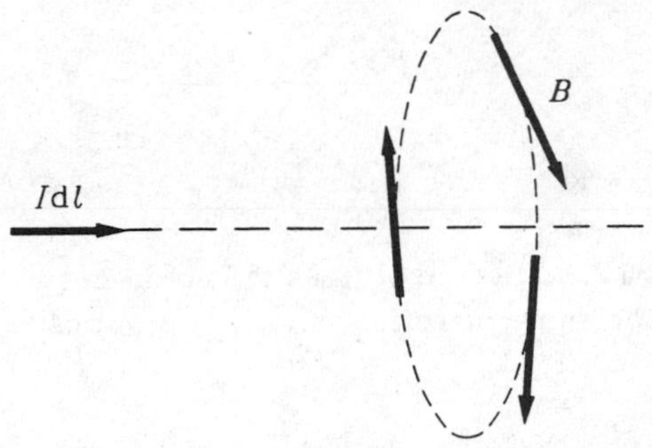

Fig. Q12.4. A line of B of a current element.

Q12.6. Describe an approximate solution of the vector integrals in Eqs. (12.4), (12.5) and (12.6). — *Hint: divide the line, volume and surface into small (but finite) elements.*

Q12.7. Assume that the lines of vector B converge to and are directed toward a point in space. Would that be a realistic magnetic field? — *(a) Yes (why?). (b) No (why?). (c) Yes, if the point is on a magnetic pole.*

Q12.8. Sketch the lines of the magnetic flux density vector for two long, parallel, straight, thin conductors with equal currents when the currents are in the (1) same and (2) opposite directions. — *The sketches are shown in Fig. Q12.8. Discuss if they are correct!*

S **Q12.9.** Prove that if we have N thin wire loops with currents I, connected in series and pressed onto one another, they can be represented as a single loop with a current NI. Is this conclusion valid at all points? — *Hint: use the Biot-Savart law.*

Answer. Consider points not very close to the N pressed loops. To determine B, we have to add up vectorially the contributions of all the N current loops. Since they are pressed tightly together and the points are not close to them, the magnetic flux density is the same as for a single loop with a current NI. If we consider points that are quite close to the N loops or inside them, there will be a difference between the two results. (Why?)

Q12.10. Starting from Eq. (12.7), write the expression for the magnetic force on a closed current loop C with current I. — *Hint: integrate the expression for the force on a single current element. Do not forget that the force is a vector!*

Q12.11. Why do we always obtain two new magnets by cutting a permanent magnet? Why do we not obtain isolated "magnetic charges"? — *(a) Our cutting tools are not fine enough. (b) Because we cannot cut an atom. (c) Because a magnet is an aggregate of small current loops, not of magnetic charges.*

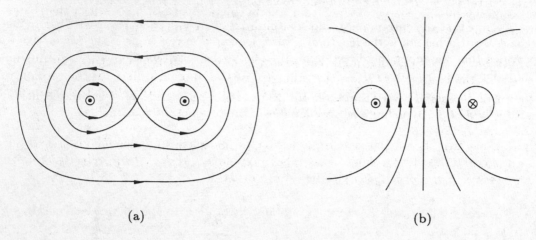

(a)   (b)

Fig. Q12.8. Sketch of lines of vector B in the cross section of two parallel conductors with currents (a) in the same direction, and (b) in opposite directions.

Q12.12. Knowing that the south pole – north pole direction of a compass needle aligns itself with the local direction of the vector B, what is the orientation of elementary current loops in the needle. — *(a) Normal to vector B. (b) Parallel to B, in the same direction. (c) Parallel to B, in the opposite direction.*

Q12.13. In which position is a planar current loop situated in a uniform magnetic field in stable equilibrium? — *Hint: see Eq. (12.9). (a) When vectors B and m have the same direction and sense. (b) When vectors B and m have the same direction, but opposite sense. (c) When vectors B and m are normal to each other.*

Q12.14. A closed surface S encloses a small conducting loop with current I. What is the magnetic flux through S? — *(a) Zero. (b) $\mu_0 I$. (c) Depends on the position of the loop, and on the exact shape of the surface.*

Q12.15. A conductor carrying a current I pierces a closed surface S. What is the magnetic flux through S? — *(a) Zero. (b) $\mu_0 I$. (c) Depends on the position of the loop, and on the exact shape of the surface.*

Q12.16. A straight conductor with a current I passes through the center of a sphere of radius R. What is the magnetic flux through the spherical surface? — *(a) Zero. (b) $\mu_0 I$. (c) $2\mu_0 I$.*

S **Q12.17.** A hemispherical surface of radius R is situated in a uniform magnetic field of flux density B. The axis of the surface makes an angle α with the vector B. Determine the magnetic flux through the surface. — *(a) It is too difficult to determine. (b) Zero. (c) $\Phi = R^2 \pi B \cos \alpha$.*

Answer. To determine the flux directly is relatively tricky. Since the flux through any close surface is zero, the flux is the same through the circular surface of radius R closing the hemispherical surface. So the flux through hemispherical surface is $\Phi = R^2 \pi B \cos \alpha$.

S **Q12.18.** Discuss the possibility of changing the kinetic energy of a charged particle by a magnetic field only. — (a) *It is possible (explain).* (b) *It is possible only if the charge velocity is parallel to vector B.* (c) *It is not possible (explain).*

Answer. The magnetic force on a moving charge, $F_\mathrm{m} = Q\boldsymbol{v} \times \boldsymbol{B}$, is normal to the instantaneous direction of motion of the charge. (Why?) So, the magnetic field can only deflect the charge, but cannot increase or decrease the magnitude of its velocity, i.e., its kinetic energy.

Q12.19. A charge Q is moving along the axis of a circular current-carrying contour normal to the plane of the contour. Discuss the influence of the magnetic field on the motion of the charge. — (a) *It deflects the charge off the axis.* (b) *It does not deflect the charge.* (c) *It deflects the charge only if it is close to the loop.*

Q12.20. An electron beam passes through a region of space undeflected. Is it certain that there is no magnetic field? Explain. — (a) *Yes (explain).* (b) *No, if the velocity is normal to vector B.* (c) *No, if the velocity is parallel to vector B.*

Q12.21. An electron beam is deflected in passing through a region of space. Does this mean that there is a magnetic field in that region? Explain. — (a) *Definitely yes.* (b) *No, because other forces can deflect the beam.* (c) *Yes, if we know that gravitation can be neglected.*

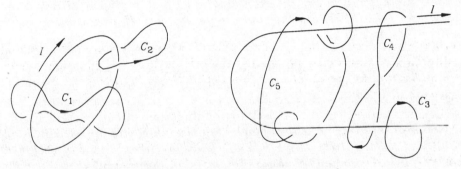

Fig. Q12.26. Contours for Ampère's law.

Q12.22. Does Ampère's law apply to a closed contour in the magnetic field of a single small charge Q moving with a velocity v? Explain. — (a) *Yes, because it is valid for any current in a vacuum.* (b) *No, because this does not represent a current.* (c) *Yes, except that it is a time-varying expression.*

S **Q12.23.** In a certain region of space the magnetic flux density vector B has the same direction at all points, but its magnitude is not constant in the direction perpendicular to its lines. Are there currents in that part of space? Explain. — (a) *There are not enough data for a conclusion.* (b) *No.* (c) *Yes.*

Answer. Assume a rectangular contour with two sides parallel to the lines of vector B. The line integral of vector B along such a contour is not zero. According to Ampère's law, there is a current through the contour.

Q12.24. An infinitely long, straight, cylindrical conductor of rectangular cross section carries a current of intensity I. Is it possible to determine the magnetic flux density inside and outside the conductor starting from Ampère's law? Explain. — (a) *Yes, Ampére's law can be used* *for all cylindrical conductors. (b) No, because the current is not distributed uniformly over its cross section. (c) No, because the structure is not sufficiently symmetrical.*

Q12.25. Can the contour in Ampère's law pass through a current-carrying conductor? Explain. — *(a) The contour must not pass through a conductor. (b) Yes, if the conductor is thin. (c) Yes, without limitations.*

Q12.26. What is the left-hand side in Ampère's law equal to for the five contours in Fig. Q12.26? — *(a) For C_1 it is zero, for C_2 it equals $-I$, for C_3 it is I, for C_4 it is $2I$, and for C_5 it is zero. (b) For C_1 it is I, for C_2 it is zero, for C_3 it is $2I$, for C_4 it is I, and for C_5 it is zero. (c) For C_1 it is zero, for C_2 it equals $-2I$, for C_3 it is $2I$, for C_4 it is I, and for C_5 it is $2I$.*

Q12.27. Compare Gauss' and Ampère's law, and explain the differences and similarities of the two. — *(a) The two laws cannot be compared at all. (b) The two laws are practically the same laws, one for the electrostatic field, the other for the magnetic field. (c) Gauss's law includes the flux of a vector function through a closed surface, Ampère's law the line integral of a vector function along a closed contour.*

PROBLEMS

P12.1. Prove that the magnetic force on a closed wire loop of any form, situated in a uniform magnetic field, is zero. — *Hint: note that the constant vector $\boldsymbol{B}$ in the integral for the force can be taken out of the integral sign.*

S **P12.2.** Prove that the moment of magnetic forces on a closed planar wire loop of arbitrary shape (Fig. P12.2), of area S and with current I, situated in a uniform magnetic field of flux density $\boldsymbol{B}$, is $\boldsymbol{M} = \boldsymbol{m} \times \boldsymbol{B}$, where $\boldsymbol{m} = IS\boldsymbol{n}$, and $\boldsymbol{n}$ is the unit vector normal to S determined according to the right-hand rule with respect to the direction of the current in the loop. — *Hint: the vector $\boldsymbol{B}$ of an arbitrary uniform magnetic field can be decomposed into a component normal to S, and that parallel to S (Fig. P12.2). Only that component of vector $\boldsymbol{B}$ which is parallel to S can result in a moment of magnetic forces on the current loop.*

Solution. Vector $\boldsymbol{B}$ of an arbitrary uniform magnetic field can be decomposed into a component normal to S, and that parallel to S. Elemental forces on current elements due to the component normal to S are in the plane of the loop. Therefore they tend to compress or extend the loop, but cannot turn it. So only that component of vector $\boldsymbol{B}$ which is parallel to S can result in a moment of magnetic forces on the current loop.

Shown in Fig. P12.2 is a loop with the indicated direction of vector $\boldsymbol{B}$ and the x axis normal to that direction. The magnitude of the magnetic force on the indicated current element is $dF_\mathrm{m} = I \, dl \, B \sin \alpha$, where α is the smaller angle between the vectors $d\boldsymbol{l}$ and $\boldsymbol{B}$. For all the elements above the x axis the forces act into the paper, and for those below the x axis they act towards the reader.

The moment of a force dF with respect to the x axis is $dM = r(x) \, dF = I r(x) \, dl \sin \alpha B = IBr(x) \, dx$. The total moment is the sum of the moments on all the current elements. Since the moment on the elements above the x axis tend to turn that part of the loop into the paper, and on the elements below the x axis towards the reader, all elemental moments tend to turn the loop about the x axis in the same direction. Therefore the total moment is the sum of all the elemental moments around the loop,

$$M = \oint_C dM = \oint_C IBr(x)\,dx = IB \oint_C r(x)\,dx = IBS,$$

where S is the area of the loop. It is a simple matter to show that the correct direction of the vector M is obtained if it is written in the form $M = m \times B$, where $S = ISn$. The vector product automatically eliminates the component of vector B normal to S, so the formula has a general value.

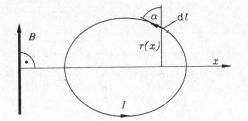

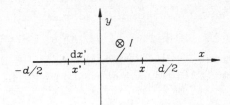

Fig. P12.2. A planar current loop.

Fig. P12.3. Cross section of current strip.

P12.3. Find the magnetic flux density vector at a point A in the plane of a straight current strip (Fig. P12.3). The strip is d wide and a current I flows through it. Assume that point A is x away from the center of the strip, where $x < d/2$. — (a) $B(x) = [(\mu_0 I)/(2\pi d)] \ln[(x + d/2)/(d/2)]$. (b) $B(x) = [(\mu_0 I)/(4\pi d)] \ln[(x + d)/d]$. (c) $B(x) = [(4\mu_0 I)/(\pi d)] \ln[(x + d)/(d/2)]$.

P12.4. A thin dielectric disk of radius $a = 10\,\text{cm}$ has a surface charge density of $\sigma = 2 \cdot 10^{-6}\,\text{C/m}^2$. Find the magnetic flux density at the center of the disk if the disk is rotating at $n = 15,000\,\text{rpm}$ around the axis perpendicular to its surface. — (a) $B = 8.275 \cdot 10^{-11}\,T$. (b) $B = 10.275 \cdot 10^{-11}\,T$. (c) $B = 6.275 \cdot 10^{-11}\,T$.

P12.5. Determine the magnetic moment of a thin triangular loop in the form of an equilateral triangle of side length a with current I. — *Hint: use the general result of problem P12.2.*

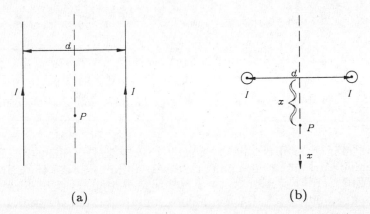

(a)

(b)

Fig. P12.6. Two wires with equal currents. (a) Front view, (b) top view.

P12.6. (1) Find the magnetic flux density vector at point P in the field of two very long straight wires with equal currents I flowing through them. Point P lies in the symmetry plane between the two wires and is x away from the plane defined by the two wires. The front view of the wires is shown in Fig. P12.6a, and the top view in Fig. P12.6b. (2) What is the magnetic flux density equal to at any point in that plane if the current in one wire is I and in the other

$-I$? — *In case (1): (a) $B(x) = \mu_0 I x / \{\pi[x^2 + (d/2)^2]\}$. (b) $B(x) = \mu_0 I d / \{\pi[x^2 + (d/2)^2]\}$.*
(c) $B(x) = 2\mu_0 I d / \{\pi[x^2 + (d/2)^2]\}$.

S **P12.7.** Determine the magnetic flux density along the axis normal to the plane of a circular loop. The loop radius is a and current intensity in it is I. — *Let the z axis be normal to the loop, with the origin at the loop center, as in Fig. P12.7. (a) $B = B_z = 2\mu_0 I a^2 / [2(a^2 + z^2)^{3/2}]$. (b) $\mu_0 I a^2 / [4(a^2 + z^2)^{3/2}]$. (c) $B = B_z = \mu_0 I a^2 / [2(a^2 + z^2)^{3/2}]$.*

Solution. Let the z axis be normal to the loop, with the origin at the loop center. The magnetic flux density due to two symmetrical current elements is along the z axis (Fig. P12.7). So, we have

$$|d\mathbf{B} + d\mathbf{B}'| = 2\frac{\mu_0 I}{4\pi}\frac{dl}{r^2}\cos\alpha = \frac{\mu_0 I}{2\pi}\frac{a\,dl}{(a^2 + z^2)^{3/2}}.$$

The total magnetic flux density is also directed along the z axis. According to the above equation, its magnitude is

$$B = B_z = \frac{\mu_0 I}{2\pi}\frac{a\pi a}{(a^2 + z^2)^{3/2}} = \frac{\mu_0 I a^2}{2(a^2 + z^2)^{3/2}}.$$

What do we obtain for $z = 0$?

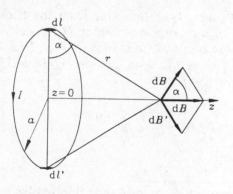

Fig. P12.7. A circular current loop.

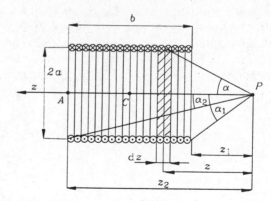

Fig. P12.8. A solenoid of circular cross section.

P12.8. We know the magnetic flux density inside a very long (theoretically infinite) thin solenoid. Usually solenoids are not long enough, so we cannot assume they are infinite. Consider a solenoid of circular cross section with a radius a, b long, and having N turns with a current I flowing through them, Fig. P12.8. (The solenoid is actually a spiral winding, but in the figure it is shown as many closely packed circular loops.)

(a) How many turns are there on a length dx of the coil? We can replace this small piece of the coil with a single circular loop with a current dI. What is dI equal to?

(b) Write the expression for the magnetic flux density $d\mathbf{B}(x)$ of one of the loops with a current dI, at any point P along the axis of the solenoid.

(c) Write the expression for the total magnetic flux density at point P resulting from all of the solenoid turns. (This is an integral.)

(d) Solve the integral. It reduces to a simpler integral if you notice that $x = a/\tan\alpha$ and $a^2 + x^2 = a^2/\sin^2\alpha$.

(e) If the solenoid is thin and long, how much larger is $\boldsymbol{B}$ at point C at the center of the solenoid than that at point A at the edge? Calculate the values of the magnetic flux density at these two points if $I = 2\,\mathrm{A}$, $b/a = 50$, $N = 1000$, and $b = 1\,\mathrm{m}$.

— *Hint: follow the steps suggested in the problem. We give only the general result of part (e). (a) The value of B_z at point C is the same as that at point A. (b) The value of B_z at point C is twice that at point A. (c) The value of B_z at point C is π times that at point A.*

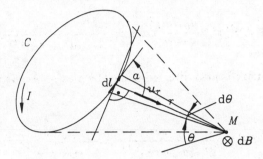

Fig. P12.9. A planar current loop.

S **P12.9.** A closed planar current loop carries a current of intensity I. Starting from the Biot-Savart law, derive the simplified integral expression for the magnitude of vector $\boldsymbol{B}$ for points in the plane of the loop. — *Hint: consider the loop and the point M indicated in Fig. P12.9. According to the Biot-Savart law, all current elements $I\,d\boldsymbol{l}$ of the loop produce vectors $d\boldsymbol{B}$ normal to the plane of the loop, i.e., to the plane of the drawing.*

$$(a)\ B = \frac{\mu_0 I}{4\pi} \oint_C \frac{d\theta}{r}. \quad (b)\ B = \frac{\mu_0 I}{2\pi} \oint_C \frac{d\theta}{r}. \quad (c)\ B = \frac{\mu_0 I}{4\pi} \oint_C \frac{d\theta}{r^2}.$$

Solution. Consider the loop in Fig. P12.9, and the point M indicated. According to the Biot-Savart law, all current elements $I\,d\boldsymbol{l}$ of the loop produce vectors $d\boldsymbol{B}$ normal to the plane of the loop (plane of the drawing). The indicated element creates the vector $d\boldsymbol{B}$ in the direction shown, of magnitude

$$dB = \frac{\mu_0 I}{4\pi} \frac{dl \sin\alpha}{r^2}.$$

From the figure we see that $dl = r\,d\theta/\cos(\pi/2 - \alpha) = r\,d\theta/\sin\alpha$ (the angles θ and α are shown in the figure). The above expression therefore simplifies to

$$dB = \frac{\mu_0 I}{4\pi} \frac{d\theta}{r}. \tag{a}$$

The total magnetic flux density due to a closed current-carrying planar loop C at any point in the plane of the loop is thus

$$B = \frac{\mu_0 I}{4\pi} \oint_C \frac{d\theta}{r}. \qquad\qquad (b)$$

P12.10. Derive the expression for magnitude of vector $\boldsymbol{B}$ of a straight current segment (Fig. P12.10). The segment is a part of a closed current loop, but only the contribution of the segment is required. — (a) $B = [\mu_0 I/(2\pi a)](\sin\theta_2 - \sin\theta_1)$. (b) $B = [\mu_0 I/(4\pi a)](\sin\theta_2 - \sin\theta_1)$. (c) $B = [\mu_0 I/(\pi a)](\sin\theta_2 - \sin\theta_1)$.

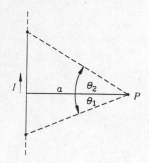

Fig. P12.10. A straight current filament.

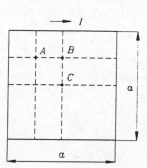

Fig. P12.11. A square current loop.

P12.11. Evaluate the magnetic flux density vector at points A, B, and C in the plane of a square current loop shown in Fig. P12.11. — *Hint: use the result of the preceding problem. Note that $\boldsymbol{B}$ is directed into the paper. For point A one of the following results is correct:* (a) $B_A = \mu_0 I(\sqrt{10} + 2\sqrt{2})/(3\pi a)$. (b) $B_A = 2\mu_0 I(\sqrt{10} + 2\sqrt{2})/(\pi a)$. (c) $B_A = 2\mu_0 I(\sqrt{10} + 2\sqrt{2})/(3\pi a)$.

P12.12. The lengths of wires used to make a square and a circular loop with equal current are the same. Calculate the magnetic flux density at the center of both loops. In which case it is greater? — (a) $B_{\text{circle}}/B_{\text{square}} = \pi^2/(8\sqrt{2}) < 1$. (b) $B_{\text{circle}}/B_{\text{square}} = (8\sqrt{2})/\pi^2 > 1$. (c) $B_{\text{circle}}/B_{\text{square}} = 1$.

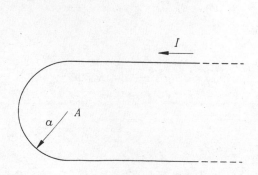

Fig. P12.13. Short-circuited two-wire line.

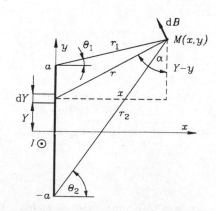

Fig. P12.15. A thin strip with current.

S **P12.13.** Evaluate the magnetic flux density at point A in Fig. P12.13. — *Directed towards the reader. (a) $B_A = \mu_0 I (2+\pi)/(4\pi a)$. (b) $B_A = \mu_0 I(2+\pi)/(4\pi a)$. (c) $B_A = \mu_0 I(2+\pi)/(4\pi a)$.*

Solution. It equals to one half the magnetic flux density at the center of a circular loop of radius a plus one half of the flux density at the midpoint of two infinite current filaments a distance $2a$ apart, both with currents I. So we have that $B_A = \mu_0 I(2+\pi)/(4\pi a)$, and is directed towards the reader.

P12.14. Evaluate the magnetic flux density at point A in the plane of a straight flat thin strip of width d with current I. Assume that the point A is at a distance x $(x > d/2)$ from the center line of the strip. Plot your result as a function of x. — *Hint: use the solution of problem P12.6 as the starting point. What should this expression physically become for $x \gg d/2$? Prove that you get what you expect.*

P12.15. Repeat problem P12.14 for a point in a cross section of the system having coordinates (x, y). Assume the origin at the strip center line, the x axis normal to the strip, and the y axis parallel to the long side of the strip cross section. Plot the magnitude of all components of the magnetic flux density vector as a function of x and y. — *Use the notation shown in Fig. P12.15.*

$$\text{(a)} \quad B_x = \frac{\mu_0 I}{\pi a} \ln \frac{r_1}{r_2}, \qquad B_y = \frac{\mu_0 I}{4\pi a}(\theta_2 - \theta_1).$$

$$\text{(b)} \quad B_x = \frac{\mu_0 I}{2\pi a} \ln \frac{r_1}{r_2}, \qquad B_y = \frac{\mu_0 I}{4\pi a}(\theta_2 - \theta_1).$$

$$\text{(c)} \quad B_x = \frac{\mu_0 I}{4\pi a} \ln \frac{r_1}{r_2}, \qquad B_y = \frac{\mu_0 I}{4\pi a}(\theta_2 - \theta_1).$$

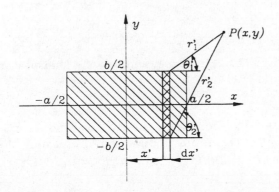

Fig. P12.16. A rectangular conductor.

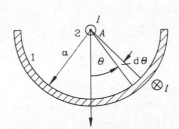

Fig. P12.19. A semicircular conductor.

P12.16. A very long rectangular conductor with current I has sides a (along the x axis) and b (along the y axis). Write the integral determining the magnetic flux density at any point of the xy plane. Do *not* attempt to solve the integral (it is tricky). — *Hint: use the result of the preceding problem and refer to Fig. P12.16.*

P12.17. Determine and plot the magnetic flux density along the axis normal to the plane of a square loop of side a carrying a current I. — *Hint: divide the loop into four straight segments and add the four vectors B at a point of the loop axis (z axis with the origin at the loop center) using the solution of problem P12.10. The resultant vector B is along* *the loop axis, and its magnitude is (a)* $B(z) = 4\mu_0 Ia/[\pi\sqrt{(a^2 + z^2)(2a^2 + 4z^2)}]$. *(b)* $B(z) = 4\mu_0 Ia/[\pi\sqrt{(a^2 + 4z^2)(2a^2 + z^2)}]$. *(c)* $B(z) = 4\mu_0 Ia/[\pi\sqrt{(a^2 + 4z^2)(2a^2 + 4z^2)}]$.

P12.18. A metal spherical shell of radius $a = 10\,\text{cm}$ is charged with a maximal charge that does not initiate the corona on the sphere surface. It rotates about the axis passing through its center with angular velocity $\omega = 50,000\,\text{rad/min}$. Determine the magnetic flux density at the center of the sphere. — *(a)* $B = 2.85\,nT$. *(b)* $B = 3.85\,nT$. *(c)* $B = 1.85\,nT$.

S **P12.19.** A very long straight conductor of semicircular cross section of radius a (Fig. P12.19) carries a current I. Determine the flux density at point A. — *(a)* $B = 2\mu_0 I/(\pi^2 a)$. *(b)* $B = \mu_0 I/(\pi^2 a)$. *(c)* $B = \mu_0 I/(2\pi^2 a)$.

Solution. The problem is solved by subdividing the semicircle into narrow strips of width $a\,d\theta$, with currents $dI = I\,d\theta/\pi$. We need to add the elemental magnetic flux density vectors at point A as vectors. The resultant vector is then found to be directed to the right, and to be of magnitude $B = \mu_0 I/(\pi^2 a)$.

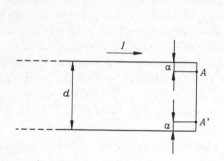

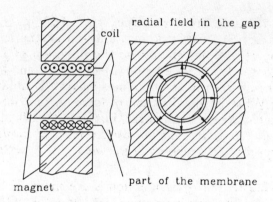

Fig. P12.21. Short-circuited two-wire line.   **Fig. P12.22.** Magnet and coil of a loudspeaker.

P12.20. Assume in Fig. P12.19 that the thin wire 2 extends along the axis of conductor 1. Wire 2 carries the same current I as conductor 1, but in the opposite direction. Determine the magnetic force per unit length on conductor 2. — *(a)* $F' = 2\mu_0 I^2/(\pi^2 a)$. *(b)* $F' = \mu_0 I^2/(\pi^2 a)$. *(c)* $F' = \mu_0 I^2/(2\pi^2 a)$.

S **P12.21.** Determine the magnetic force on the segment $A - A'$ of the two-wire-line short circuit shown in Fig. P12.21. — *Hint: note that the magnetic flux density due to both conductors along the short circuit is the same, equal to one half of that due to an infinite conductor.* (a) $F = (\mu_0 I^2/\pi)\,\ln[(d - a)/a]$. (b) $F = (2\mu_0 I^2/\pi)\,\ln[(d - a)/a]$. (c) $F = [(\mu_0 I^2)/(2\pi)]\,\ln[(d - a)/a]$.

Solution. Since only the total force is required, it can be calculated simply by the following reasoning. The magnetic flux density due to both conductors along the short circuit is the same, equal to one half of that due to an infinite conductor, except that one is the largest near the top, and the other near the bottom conductor. The total force on the short circuit is thus the same as if it were situated in the field of an infinite conductor (one or the other extended to infinity). So we have that the force is directed to the right, and is of intensity (x is the distance of the point from that infinite conductor)

$$F = \int_a^{d-a} \frac{\mu_0 I^2 \, dx}{2\pi x} = \frac{\mu_0 I^2}{2\pi} \ln \frac{d-a}{a}.$$

P12.22. Shown in Fig. P12.22 is a sketch of a permanent magnet used in loudspeakers. The lines of the magnetic flux density vector are radial, and at the position of the coil it has a magnitude $B = 1\,\mathrm{T}$. Determine the magnetic force on the coil (which is glued to the loudspeaker membrane) at the instant when the current in the coil is $I = 0.15\,\mathrm{A}$, in the indicated direction. The number of turns of the coil is $N = 10$, and its radius $a = 0.5\,\mathrm{cm}$. — *The force is towards the left in Fig. P12.22 (left part). (a)* $F = 0.027\,N$. *(b)* $F = 0.037\,N$. *(c)* $F = 0.047\,N$.

P12.23. Prove that the magnetic force on a *segment* of a closed current loop with current I, situated in a uniform magnetic field of flux density B, does not depend on the segment shape but only on the position of its two end points. — *Hint: note that, since the magnetic field is uniform, vector B can be taken outside the integral representing the total force.*

P12.24. A circular current loop of radius a and with current I is cut into halves that are in contact. It is situated in a uniform magnetic field of flux density B normal to the plane of the loop. (1) What should be the direction of B with respect to that of the current in the loop in order that the magnetic force press the loop halves one onto the other? (2) What is the force on each of the two loop halves? Evaluate the force for $a = 10\,\mathrm{cm}$, $I = 2\,\mathrm{A}$, and $B = 1\,\mathrm{T}$. (3) What is the direction of force on the two halves of the loop due to the current in the loop itself? (Neglect this force, but note that it always exists.) — *B should be normal to the plane of the contour, and the direction of B should be defined according to the* left-hand *rule with respect to the current in the loop. (a)* $F = 2aIB$. *(b)* $F = aIB$. *(c)* $F = aIB/2$.

S **P12.25.** Three circular loops are made of three equal pieces of wire of length b, one with a single turn, one with two turns, and one with three turns of wire. If the same current I exists in the loops and they are situated in a uniform magnetic field of flux density B, determine the maximal moment of magnetic forces on the three loops. Then solve for the moments if $I = 5\,\mathrm{A}$, $b = 50\,\mathrm{cm}$, and $B = 1\,\mathrm{T}$. — *(a)* $M_1 = 0.995\,Nm$, $M_2 = 0.497\,Nm$, $M_3 = 0,332\,Nm$. *(b)* $M_1 = 0.0995\,Nm$, $M_2 = 0.0497\,Nm$, $M_3 = 0,00332\,Nm$. *(c)* $M_1 = 0.00995\,Nm$, $M_2 = 0.00497\,Nm$, $M_3 = 0,0332\,Nm$.

Solution. The moment of magnetic forces is given by $M = m \times B = INS \times B$, where S is the area of one turn of the loop, and N the number of turns. The maximal moment of magnetic forces is thus $M = INSB$. We need to determine the areas of the three loops.

The area of the loop with one turn is $S_1 = (b/2\pi)^2 \pi = b^2/(4\pi)$, that of two turns is $S_2 = b^2/(16\pi)$, and that with three turns is $S_3 = b^2/(36\pi)$. The maximal moment of magnetic forces on the three loops is $M_1 = 0.0995\,\mathrm{Nm}$, $M_2 = 0.0497\,\mathrm{Nm}$, and $M_3 = 0,0332\,\mathrm{Nm}$.

P12.26. Determine the moment of magnetic forces acting on a rectangular loop of sides a and b, and with current of intensity I, situated in a uniform magnetic field of flux density B. Side b of the loop is normal to the lines of B, and side a parallel with them. — *(a)* $M = a(a+b)B$. *(b)* $M = a(b-a)B$. *(c)* $M = abB$.

S **P12.27.** Two thin, parallel, coaxial circular loops of radius a are a distance a apart. Each loop carries a current I. Prove that at the midpoint between the loops, on their common axis, the first three derivatives of the axial magnetic flux density with respect to a coordinate along the axis are zero. (This means that the field around that point is highly uniform. Two such coils are known as the *Helmholtz coils*.) — *Hint: the vector $\boldsymbol{B}$ is along the axis of the loops. Adopt the axis to be the z axis, with the origin at the center of one loop. Find the total $B_z(z)$ and prove that the first, second, and even the third derivative of $B_z(z)$ is zero for $z = a/2$.*

Solution. From the solution to problem P12.7, the total magnetic flux density vector is along the axis of the two loops. Let us adopt it to be the z axis, with the origin at the center of one loop. Its magnitude is given by [note that the distance from the center of one loop is z, and from the other $(a - z)$]

$$B_z(z) = \frac{\mu_0 I a^2}{2} \left\{ \frac{1}{(a^2 + z^2)^{3/2}} - \frac{1}{[a^2 + (a - z)^2]^{3/2}} \right\}.$$

It is left to the reader to prove that the first, second, and even the third derivative of $B_z(z)$ is zero for $z = a/2$.

P12.28. Write the expression for the vector $\boldsymbol{B}$ inside a long circular conductor of radius a carrying a current I. To that end, use the current density vector $\boldsymbol{J}$ inside the conductor, and the vector $\boldsymbol{r}$ representing the distance of the point considered inside the conductor to the conductor axis. — *(a) $\boldsymbol{B}(\boldsymbol{r}) = -\mu_0 \boldsymbol{J} \times \boldsymbol{r}/2$. (b) $\boldsymbol{B}(\boldsymbol{r}) = \mu_0 \boldsymbol{J} \times \boldsymbol{r}/2$. (c) $\boldsymbol{B}(\boldsymbol{r}) = \mu_0 \boldsymbol{J} \times \boldsymbol{r}$.*

P12.29. A very long cylindrical conductor of circular cross section of radius a has a hole of radius b. The axis of the hole is a distance d $(d + b < a)$ from the conductor axis. Using the principle of superposition and the expression for the magnetic flux density vector inside a round conductor from the preceding problem, prove that the magnetic field in the cavity is uniform. — *Hint: note that the magnetic flux in the cavity can be obtained as that in a solid conductor of radius a, minus that in a solid conductor of radius b (centered off axis as specified), both with current densities $\boldsymbol{J}$. and use the result of the preceding problem. (a) $\boldsymbol{B} = \mu_0 \boldsymbol{J} \times \boldsymbol{d}/2$. (b) $\boldsymbol{B} = -\mu_0 \boldsymbol{J} \times \boldsymbol{d}/2$. (c) $\boldsymbol{B} = \mu_0 \boldsymbol{J} \times \boldsymbol{d}$. $\boldsymbol{d} = \boldsymbol{r} - \boldsymbol{r}'$ is the position vector of the cavity axis with respect to the cylinder axis.*

*P12.30. Prove that the divergence of the magnetic flux density vector given by the Biot-Savart law is zero. — *Hint: take the divergence of the left-hand and right-hand sides in the Biot-Savart law in Eq. (12.5). Note that the divergence operator acts on the field coordinates, while the integral is over the source coordinates, so that the divergence operator can be taken inside the integral to act on the integrand only.*

S **P12.31.** A straight, very long, thin conductor has a charge Q' per unit length. It also carries a current of intensity I. A charge Q is moving with a velocity v parallel to the wire, at a distance d from it, unaffected by the simultaneous action of both the electric and magnetic force. Determine the velocity v of the charge, assuming the necessary correct direction of the current and the sign of the charge on the conductor. — *Assume that $Q > 0$ and that the current in the conductor is in the same direction as the charge motion. (a) $v = Q'/(\epsilon_0 \mu_0 I)$. (b) $v = Q'/(2\epsilon_0 \mu_0 I)$. (c) $v = Q'/(4\epsilon_0 \mu_0 I)$.*

Solution. Assume that $Q > 0$ and that the current in the conductor is in the same direction as the charge motion. The magnetic force on the charge is then attracting the charge towards the conductor, and is of intensity $F_m = \mu_0 v Q I/(2\pi d)$. In order that the total force on the charge be zero, the charge per unit length of the conductor, Q', must be positive. If the magnetic and electric forces on the charge are of equal magnitude, we have

$$F_m = \frac{\mu_0 v Q I}{2\pi d} = F_e = \frac{QQ'}{2\pi\epsilon_0 d},$$

from where the velocity of the charge is $v = Q'/(\epsilon_0 \mu_0 I)$.

P12.32. Starting from the magnetic force between two current elements, derive the expression for the magnetic force between two moving charges. — *Hint: note that the product $I d\mathbf{l}$ for a current element can be written in the form $NQv d v$, where N is the number of charge carriers per unit volume, $\mathbf{v}$ their drift velocity, and dv the elemental volume with current.*

P12.33. Assuming that the expression for the magnetic force between two moving charges from the preceding problem is true, compare the maximal possible magnetic force between the charges with the Coulomb force between them. The charges are moving with equal velocities v and are at a distance r. — *(a) $F_m/F_e = \epsilon_0 \mu_0 v^2$. (b) $F_m/F_e = 4\epsilon_0\mu_0 v^2$. (c) $F_m/F_e = \epsilon_0\mu_0 v^2/4$.*

P12.34. A copper wire of circular cross section and radius $a = 1\,\mathrm{mm}$ carries a current of $I = 50\,\mathrm{A}$. This is the largest current that can flow through the wire without damaging the conductor material. Plot the magnitude of the magnetic flux density vector as a function of distance from the center of the wire. Calculate the magnetic flux density at the surface of this wire. — *(a) $B = 0.01\,T$. (b) $B = 0.001\,T$. (c) $B = 0.0001\,T$.*

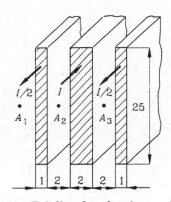

Fig. P12.35. DC line for aluminum electrolysis.

P12.35. A plant for aluminum electrolysis uses a dc current of $15\,\mathrm{kA}$ flowing through a line which consists of three metal plate electrodes, as in Fig. P12.35. All the dimensions in the figure are given in centimeters. Find the approximate magnetic flux density at points A_1, A_2, and A_3 shown in the figure. — *(a) $B_{A1} \simeq 0$, $B_{A2} = 17.7\,mT$, upward, $B_{A3} = 17.7\,mT$, downward. (b) $B_{A1} \simeq 0$, $B_{A2} = 27.7\,mT$, upward, $B_{A3} = 27.7\,mT$, downward. (c) $B_{A1} \simeq 0$, $B_{A2} = 37.7\,mT$, upward, $B_{A3} = 37.7\,mT$, downward.*

P12.36. A very long cylinder of radius a has a volume charge of density ρ. Find the expression for the magnetic flux density vector inside as well as outside of the cylinder if the cylinder is rotating around its axis with an angular velocity ω. Plot your results. — *Note that if the cylinder is long, there is practically no field outside the cylinder, and inside it is axial. (a) $B(r) = \mu_0\rho\omega(a^2 - r^2)/2$. (b) $B(r) = \mu_0\rho\omega(a^2 - r^2)/4$. (c) $B(r) = \mu_0\rho\omega(a^2 - r^2)/8$.*

P12.37. Find the magnetic flux density between and outside large current sheets shown in Fig. P12.37. — *Hint: use Ampére's law to find the magnetic field of a single current sheet, and then use superposition.*

S **P12.38.** A current $I = 0.5$ A flows through the torus winding shown in Fig. P12.38. Find the magnetic flux density at points A_1, A_2 and A_3 inside the torus. There are $N = 2500$ turns, $a = 5$ cm, $b = 10$ cm, and $h = 4$ cm. — *(a)* $B_{A_1} = B(a) = 10\,mT$, $B_{A_2} = B[(a+b)/2] = 6.66\,mT$, $B_{A_3} = B(b) = 5\,mT$. *(b)* $B_{A_1} = 5\,mT$, $B_{A_2} = 3.33\,mT$, $B_{A_3} = 2.5\,mT$. *(c)* $B_{A_1} = 2.5\,mT$, $B_{A_2} = 1.67\,mT$, $B_{A_3} = 1.25\,mT$.

Solution. The problem is solved by the application of the integral form of Ampère's law. (Why can we solve it in this way?) Outside the torus there is no field. (Is this statement strictly true? Explain.) If we apply Ampère's law to a circular contour of radius r ($a < r < b$), coaxial with the torus and inside it, we get

$$2\pi r B(r) = \mu_0 NI, \quad \text{from which} \quad B(r) = \mu_0 \frac{NI}{2\pi r}.$$

For given numerical values, we obtain $B_{A_1} = B(a) = 5$ mT, $B_{A_2} = B[(a+b)/2] = 3.33$ mT , $B_{A_3} = B(b) = 2.5$ mT.

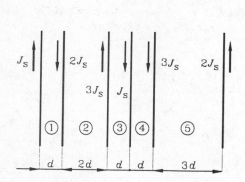

Fig. P12.37. Parallel current sheets.

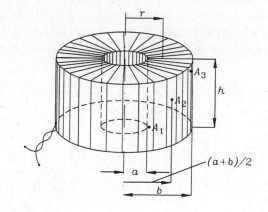

Fig. P12.38. A densely wound thick toroidal coil.

P12.39. Find the dimensions and required number of turns for a torus such as the one in the previous problem so that the following design parameters are satisfied: (1) the magnetic flux density in the middle of the torus cross section is 6 mT; (2) the cross section of the core has dimensions $b - a = 10$ cm and $h = 10$ cm; and (3) the magnetic flux density does not vary by more than 3% from the value in the middle of the cross section. Assume you have at your disposal an insulated copper wire with a 1.5-mm diameter which can tolerate a maximum current of $I_{max} = 7.5$ A. If it is not possible to design the winding as a single-layer coil, design a multi-layer winding. Note: many possible designs meet the criteria; chose the one that uses the least amount of wire, i.e., that has the lowest cost. — *Hint: start with condition (3). Note that the solution is not unique, so that the designer needs to make the final decision.*

P12.40. Find and plot the magnetic flux density vector due to a current I flowing through a hollow cylindrical conductor of inner radius a and outer radius b. — *Hint: note that the lines of vector B are circles, that B is of constant magnitude along such a circle, and apply the Ampère law.*

13. Magnetic Field in Materials

• In magnetic terms, atoms and molecules inside matter resemble tiny current loops. If a piece of matter is situated in a magnetic field, the moment of magnetic forces partly aligns these loops, and we say that the substance is *magnetized*. The magnetic field produced by the substance is due to these aligned current loops, known as *Ampère's currents*. A substance in the magnetic field can therefore be visualized as a large set of oriented elementary current loops situated in a vacuum. These oriented loops can be replaced by equivalent *macroscopic* currents *situated in a vacuum*, known as the *magnetization currents*.

• An elementary current loop is first characterized by a *magnetic moment*, $m = I\,S$, where I is the loop current and S its vector area. Next the *magnetization vector*, M, is defined, as

$$M = \frac{(\sum m)_{\text{in } dv}}{dv} = N m \qquad \text{(A/m)}, \qquad (13.1)$$

where N is the number of Ampère's currents per unit volume.

• The Ampère currents can be considered to be situated in a vacuum. Consequently, they can be incorporated in Ampère's law (which is valid for currents in a vacuum):

$$\oint_C B \cdot dl = \mu_0 \left(\int_S J \cdot dS + \oint_C M \cdot dl \right), \qquad (13.2)$$

or

$$\oint_C H \cdot dl = \int_S J \cdot dS, \qquad (13.3)$$

where

$$H = B/\mu_0 - M \qquad \text{(A/m)} \qquad (13.4)$$

is known as the *magnetic field intensity vector*. For *linear magnetic materials*

$$M = \chi_m H \qquad (\chi_m - \text{dimensionless}, \ M - \text{A/m}), \qquad (13.5)$$

where χ_m is the *magnetic susceptibility*. Thus, for linear materials,

$$B = \mu_0(1 + \chi_m)H = \mu_0 \mu_r H = \mu H \qquad (\mu_r - \text{dimensionless}, \ \mu - \text{H/m}). \qquad (13.6)$$

The constant μ_r is the *relative permeability*, and μ the *permeability* of the material. If Eq. (13.6) holds, the material is *linear*, else it is *nonlinear*. If μ is the same at all points, the material is said to be *homogeneous*, otherwise it is *inhomogeneous*. Linear magnetic materials can be *diamagnetic* ($\chi_m < 0$, i.e., $\mu_r < 1$), or *paramagnetic* ($\chi_m > 0$, i.e., $\mu_r > 1$).

• Ampère's law in Eq. (13.3) can be transformed into a differential equation, i.e., its differential form,

$$\text{curl}\,H = \nabla \times H = J, \tag{13.7}$$

valid for time-invariant currents.

• The volume density of magnetization current is given by

$$J_m = \nabla \times M \qquad (\text{A/m}^2). \tag{13.8}$$

It is zero in homogeneous materials with no macroscopic currents. On a boundary between two magnetized materials, the surface magnetization current density is given by

$$J_{ms} = n \times (M_1 - M_2) \qquad (\text{A/m}), \tag{13.9}$$

where n is the unit vector normal to the boundary, directed into medium 1.

• The most important boundary conditions at the interface of two magnetic materials are

$$H_{1\text{tang}} = H_{2\text{tang}}, \tag{13.10}$$

(no surface currents on the boundary), and

$$B_{1\text{normal}} = B_{2\text{normal}}. \tag{13.11}$$

In linear media these equations become

$$\frac{B_{1\text{tang}}}{\mu_1} = \frac{B_{2\text{tang}}}{\mu_2}, \qquad \mu_1 H_{1\text{normal}} = \mu_2 H_{2\text{normal}}. \tag{13.12}$$

• Diamagnetic and paramagnetic materials are linear, and have $\mu_r \simeq 1$ (less than one for diamagnetic materials, greater than one for paramagnetic materials). The most important magnetic materials in electrical engineering are known as *ferromagnetic materials*, nonlinear materials with very large value of relative permeability. In ferromagnetic materials, groups of atoms (*Weiss' domains*) are formed as small saturated magnets. Magnetization of ferromagnetic materials is obtained by aligning these domains, which is accompanied by *hysteresis losses*. Above the *Curie temperature*, a ferromagnetic material becomes paramagnetic. *Ferrites* are materials that have neighboring Weiss domains of different sizes, oriented in opposite directions. Therefore they have less pronounced magnetic properties than ferromagnetic materials, but many of them have high electrical resistivity, which makes them important for high-frequency applications. (High resistivity makes so-called skin effect less pronounced and reduces eddy-current losses at high frequencies.)

• Ferromagnetic materials are characterized by *magnetization curves*, obtained by measurements. These are curves of B versus H. If the material was previously unmagnetized, the *initial magnetization curve* is obtained. If H is periodic, a *hysteresis loop* is obtained instead. The curve connecting the tips of hysteresis loops corresponding to different amplitudes of H is the *normal magnetization curve*.

• Ferromagnetic materials are characterized by several "permeabilities". The ratio B/H along an initial magnetization curve at $H = 0$ is the *initial permeability*, and that along a normal magnetization curve is the *normal permeability*. Also used are complex permeability, differential permeability, etc.

• *Magnetic circuits* are structures designed to channel the magnetic flux, and are used in a wide variety of devices (cores of transformers, motors, generators, relays, etc.) An approximate solution of magnetic circuits is based on assuming them to be thin (with respect to the length of the branches), and linear. Such magnetic circuits are analyzed using Kirchhoff's laws for magnetic circuits, which are analogous to these laws for electric circuits. In these equations, the electromotive force is replaced by the *magnetomotive force* (the product NI of the number of turns in a coil and its current), the current by the magnetic flux, and the resistance by the "magnetic resistance", or *reluctance*. For a branch of length l, cross-section area S and permeability μ, the reluctance is given by

$$R_\mathrm{m} = \frac{l}{\mu S} \qquad (1/\mathrm{H}).\tag{13.13}$$

Real magnetic circuits are both nonlinear and are not thin. Their analysis is quite complicated, especially if the circuit is not quite simple.

QUESTIONS

Q13.1. Are any conventions implicit in the definition of the magnetic moment of a current loop? — *(a) There are no conventions at all. (b) The direction around the loop is adopted in the direction of the current. (c) In addition to (b), the unit vector normal to the surface and the direction around the loop are connected through the right-hand rule.*

Q13.2. A magnetized body is introduced into a uniform magnetic field. Is there a force on the body? Is there a moment of magnetic forces on the body? Explain. — *(a) There is a force in the direction of vector B, there is a moment of magnetic forces. (b) There is no force, there is a moment. (c) There is a force in the direction opposite to that of vector B, there is a moment.*

S **Q13.3.** A small body made of soft iron is placed on a table. Also on the table is a permanent magnet. If the body is pushed toward the magnet, ultimately the magnet will pull the body toward itself, so that the body will acquire certain kinetic energy before it hits the magnet. Where did this energy come from? — *(a) From our pushing the body. (b) From the magnet itself. (c) The small body had a potential energy in the field of the permanent magnet.*

Answer. A small body had a potential energy in the field of the permanent magnet. This potential energy was used to accelerate the body.

Q13.4. Prove that the units for B and $\mu_0 M$ are the same. — *Hint: use the Biot-Savart law to express the tesla in terms of the unit for μ_0.*

Q13.5. The source of the magnetic field is a permanent magnet of magnetization M. What is the line integral of the vector H around a contour which passes through the magnet? — *(a) Depends on the shape of the magnet and its magnetization. (b) Depends on the direction of the path with respect to the magnetization vector. (c) Zero (explain).*

S **Q13.6.** The magnetic core of a thin toroidal coil is magnetized to saturation, and then the current in the coil is switched off. The remanent (i.e., remaining) flux density in the core is B_r. Determine the magnetization vector and the magnetic field strength vector in the core. — *(a) $M = B_r$. (b) $M = \mu_0 B_r$. (c) $M = B_r/\mu_0$.*

Answer. Since $B = \mu_0 H + M$, and $H = 0$ in the core (remanent flux density exists for $H = 0$), $M = B_r$.

Q13.7. Is there a magnetic field in the air around the core in question Q13.6? Explain. — *(a) Certainly, because the core is magnetized. (b) No, since there is no leakage flux. (c) Yes, but only if the permeability of the core is not high.*

Q13.8. Why is the equation

$$\int_S \nabla \times H \cdot dS = \int_S J \cdot dS$$

valid for any contour C and any surface bounded by C? — *(a) Because we adopt that, at all points, curl $H = J$. (b) It is valid for any C and any surface only under certain conditions (which ones?). (c) Because it was derived using the Stokes's theorem.*

Q13.9. Why is the reference direction of the vector J_{ms} in Fig. Q13.9 into the paper? — *Hint: recall the conventions implicit in Eq. (13.9).*

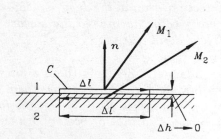

Fig. Q13.9. Boundary of magnetized material.

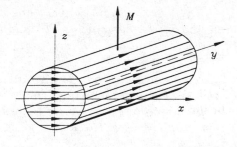

Fig. Q13.11. Surface magnetization currents.

Q13.10. Suppose that all atomic currents contained in the page you are reading can be oriented so that m is toward you. What is then their macroscopic resultant, and what is (qualitatively) the magnetic field of such a "magnetic sheet"? — *(a) As that of two layers of current on the two sheet sides. (b) As that of a circulating current around the sheet circumference. (c) The sheet is thin, and there is no magnetic field.*

Q13.11. What are the macroscopic resultants of the microscopic currents of a short, circular, cylindrical piece of magnetized matter with uniform magnetization M at all points, if: (1) M

is parallel to the cylinder axis, or (2) *M* is perpendicular to the axis? — *(a) In case 1, surface currents on the two cylinder bases. In case 2 the same, but of different distribution. (b) In both cases, a volume distribution of magnetization currents. (c) In case 1, a current sheet around the curved surface, equivalent to a short solenoid. In case 2, surface magnetization currents are like those in Fig. Q13.11.*

Q13.12. Sketch roughly the lines of vectors *M*, *B*, and *H* in the two cases in question Q13.11. — *Hint: discuss if the sketches in Fig. Q13.12 are correct.*

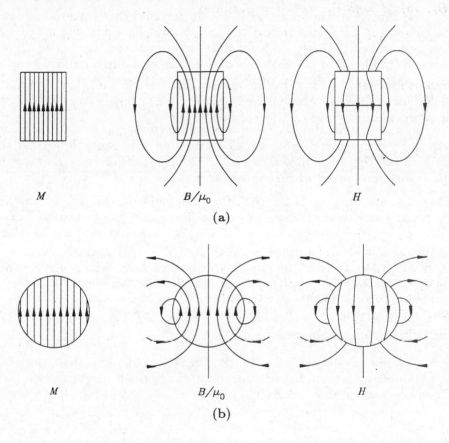

Fig. Q13.12. Sketch of lines of vectors *M*, *B*, and *H* for the two cases of magnetized short cylinder described in question Q13.11.

Q13.13. A ferromagnetic cube is magnetized uniformly over its volume. The magnetization vector is perpendicular to two sides of the cube. What is this cube equivalent to in terms of the magnetic field it produces? — *(a) To a short solenoid of square cross section. (b) To two current sheets (over the two mentioned cube sides). (c) To a volume distribution of currents.*

Q13.14. Is the north magnetic pole of the earth close to its geographical North Pole? Explain. (See Chapter 17.) — *Hint: recall that the magnetic needle turns its north pole towards the North.*

S **Q13.15.** If a high-velocity charged elementary particle pierces a toroidal core in which there is only remanent flux density, and *H* = 0, will the particle be deflected by the magnetic field?

Explain. — *(a) No, because in the core $H = 0$. (b) No, because in the core $B = 0$. (c) Yes, because in the core B is not zero, although H is.*

Answer. It will be deflected, because the force on the particle is $F = Qv \times B$, and $B \neq 0$.

Q13.16. Sketch the initial magnetization curve corresponding to a change of H from zero to $-H_m$. — *Hint: assume that H starts from zero and becomes negative, and think about what the variation of B must be like.*

Q13.17. Suppose that the magnetization of a thin toroidal core corresponds to the point B_r (remanent flux density). The coil around the core is removed, and the magnetic flux density uniformly decreased to zero by some appropriate *mechanical or thermal treatment*. How does the point in the B-H plane go to zero? — *(a) Backwards along the magnetization curve. (b) Continues to a point on the $-H$ axis, and then along that axis to the origin. (c) Along the B axis to the origin.*

S **Q13.18.** The initial magnetization curve of a certain ferromagnetic material is determined for a thin toroidal core. Explain the process of determining the magnetic flux in the core of the same material, but of the form shown in Fig. P13.5a. — *Hint: note that the magnetic field intensity differs from one point of such a core to another, as $H(r) = NI/(2\pi r)$.*

Answer. The magnetic field intensity differs from one point of the core to another, as $H(r) = NI/(2\pi r)$. We need to take the corresponding value of B from the magnetization curve, to multiply it by $h\,dr$ (the area of a small rectangular surface of the core cross section), and to integrate from a to b.

Q13.19. Make a rough sketch of the curve in the B-H plane obtained if H is increased to H_m, then decreased to zero, then again increased to H_m and decreased to zero, and so on. — *(a) A straight line segment. (b) A flat hysteresis loop. (c) Even a rough sketch is not possible without measurements.*

Q13.20. Is it possible to obtain a higher remanent flux density than that obtained when saturation is attained and then H reduced to zero? Explain. — *(a) Yes (how?). (b) No (why?). (c) For some materials (describe).*

S **Q13.21.** What do you expect would happen if a thin slice is cut out of a ferromagnetic toroid with remanent flux density in it? Explain. — *(a) Nothing. (b) The operating point will move downwards along the B axis. (c) The operating point will move to the left and downward.*

Answer. The simplest way to visualize the situation is to imagine that the slice was not cut out, but that, instead, a surface current of equal magnitude to the magnetization surface current, but opposite to it, is "glued" over the slice. This amounts to the same, but is easier to understand. The working point in the $H - B$ diagram would move from the point $(0, B_r)$ to the left and downward, because we have a "coil" with negative current.

Q13.22. A rod of ferromagnetic material can be magnetized in various ways. If a magnetized rod attracts most ferromagnetic powder (e.g., iron filings) near its ends, and very little in its middle region, how is it magnetized? — *(a) Not enough data for a conclusion. (b) Normal to its axis. (c) Along its axis.*

Q13.23. If a small diamagnetic body is close to a strong permanent magnet, does the magnet attract or repel it? Explain. — *(a) It repels it. (b) It attracts it. (c) It does not exert a force on it.*

Q13.24. Answer question Q13.23 for a small paramagnetic body. — *The answers are the same as for the preceding question.*

Q13.25. While the core in question Q13.17 is still magnetized, if just one part of the core is heated above the Curie temperature, will there be a magnetic field in the air? If you think there will be, what happens when the heated part has cooled down? — *(a) No. (b) Yes, but will practically disappear when it cools down (why?). (c) It depends how large the heated part is.*

S **Q13.26.** Assuming that you use a large number of small current loops, explain how you can make a model of (1) a paramagnetic material, and (2) a ferromagnetic material? — *(a) It is not possible to make models in this way. (b) Just orient the loops in both cases. (c) In case 1, suspend the loops on springs individually, and in random directions. In case 2, make groups of many oriented current loops (with the magnetic moment, m, of loops in a group in the same directions). Suspend these groups of loops on springs, and allow these groups of loops to turn with friction.*

Answer. (1) Suspend the loops individually, and in random directions, on springs that allow them to turn only to some extent in response to the applied magnetic field. (2) Suspend many *large groups* of strictly oriented current loops in random directions, on springs that allow them to turn *with friction* in response to the applied magnetic field.

Q13.27. If the current in the coil wound around a ferromagnetic core is sinusoidal, is the magnetic flux in the core also sinusoidal? Explain. — *(a) Depends on the amplitude of current. (b) Yes. (c) No.*

Q13.28. Analyze similarities and differences between Kirchhoff's laws for dc electric circuits and magnetic circuits. — *Hint: analyze possible approximations in the two sets of laws.*

Q13.29. How do you determine the direction of the magnetomotive force in a magnetic circuit? — *(a) The magnetomotive force does not have a direction. (b) According to the right-hand rule with respect to the current in the winding. (c) According to the left-hand rule.*

S **Q13.30.** A thin magnetic circuit is made of a ferromagnetic material with an initial magnetization curve that can be approximated by the expression $B(H) = B_0 H/(H_0 + H)$, where B_0 and H_0 are constants. If the magnetic field strength, H, in the circuit is much smaller than H_0, can the circuit be considered as linear? What in that case is the permeability of the material? What is the physical meaning of the constant B_0? — *(a) No, it cannot be considered as linear. (b) Depends on the values of the parameters. (c) Yes, and the permeability is $\mu = B_0/H_0$.*

Answer. It can, for then $B(H) = B_0 H/(H_0 + H) \simeq (B_0/H_0)H$. The permeability of the material is thus $\mu = B_0/H_0$. If $H \gg H_0$, then $B(H) \simeq B_0$, so that B_0 is the magnetic flux density corresponding to saturation.

Q13.31. Why can't we have a magnetic circuit with no leakage flux (stray field in the air surrounding the magnetic circuit)? — *(a) Because μ cannot be large enough. (b) Because μ_0 is too large. (c) A combination of the two preceding answers.*

Q13.32. Is it possible to construct a magnetic circuit closely analogous to a dc electric circuit, if the latter is situated (1) in a vacuum, or (2) in an imperfect dielectric? Explain. — *(a) No, in both cases. (b) Yes in case 1, no in case 2. (c) No in case 1, yes in case 2.*

Q13.33. One half of the length of a thin toroidal coil is filled with a ferromagnetic material, and the other half with some paramagnetic material. Can the problem be analyzed as a magnetic circuit? Explain. — *(a) No. (b) Yes. (c) Depends on the permeability of the paramagnetic material.*

PROBLEMS

S **P13.1.** The magnetic moment of the earth is about $8 \cdot 10^{22}$ Am^2. Imagine that there is a giant loop around the earth's equator. How large does the current in the loop have to be to result in the same magnetic moment? Would it be theoretically possible to cancel the magnetic field of the earth with such a current loop (1) on its surface, or (2) at far points? The radius of the earth is approximately 6370 km. — *(a)* $I = 6.28 \cdot 10^6$ A. *(b)* $I = 6.28 \cdot 10^7$ A. *(c)* $I = 6.28 \cdot 10^8$ A.

Solution. The magnitude of the magnetic moment of the contour is given by

$$m = I\pi R^2,$$

where R is the radius of the earth, so that the loop current has to be as large as $I = 6.28 \cdot 10^8$ A! With such a loop it could not be possible, not even theoretically, to cancel the magnetic field of the earth on its surface, because this field is due to currents distributed inside the earth volume, and not on the earth equator. At far points, however, the magnetic field of an arbitrary current distribution depends on the magnetic moment of the equivalent contour only, and the cancellation of the magnetic field of the earth is theoretically possible.

P13.2. The number of iron atoms in one cubic centimeter is approximately $8.4 \cdot 10^{22}$, and the product of μ_0 and the maximum possible magnetization (corresponding to "saturation") is $\mu_0 M_{\text{sat}} = 2.15$ T. Calculate the magnetic moment of an iron atom. — *(a)* $m = 1.04 \cdot 10^{-21}$ Am^2. *(b)* $m = 4.04 \cdot 10^{-20}$ Am^2. *(c)* $m = 2.04 \cdot 10^{-23}$ Am^2.

P13.3. A thin toroid is uniformly magnetized along its length with a magnetization vector of magnitude M. No free currents are present. Noting that the lines of M, B, and H inside the toroid are circles by symmetry, determine the magnitude of B and prove that $H = 0$. — *Hint: use Ampère's law to prove that $H = 0$.*

S **P13.4.** A straight, long copper conductor of radius a is covered with a layer of iron of thickness d. A current of intensity I exists in this composite wire. Assuming that the iron permeability is μ, determine the magnetic field, the magnetic flux density, and the magnetization in copper and iron parts of the wire. Note that the current density in the copper and iron parts of the wire is not the same. — *Hint: assume that σ_{Cu} is the conductivity of copper, and σ_{Fe} that of iron. Note that the electric field intensities in copper and iron are the same (due to continuity of E_{tang} on the boundary).*

Solution. Let σ_{Cu} be the conductivity of copper, and σ_{Fe} that of iron. The electric field intensities in copper and iron are the same (due to continuity of E_{tang} on the boundary). Hence we have that

$$\frac{J_{\text{Cu}}}{J_{\text{Fe}}} = \frac{\sigma_{\text{Cu}}}{\sigma_{\text{Fe}}}.$$

The total current in the wire is given by

$$I = J_{Cu}\pi a^2 + J_{Fe}\pi(d^2 + 2ad),$$

so that $J_{Cu} = I/[\pi a^2 + \pi(d^2 + 2ad)\sigma_{Fe}/\sigma_{Cu}]$ and $J_{Fe} = J_{Cu}\sigma_{Fe}/\sigma_{Cu}$.

Due to symmetry, vector $\boldsymbol{H}$ has a circular component only, and its intensity depends only to the distance r from the wire axis. By applying the generalized Ampère's law to the contour in the form of a circle of radius r, we obtain

$$H = \frac{J_{Cu}r}{2} \quad \text{(in copper)}, \qquad H = \frac{J_{Cu}a^2 + J_{Fe}(r^2 - a^2)}{2r} \quad \text{(in iron)}.$$

In copper, $\boldsymbol{B} = \mu_0\boldsymbol{H}$ and $\boldsymbol{M} = 0$. In iron, $\boldsymbol{B} = \mu\boldsymbol{H}$ and $\boldsymbol{M} = (\mu/\mu_0 - 1)\boldsymbol{H}$.

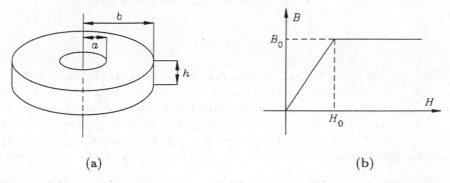

Fig. P13.5. (a) A ferromagnetic core, and (b) its idealized initial magnetization curve.

P13.5. The ferromagnetic toroidal core sketched in Fig. P13.5a has an idealized initial magnetization curve as shown in Fig. P13.5b. Determine the magnetic field strength, the magnetic flux density, and the magnetization at all points of the core, if the core is wound uniformly with $N = 628$ turns of wire with current of intensity (1) 0.5 A, (2) 0.75 A, and (3) 1 A. The core dimensions are $a = 5\,\text{cm}$, $b = 10\,\text{cm}$, and $h = 5\,\text{cm}$, and the constants of the magnetization curve are $H_0 = 1000\,\text{A/m}$, and $B_0 = 2\,\text{T}$. For the three cases determine the magnetic flux through the core cross section. Assume that the core was not magnetized prior to turning on the current in the winding. — *Hint: use Ampère's law to find H, and note that in the three cases different parts of the magnetization curve are valid. The magnetic flux through the core cross section for the three cases is: (a) $\Phi_1 = 0.466\,mWb$, $\Phi_2 = 3.658\,mWb$, $\Phi_3 = 4\,mWb$. (b) $\Phi_1 = 3.466\,mWb$, $\Phi_2 = 4.658\,mWb$, $\Phi_3 = 5\,mWb$. (c) $\Phi_1 = 6.466\,mWb$, $\Phi_2 = 2.658\,mWb$, $\Phi_3 = 2.1\,mWb$.*

P13.6. A straight conductor of circular cross section of radius a and permeability μ carries a current I. A coaxial conducting tube of inner radius b ($b > a$) and outer radius c, with no current, also has a permeability μ. Determine the magnetic field intensity, magnetic flux density, and magnetization at all points. Determine the volume and surface densities of macroscopic currents equivalent to Ampère currents. — *Hint: use Ampère's law to find H, and hence B and M, and Eqs. (13.8) and (13.9) to find the magnetization currents.*

P13.7. Repeat problem P13.6 if the conductor and the tube are of permeability $\mu(H) = \mu_0 H/H_0$, where H_0 is a constant. — *Hint: note that in this case $B = \mu_0 H^2/H_0$ and $M =$*

$(H/H_0 - 1)H$ *in the conductor and in the tube, and that H is the same as in the preceding problem.*

P13.8. Repeat problems P13.6 and P13.7 assuming that the tube carries a current $-I$, so that the conductor and the tube make a coaxial cable. — *Hint: the only difference is that H in the tube is determined also by the current in the tube.*

P13.9. A ferromagnetic sphere of radius a is magnetized uniformly with a magnetization vector M. Determine the density of magnetization surface currents equivalent to the magnetized sphere. — *Hint: use Eq. (13.9). The surface currents are circulating currents, as in Fig. P13.9.*

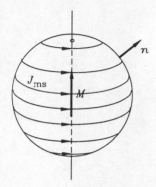

Fig. P13.9. A magnetized sphere.

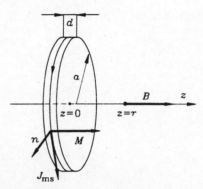

Fig. P13.10. A magnetized thin disk.

S **P13.10.** A thin circular ferromagnetic disk of radius $a = 2\,\text{cm}$ and thickness $d = 2\,\text{mm}$ is uniformly magnetized normal to its bases. The vector $\mu_0 M$ is of magnitude 0.1 T. Determine the magnetic flux density vector on the disk axis normal to its bases, at a distance $r = 2\,\text{cm}$ from the center of the disk. — *Hint: there are no volume magnetization currents, and surface magnetization currents are obtained from Eq. (13.9), Fig. P13.10.*

Solution. Since M is constant (the disk is uniformly magnetized), $J_m = \nabla \times M = 0$, i.e., there are no volume magnetization currents inside the disk. The distribution of the surface magnetization currents can be determined on the basis of Eq.(13.16). On the two cylinder bases, $J_{ms} = 0$, since vectors n and M are parallel. On the cylinder curved surface, however, there are circulating surface magnetization currents, of density $J_{ms} = M$ and with direction indicated in Fig. P13.10. The magnetized disk being thin ($d \ll a$), it can be replaced by a circular loop of radius a, situated in a vacuum. The loop current is

$$I = J_{ms}d = Md.$$

The magnetic flux density at the point specified in the text of the problem is hence (see problem P12.7)

$$B = \frac{\mu_0 M d a^2}{2(a^2 + r^2)^{3/2}} = 1.77\,\text{mT}.$$

The vector B is shown in Fig. P13.10.

P13.11. A thin ferromagnetic toroidal core was magnetized to saturation, and then the current in the winding wound about the core was turned off. The remanent flux density of the core material is $B_r = 1.4\,\text{T}$. Determine the surface current density on the core equivalent to the Ampère currents. If the mean radius of the core is $R = 5\,\text{cm}$, and the winding has $N = 500$ turns of wire, find the current in the winding corresponding to this equivalent surface current. If the cross section area of the core is $S = 1\,\text{cm}^2$, find the magnetic flux in the core. — *The magnetic flux in the core is (a)* $\Phi = 85\,\mu Wb$. *(b)* $\Phi = 140\,\mu Wb$. *(c)* $\Phi = 23\,\mu Wb$.

P13.12. A round ferrite rod of radius $a = 0.5\,\text{cm}$ and length $b = 10\,\text{cm}$ is magnetized uniformly over its volume. The vector $\mu_0 M$ is in the direction of the rod axis, of magnitude $0.07\,\text{T}$. Determine the magnetic flux density at the center of one of the rod bases. Is it important whether the point is inside the rod, outside the rod, or on the very surface of the rod? — *(a)* $B = 15\,mT$. *(b)* $B = 25\,mT$. *(c)* $B = 35\,mT$.

P13.13. Shown in Fig. P13.13 is a stripline with ferrite dielectric. Since $a \gg d$, the magnetic field outside the strips can be neglected. Under this assumption, find the magnetic field intensity between the strips if the current in the strips is I. If the space between the strips is filled with a ferrite of relative permeability μ_r that can be considered constant, determine the magnetic flux density and magnetization in the ferrite, and the density of surface currents equivalent to the Ampère currents in the ferrite. — *Hint: use the Ampère law to find H, and hence B, M, and J_{ms}. (a)* $J_{ms} = (\mu_r - 1)I/(2a)$. *(b)* $J_{ms} = (\mu_r - 1)I/a$. *(c)* $J_{ms} = 2(\mu_r - 1)I/a$.

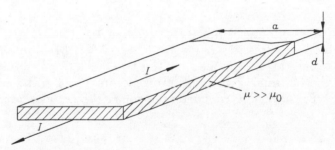

Fig. P13.13. A stripline with ferrite dielectric.

S **P13.14.** The ferromagnetic cube shown in Fig. P13.14a is magnetized in the direction of the z axis so that the magnitude of the magnetization vector is $M_z(x) = M_0 x/a$. Find the density of volume currents equivalent to the Ampère currents inside the cube, as well as the surface density of these currents over all cube sides. Follow the surface currents and note that in part they close through the magnetized material. — *Hint: the volume and surface currents in a cross section of the cube perpendicular to the z axis are as in Fig. P13.14b.*

Solution. Having in mind the expression for the curl in rectangular coordinates, the density of volume magnetization currents inside the cube is given by

$$\boldsymbol{J}_m = \nabla \times M = -\frac{\partial M_z}{\partial x}\,\boldsymbol{u}_y = -\frac{M_0}{a}\,\boldsymbol{u}_y.$$

According to Eq. (13.16), there are no resultant surface currents on the upper, lower and back side of the cube, while on the left, right and front cube side, the surface magnetization current density is $\boldsymbol{J}_{ms1} = M_0(x/a)\,\boldsymbol{u}_x$, $\boldsymbol{J}_{ms2} = -M_0(x/a)\,\boldsymbol{u}_x$, and $\boldsymbol{J}_{ms3} = M_0\,\boldsymbol{i}_u$, respectively.

By analysing the above expressions we can conclude that, indeed, the surface magnetization currents close in part through the cube, by forming the current contour integral with the volume magnetization currents. This is shown graphically in a cross section of the cube perpendicular to the z axis in Fig. P13.14b.

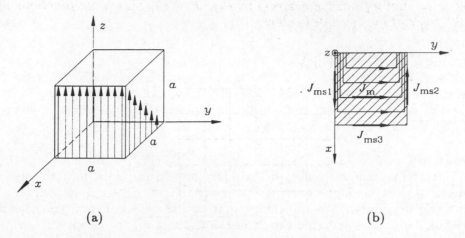

(a)   (b)

Fig. P13.14. (a) A magnetized ferromagnetic cube, and (b) distribution of volume and surface magnetization currents inside and over the cube.

P13.15. Prove that on the boundary surface of two magnetized materials the surface magnetization current is given by the expression $J_{ms} = n \times (M_1 - M_2)$. M_1 and M_2 are magnetization vectors in the two materials at close points on the two sides of the boundary, and n is the unit vector normal to the boundary, directed into medium 1. — *Hint: interpret properly the Ampère law in Eq. (13.2), and find the current through the rectangular contour sketched in Fig. Q13.9.*

P13.16. At a point of a boundary surface between air and a ferromagnetic material of permeability $\mu \gg \mu_0$ the lines of vector B are not normal to the boundary surface. Prove that the magnitude of the magnetic flux density vector in the ferromagnetic material is then much greater than that in air. — *Hint: use boundary conditions in Eqs. (13.12).*

P13.17. A current loop is in air above a ferromagnetic half-space. Prove that the field in the air due to the Ampère currents in the half-space is very nearly the same as that due to a loop below the boundary surface symmetrical to the original loop, carrying the current of the same intensity *and direction*, with the magnetic material removed. (This is the image method for ferromagnetic materials.) — *Hint: prove that the magnetic field of the original loop and its image satisfy the boundary conditions on the interface.*

P13.18. Inside a uniformly magnetized material, of relative permeability μ_r, are two cavities. One is a needlelike cavity in the direction of the vector B. The other is a thin-disk cavity, normal to that vector. Determine the ratio of magnitudes of the magnetic flux density in the two cavities and that in the surrounding material. Using these results, estimate the greatest possible theoretical possibility of "magnetic shielding" from time-invariant external magnetic field. (We shall see that the shielding effect is greatly increased for time-varying fields.) — *Hint: use boundary conditions in Eqs. (13.12).*

S **P13.19.** Sketched in Fig. P13.19 is the normal magnetization curve of a ferromagnetic material. Using this diagram, estimate the relative normal and differential permeability, and plot their dependence on the magnetic field strength. What are the initial and maximal permeabilities of the material? — *Hint: check if in the table below the values of normal relative permeability, $\mu_{rn} = (B/H)/\mu_0$, and differential relative permeability, $\mu_{rd} = (dB/dH)/\mu_0$, agree with your estimates from the diagram in Fig. P13.19. Sketch the functions $\mu_{rn}(H)$ and $\mu_{rd}(H)$ on the basis of the tabulated values.*

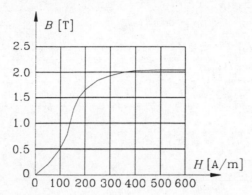

Fig. P13.19. A normal magnetization curve.

Solution. In the table below are given a few values of the normal relative permeability, $\mu_{rn} = (B/H)/\mu_0$, and the differential relative permeability, $\mu_{rd} = (dB/dH)/\mu_0$, which are estimated from the diagram in Fig. P13.19. It is left to the reader to sketch the functions $\mu_{rn}(H)$ and $\mu_{rd}(H)$ on the basis of the tabulated values.

H (A/m)	0	100	200	300	400	500
μ_{rn}	2600	3800	6500	5100	4100	3300
μ_{rd}	2600	5600	2900	1200	300	60

The initial relative permeability is $\mu_{ri} = 2600$, while the maximal normal and differential relative permeabilities are $(\mu_{rn})_{max} = 6900$ for $H = 180$ A/m and $(\mu_{rd})_{max} = 11000$ for $H = 150$ A/m.

P13.20. Approximate the normal magnetization curve in Fig. P13.19 in the range $0 \le H \le 300$ A/m by a straight line segment, and estimate the largest deviation of the normal relative permeability in this range from the relative permeability of such a hypothetical linear material. — *(a) about 800 and −500. (b) about 1800 and −2500. (c) about 8000 and −5000.*

P13.21. Figs. P13.21a and P13.21b show two hysteresis loops corresponding to sinusoidal variation of the magnetic field strength in two ferromagnetic cores between $-H_m$ and $+H_m$. Plot the time dependence of the magnetic flux density in the core. Does the magnetic flux also have a sinusoidal time dependence? — *Hint: check if Fig. P13.21c shows the dependence on time you expect of the magnetic field intensity, $H(t)$, and the corresponding magnetic flux densities, $B_a(t)$ and $B_b(t)$.*

P13.22. If in Figs. P13.21a and b the magnetic flux density varies sinusoidally in time, sketch the time dependence of the magnetic field strength in the core. Is it also sinusoidal? If the hysteresis loops were obtained by measurements with sinusoidal magnetic field strength, is it absolutely correct to use such loops in this case? — *Hint: check if the variations of H in time, obtained from the hysteresis loops in Figs. P13.21a and P13.21b and sketched in Fig. P13.22, are what you expect.*

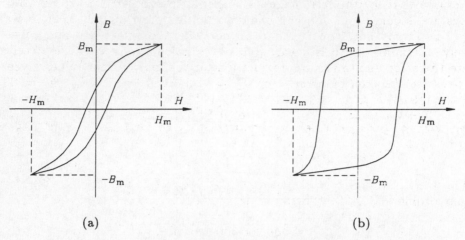

(a)   (b)

Fig. P13.21. Hysteresis loops for two ferromagnetic materials.

S **P13.23.** The initial magnetization curve (first part of hysteresis curve) $B(H)$ of a ferromagnetic material used for a transformer was measured and it was found that it can be approximated by a function of the form $B(H) = B_0 H/(H_0 + H)$, where the coefficients are $B_0 = 1.37$ T and $H_0 = 64$ A/m. Then a thin torus with mean radius $R = 10$ cm and a cross-section of $S = 1$ cm^2 is made out of this ferromagnetic material, and $N = 500$ turns are densely wound around it. Find $B(H)$ and the flux through the magnetic circuit as a function of current intensity I through the winding. Find the flux for (1) $I = 0.25$ A, (2) $I = 0.5$ A, (3) $I = 0.75$ A, and (4) $I = 1$ A. — *Hint: use Ampère's law to find H in the core, then B from the given magnetization curve.* (a) $\Phi_1 = 204\,\mu Wb$, $\Phi_2 = 218\,\mu Wb$, $\Phi_3 = 224\,\mu Wb$, $\Phi_4 = 227\,\mu Wb$. (b) $\Phi_1 = 94\,\mu Wb$, $\Phi_2 = 98\,\mu Wb$, $\Phi_3 = 94\,\mu Wb$, $\Phi_4 = 97\,\mu Wb$. (c) $\Phi_1 = 104\,\mu Wb$, $\Phi_2 = 118\,\mu Wb$, $\Phi_3 = 124\,\mu Wb$, $\Phi_4 = 127\,\mu Wb$.

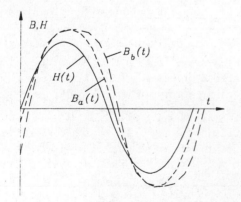 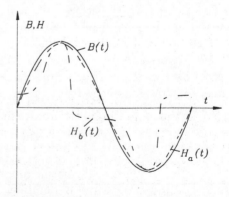

Fig. P13.21c. $B(t)$ for sinusoidal $H(t)$.   **Fig. P13.22.** $H(t)$ for sinusoidal $B(t)$.

Solution. The magnetic field intensity in the core is given by $H = NI/(2\pi R)$ (according to Ampère's law). The magnetic flux density can then be determined by the function approximating the initial magnetization curve, $B = B(H)$, and, finally, the magnetic flux through the magnetic circuit found as $\Phi = BS$. For the four specified values of I, we obtain: (1) $\Phi = 104\,\mu Wb$, (2) $\Phi = 118\,\mu Wb$, (3) $\Phi = 124\,\mu Wb$, and (4) $\Phi = 127\,\mu Wb$.

P13.24. Assume that for the ferromagnetic material in problem P13.23 you did not have a measured hysteresis curve, but you had one data point: for $H = 1000\,A/m$, B was measured to be $B = 2\,T$. From that, you can find an equivalent permeability and solve the circuit approximately, assuming that it is linear. Repeat the calculations from the preceding problem and calculate the error due to this approximation for the four current values given in problem 13.23. — *Let δ be the relative error. (a) $\Phi_1 = 40\,\mu Wb$ ($\delta = 61.5\,\%$), $\Phi_2 = 80\,\mu Wb$ ($\delta = 32.2\,\%$), $\Phi_3 = 119\,\mu Wb$ ($\delta = 4\,\%$), $\Phi_4 = 159\,\mu Wb$ ($\delta = 25.2\,\%$). (b) $\Phi_1 = 20\,\mu Wb$ ($\delta = 31.5\,\%$), $\Phi_2 = 40\,\mu Wb$ ($\delta = 12.2\,\%$), $\Phi_3 = 60\,\mu Wb$ ($\delta = 2\,\%$), $\Phi_4 = 80\,\mu Wb$ ($\delta = 15.2\,\%$). (c) $\Phi_1 = 60\,\mu Wb$ ($\delta = 31.5\,\%$), $\Phi_2 = 120\,\mu Wb$ ($\delta = 52.2\,\%$), $\Phi_3 = 160\,\mu Wb$ ($\delta = 45\,\%$), $\Phi_4 = 180\,\mu Wb$ ($\delta = 55.2\,\%$).*

P13.25. The thick toroidal core sketched in Fig. P13.25 is made out of the ferromagnetic material from problem P13.23. There are $N = 200$ turns wound around the core, and the core dimensions are $a = 3\,cm$, $b = 6\,cm$, and $h = 3\,cm$. Find the magnetic flux through the core for $I = 0.2\,A$ and $I = 1\,A$ in two different ways: (1) using the mean radius; and (2) by dividing the core into 5 layers and finding the mean magnetic field in each of the layers. — *(a) $\Phi_1 = 0.85\,(1.13)\,mWb$, the same for case 2. (b) $\Phi_1 = 0.54\,(0.78)\,mWb$, $\Phi_2 = 0.62\,(0.88)\,mWb$. (c) $\Phi_1 = 0.27\,(0.43)\,mWb$, $\Phi_2 = 0.57\,(0.76)\,mWb$.*

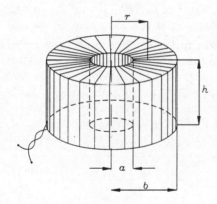

Fig. P13.25. A thick toroidal coil.

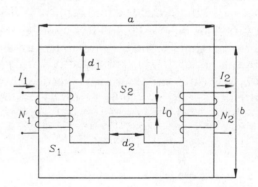

Fig. P13.26. A magnetic circuit with air gap.

S **P13.26.** Find the number of turns $N_1 = N_2 = N$ for the magnetic circuit shown in Fig. P13.26 so that the magnetic flux density in the air gap is $B_0 = 1\,T$ when $I_1 = I_2 = I = 5\,A$. The core is made of the same ferromagnetic material as in problem P13.23. Solve the problem in two ways: (1) taking the magnetic resistance of the core into account; and (2) neglecting the magnetic resistance of the core. Use the following values: $a = 10\,cm$, $b = 6\,cm$, $d_1 = d_2 = 2\,cm$, $S_1 = S_2 = S = 4\,cm^2$, and $l_0 = 1\,mm$. What is the percentage difference between the answers in (1) and (2)? — *(a) $N = 263$, error in neglecting the reluctance of the core is about 5%. (b) $N = 163$, error in neglecting the reluctance of the core is about 2.5%. (c) $N = 363$, error in neglecting the reluctance of the core is about 7.5%.*

Solution. (1) The magnetic flux density in the central part of the core is $B_2 = B_0 = 1\,T$. By symmetry, the magnetic flux densities in the left and right branches of the magnetic circuit are the same. Using the first Kirchhoff's law for magnetic circuits, we obtain $B_1 = 0.5$ T. The magnetic field intensity in the core is given by

$$H = \frac{H_0 B}{B_0 - B},$$

which results in $H_1 = 36.78$ A/m and $H_2 = 173$ A/m. The length of the left (or right) branch is $l_1 = 11.14$ cm, and that of the central branch $l_2 = 3.9$ cm. The second Kirchhoff's law for magnetic circuits finally yields

$$NI = H_1 l_1 + H_2 l_2 + H_0 l_0,$$

where $H_0 = B_0/\mu_0$, so that $N = 163$.

(2) If we neglect the magnetic resistance of the core, i.e., if we neglect $H_1 l_1 + H_2 l_2$ with respect to $H_0 l_0$, the result is $N = 159$. The error is about 2.5 %.

P13.27. The magnetization curve of a ferromagnetic material used for a magnetic circuit can be approximated by $B(H) = 2H/(400 + H)$, where B is in T and H is in A/m. The magnetic circuit has a cross-sectional area of $S = 2$ cm^2, a mean length of $l = 50$ cm, and $N = 200$ turns with $I = 2$ A flowing through them. The circuit has an air gap $l_0 = 1$ mm long. Find the magnetic flux density vector in the air gap. — (a) $B = 0.433$ T. (b) $B = 0.833$ T. (c) $B = 0.633$ T.

P13.28. The magnetic circuit shown in Fig. P13.28 is made out of the same ferromagnetic material as the one in the previous problem. The dimensions of the circuit are $a = 6$ cm, $b = 4$ cm, $d = 1$ cm, $S_1 = S_2 = S = 1$ cm^2, $N_1 = 50$, $N_2 = 80$, and $N_3 = 40$. With $I_2 = I_3 = 0$, find the value of I_1 needed to produce a magnetic flux of $50 \,\mu$Wb in branch 3 of the circuit. — (a) $I_1 = 0.66$ A. (b) $I_1 = 1.66$ A. (c) $I_1 = 2.66$ A.

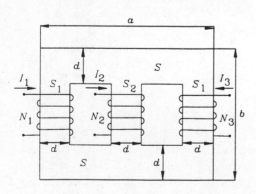

Fig. P13.28. A magnetic circuit.

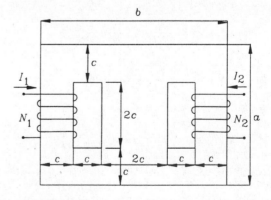

Fig. P13.29. A linear magnetic circuit.

P13.29. A linear magnetic circuit is shown in Fig. P13.29. The first winding has $N_1 = 100$ turns, and the second one $N_2 = 48$. Find the magnetic flux in all the branches of the circuit if the currents in the windings are (1) $I_1 = 10$ mA, $I_2 = 10$ mA; (2) $I_1 = 20$ mA, $I_2 = 0$ mA; (3) $I_1 = -10$ mA, $I_2 = 10$ mA. The magnetic material of the core has $\mu_r = 4000$, the dimensions

of the core are $a = 4\,cm$, $b = 6\,cm$, $c = 1\,cm$, and the thickness of the core is $d = 1\,cm$. —
Hint: designate the left, right and central branches by 1, 2 and 3, and orient them upward, downward and downward, respectively. Possible results for case 3 are: (a) $\Phi_1 = -7.25\,\mu Wb$, $\Phi_2 = 2.79\,\mu Wb$, $\Phi_3 = -14.04\,\mu Wb$. *(b)* $\Phi_1 = -3.25\,\mu Wb$, $\Phi_2 = 3.79\,\mu Wb$, $\Phi_3 = -9.04\,\mu Wb$. *(c)* $\Phi_1 = -5.25\,\mu Wb$, $\Phi_2 = 1.79\,\mu Wb$, $\Phi_3 = -7.04\,\mu Wb$.

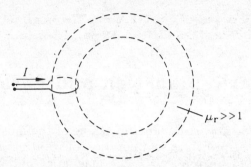

Fig. P13.30. A single loop on toroidal core.

P13.30. Shown in Fig. P13.30 is a single current loop on a toroidal core (indicated in dashed lines) of very high permeability. Assume that the core can be obtained by a gradual increase of the number of the Ampère currents, from zero to the final number per unit volume. Follow the process of creating the magnetic field in the core as the core becomes "denser". If your reasoning is correct, you should come to the answer to an important question: which is the physical mechanism of channeling the magnetic flux by ferromagnetic cores? — *Hint: when the number of Ampère currents is small, they are oriented by the magnetic field of the loop (primary field) only. By increasing the number of Ampère currents, the resultant field due to them (secondary field) is also increased, and the currents are oriented in response to the total magnetic field.*

Part 3: Slowly Time-Varying Electromagnetic Field

14. Electromagnetic Induction and Faraday's Law

- The *Faraday law of electromagnetic induction* is one of the most important laws of electromagnetism. The fundamental principles of operation of rotating generators, telephone, radio and television, etc., are based on that law. The magnetic coupling is also the result of the electromagnetic induction.

- If a charged particle is *accelerated*, an additional force to the Coulomb and magnetic forces acts on other charged particles, stationary or moving. This additional force is also much smaller than Coulomb's force, but time-varying currents in conductors involve a vast number of accelerated charges, and produce easily measurable effects. It is of the same *form* as the electric force ($F = QE$) but the electric field vector E in this case has quite different properties than the electric field vector of static charges. It is called *the induced electric field strength*. Most frequently, a charge is situated simultaneously in both a Coulomb-type and an induced electric field, so that the total electric force on the charge is given by

$$F = Q(E_{st} + E_{ind}). \qquad (14.1)$$

- If a charged particle is moving with a velocity v with respect to the source of the magnetic field, the induced electric field is obtained as

$$E_{ind} = v \times B \qquad (\text{V/m}). \qquad (14.2)$$

- A time-varying current distribution of density J produces an induced electric field given by the equation

$$E_{ind} = -\frac{\partial}{\partial t} \left(\frac{\mu_0}{4\pi} \int_v \frac{J dv}{r} \right) \qquad (\text{V/m}). \qquad (14.3)$$

In this equation, r is the distance of the volume element dv from the point where the induced field is being determined. In the case of currents over surfaces, $J dv$ in Eq. (14.3) should be substituted by $J_{rms} dS$, and in the case of a thin wire by $i dl$.

- Faraday's law is an equation giving the *total* electromotive force (emf) induced in a closed loop due to the induced electric field. More precisely, it gives the emf of the Thévenin generator equivalent to all the elemental generators due to electromagnetic induction acting along the loop. It can be expressed in several forms. If a closed conductive contour C is moving in a time constant magnetic field with a velocity v, it is given by

$$e = \oint_C (v \times B) \cdot dl \qquad \text{(V)}. \qquad (14.4)$$

The most general expression for the induced emf is

$$e = \oint_C E_{\text{ind}} \cdot dl = -\frac{d\Phi_{\text{through } C \text{ in } dt}}{dt} \qquad \text{(V)}. \qquad (14.5)$$

- A qualitative law describing the sense of the induced emf around a closed loop, the *Lentz law*, states that the induced current in a loop tends to reduce the time-variation of the magnetic flux by its own time-varying magnetic field.

- The induced currents inside conductive bodies (not wires) that are a result of the induced electric field are called *eddy currents*. Due to eddy currents, the total current in a conductor is the largest on the conductor surface. Therefore Joule's losses are increased, and the flux inside a solid ferromagnetic core is reduced greatly for time-varying fields, so that laminated cores must be used for alternating-current machinery. In some instances, eddy currents are produced on purpose, e.g., in so-called induction furnaces for melting metals.

- Substances with zero resistivity at very low temperatures are known as *superconductors*. It is impossible to change the magnetic flux through a superconducting loop by means of electromagnetic induction.

- The most general definition of voltage between two points in an electromagnetic field is defined as

$$V_{AB} = \int_A^B E_{\text{total}} \cdot dl \qquad \text{(V)}. \qquad (14.6)$$

Due to the induced electric field contained in E_{total}, the voltage between two points *depends on the path between them.*

QUESTIONS

S **Q14.1.** An observer in a coordinate system with the source of a time-invariant magnetic field observes a force $F = Qv \times B$ on a charge Q moving with a uniform velocity v with respect to his coordinate system. What is the force on Q observed by an observer moving with the charge? What is his interpretation of the velocity v? — *(a) $-F$, due to an electric field. (b) F, due to an electric field. (c) F, due to a magnetic field.*

Answer. The force is the same, but is interpreted as due to an *electric field*, E (which is equal to $v \times B$). In the observer's interpretation, v is the *negative* of the velocity of the first coordinate system with respect to him/her.

Q14.2. Three point charges are stationary in a coordinate system of the first observer. A second observer, in his coordinate system, moves with respect to the first with a uniform velocity. What kinds of fields are observed by the first observer, and what by the second? — *(a) Both observe an electric and a magnetic field. (b) The first observes a magnetic field, the second an electric field. (c) The first observes an electric field, the second both an electric field and a magnetic field.*

Q14.3. A small uncharged conducting sphere is moving in the field of a permanent magnet. Are there induced charges on the sphere surface? If they exist, how do an observer moving with the charge and an observer on the magnet explain their existence? — *(a) No, because they are moving only in a magnetic field. (b) Yes (what are the explanations of the two observers?). (c) No, because the magnetic field is time-constant (why?).*

Q14.4. A straight metal rod moves with a constant velocity v in a uniform magnetic field of magnetic induction B. The rod is normal to B, and v is normal to both the rod and to B. Sketch the distribution of the induced charges on the rod. What is the *electric* field of these charges inside the rod equal to? — *(a) Charges of opposite sign are induced as if the rod is in a uniform axial electric field. (b) There are no induced charges on the rod. (c) Charges of opposite sign are induced on the two sides of the rod curved surface.*

Q14.5. A small dielectric sphere moves in a uniform magnetic field. Is the sphere polarized? Explain. — *(a) Yes (why?). (b) No, because the field is uniform. (c) No, because the field is time-constant.*

S **Q14.6.** A charge Q is located close to a toroidal coil with time-varying current. Is there a force on the charge? Explain. — *(a) No, there is no electric or magnetic field outside the coil. (b) Yes, due to the magnetic field. (c) Yes, due the electric field.*

Answer. There is a force, since there is a time-varying electric field due to the time-varying current in the coil.

Q14.7. A wire of length l is situated in a magnetic field of flux density B parallel to the wire. Is an electromotive force induced in the wire if it is moved (1) along the lines of B, and (2) transverse to the lines of B? Explain. — *(a) In case 1 yes, in case 2 no. (b) In case 1 no, in case 2 yes. (c) No in both cases.*

Q14.8. Strictly speaking, do currents in branches of an ac electrical circuit depend on the circuit shape? Explain. — *(a) No, because this is the circuit-theory assumption. (b) No, because there is no induced electric field along circuit branches. (c) Yes, because there is an induced electric field along circuit branches which depends on the circuit shape.*

Q14.9. Does the shape of a dc circuit influence the currents in its branches? Explain. — *The answers are the same as in the preceding question.*

Q14.10. A circular metal ring carries a time-varying current, which produces a time-varying induced electric field. If the ring is set in oscillatory motion about the axis normal to its plane, will the induced electric field be changed? Explain your answer. — *(a) Yes, because the electrons will be additionally accelerated. (b) No, because this does not involve a motion of charges. (c) No, because the mechanical motion includes not only free charge carriers, but also all other charges inside matter.*

S **Q14.11.** A device for accelerating electrons known as the *betatron* (Fig. Q14.11) consists of a powerful electromagnet and an evacuated tube which is bent into a circle. Electrons are accelerated in the tube when the magnetic flux in the core is forced to rise approximately as a linear function of time. What is the physical mechanism of accelerating the electrons in the betatron? — *(a) By the electric field of static charges (explain). (b) By the magnetic field (explain). (c) By the induced electric field (explain).*

Answer. During the time interval in which the magnetic flux in the core rises, its magnetization and magnetization currents increase, and an induced electric field is obtained. Due to symmetry the lines of this field are circles centered at the electromagnet axis, and this field accelerates the electrons. Since they are moving in the magnetic field that exists between the magnet poles, their trajectory is bent in approximately a circle.

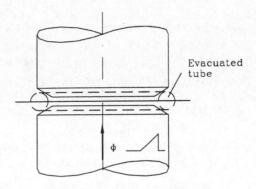

Fig. Q14.11. Sketch of a betatron.

Q14.12. Is it physically sound to speak about a partial electromotive force induced in a segment of one loop by the current in a segment of another (or even of the same) loop? Explain. — *Hint: recall the definition of the induced electric field. If a loop element is* d*l*, *what is the physical meaning of the product* $E_{\text{induced}} \cdot dl$?

Q14.13. A vertical conducting sheet (say, of aluminum) is permitted to fall under the action of gravity between the poles of a powerful permanent magnet. Is the motion of the sheet affected by the presence of the magnet? Explain. — *(a) It is slowed down. (b) It is not affected. (c) It is faster than without magnet.*

Fig. Q14.16. Two coupled coils.   **Fig. Q14.17.** A permanent magnet approaching a loop.

S **Q14.14.** A long solenoid wound on a Styrofoam core carries a time-varying current. It is encircled by three loops, one of copper, one of a resistive alloy, and the third of a bent moist filament (a poor conductor). In which loop is the induced electromotive force the greatest? — *(a) In the copper loop. (b) In the moist filament. (c) It is the same in all three.*

Answer. The electromotive force is the same in all of them, but the resulting induced current is the greatest in the copper loop.

Q14.15. What becomes different in question Q14.14 if the solenoid is wound onto a ferromagnetic core? — *(a) The answer is not changed, but the emf is greater. (b) Nothing at all is different. (c) The answer is not changed, but the emf is smaller.*

Q14.16. Assume that the current $i(t)$ in circular loop 1 in Fig. Q14.16 in a certain time interval increases linearly in time. Will there be a current in the closed conducting loop 2? If you think that there will be, what is its direction? — *(a) There is no current in loop 2. (b) There is current in loop 2, in the same direction as in loop 1. (c) There is current in loop 2, in the opposite direction to that in loop 1.*

Q14.17. What is the direction of the current induced in the loop sketched in Fig. Q14.17? Explain. — *(a) Depends on the velocity of the magnet. (b) In the clockwise direction. (c) In the counterclockwise direction.*

S **Q14.18.** The coil in Fig. Q14.18 consists of N densely wound turns of thin wire. What is the voltage of the generator when compared with the case of a single turn? Explain using both the concept of magnetic flux and that of the induced electric field. — *(a) It is N times greater. (b) It is N^2 times greater. (c) It is $\sqrt{N}$ times greater.*

Answer. It is N times greater. In terms of the magnetic flux, the flux through N turns is N times that through a single turn. In terms of the induced electric field, we need to integrate the product $\mathbf{E}_{\text{ind}} \cdot d\mathbf{l}$ along all the N turns, which amounts to N times the integral along a single turn.

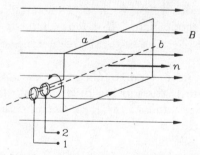

Fig. Q14.18. A coil rotating in magnetic field.

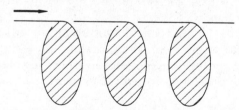

Fig. Q14.20. Three series loops.

Q14.19. A solenoid is wound onto a long, cylindrical permanent magnet. A voltmeter is connected to one end of the solenoid, and to a sliding contact that moves and makes a contact with a larger or smaller number of turns of the solenoid. Thus the magnetic flux in the closed loop of the voltmeter will vary in time. Does the voltmeter detect a time-varying voltage (assuming that it is sensitive enough to do so)? Explain. — *(a) It does detect a voltage. (b) It does not detect a voltage. (c) Depends on the velocity of the sliding contact.*

Q14.20. What is the direction of the reference unit vectors normal to the three loops in Fig. Q14.20? — *(a) In all three from left to right. (b) In all three from right to left. (c) The loops not being completely closed, it cannot be defined.*

Q14.21. A cylindrical permanent magnet falls without friction through a vertical evacuated metal tube. Is the fall accelerated? If not, what determines the velocity of the magnet? Explain. *(a) It falls as if there were no tube, because there is no friction. (b) It falls as if there were no tube, because there is no friction and no air in the tube. (c) After some time it reaches a constant velocity, due to the magnetic force of currents induced in the tube.*

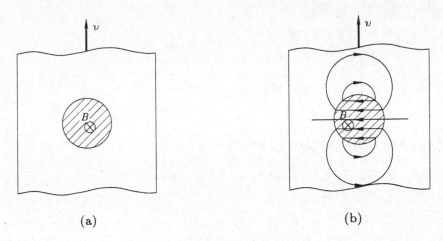

Fig. Q14.23. (a) A strip moving in a magnetic field. (b) Sketch of lines of induced currents in the strip.

S **Q14.22.** A strong permanent magnet is brought near the end plate of the pendulum of a wall timepiece. Does this influence the period of the pendulum? If it does, is the pendulum accelerated or slowed down? Does this depend on the type of the magnetic pole of the magnet closer to the pendulum plate? Explain. — *(a) Depends on the pole closer to the pendulum. (b) The pendulum is accelerated. (c) The pendulum is slowed down.*

Answer. Eddy currents are induced in the pendulum, and the magnetic force on them tends to slow down the pendulum. This effect does not depend on the magnetic pole of the magnet, only on the intensity of the magnetic field.

Q14.23. Fig. Q14.23a shows a sketch of a flat strip moving between the poles of a permanent magnet. Sketch the lines of the induced current in the strip. — *Hint: check if the lines sketched in Fig. Q14.2b are as you expected.*

Q14.24. The strip in question Q14.23 has in case (1) longitudinal, and in case (2) transverse slots with respect to the direction of motion. In which case are induced currents greater? Explain. — *(a) In case 1. (b) The same in both cases. (c) In case 2.*

Q14.25. Explain in detail how the right-hand side in the Faraday law for a rigid contour moving in a magnetic field,

$$e = \oint_C \boldsymbol{E}_{\mathrm{ind}} \cdot \mathrm{d}\boldsymbol{l} = \oint_C (\boldsymbol{v} \times \boldsymbol{B}) \cdot \mathrm{d}\boldsymbol{l} = \oint_C \left(\frac{\mathrm{d}\boldsymbol{s}}{\mathrm{d}t} \times \boldsymbol{B} \right) \cdot \mathrm{d}\boldsymbol{l} = \frac{\mathrm{d}}{\mathrm{d}t} \oint_C (\mathrm{d}\boldsymbol{s} \times \boldsymbol{B}) \cdot \mathrm{d}\boldsymbol{l},$$

is obtained from the middle expression in it. — *Hint: discuss why the factor* d/dt *can be taken out of the integral.*

S   **Q14.26.** The magnetic flux through a contour C at time t is Φ_1, and at time $t + \Delta t$ it is Φ_2. Is the time increment of the flux through C ($\Phi_2 - \Phi_1$), or ($\Phi_1 - \Phi_2$)? Explain. — *Hint: recall the definition of the increment of a function of one variable.*

Answer. It is ($\Phi_2 - \Phi_1$). Recall that the increment of a function is always defined as the difference of its value for an increased independent variable, and its value before this increase.

Q14.27. Is the *distribution* of the induced electromotive force around a contour seen from the right-hand side in Eq. (14.5)? Is it seen from the middle expression in that equation? Is it seen from the right-hand side in the Faraday law in the text of the question Q14.25? — *Hint: trace where in these equations induced elemental electromotive forces are integrated along the contour.*

S   **Q14.28.** Imagine an electric circuit with several loops situated in a slowly time-varying magnetic (and induced electric) field. Can you analyze such a circuit by circuit-theory methods? If you think you can, explain in detail how you would do it. — *Hint: there is a distributed electromotive force in all such loops. How it can be incorporated in the circuit-theory equations?*

Answer. There is a distributed electromotive force in all such loops. They need to be incorporated in the circuit-theory solution as equivalent Thévenin generators.

Q14.29. Does it make any sense at all to speak about the electromotive force induced in an *open* loop? If you think that this makes sense, explain what happens. *Hint: recall that even for a closed loop the induced emf is, in fact, a sum of elemental emf's. Will there be an induced current in an open loop?*

Q14.30. Explain in detail what a positive and what a negative electromotive force in Eq. (14.5) mean. — *(a) A negative emf acts in the direction of the contour C. (b) A positive emf acts in the direction opposite to that of the contour. (c) A positive emf acts in the direction of the contour.*

S   **Q14.31.** Why is the reduction of eddy current losses possible only if the vector $\mathbf{B}$ is parallel to a thin ferromagnetic sheet? — *(a) Because of boundary conditions for vector $\mathbf{B}$. (b) Because only then the flux through a contour in the sheet is small. (c) Because then the magnetic field in the sheet is smaller.*

Answer. If vector $\mathbf{B}$ is parallel to a thin ferromagnetic sheet, possible current loops have small areas, the enclosed magnetic flux is small, and the induced electromotive force in them is also small. If it is normal to the sheet, large current loops (as large as the sheet itself) are possible, enclosing a large flux.

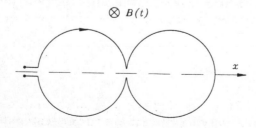

Fig. Q14.32. A loop in the form of an 8.

Q14.32. What is the induced electromotive force in the loop shown in Fig. Q14.32, if the magnetic field is time-varying? If the right half of the loop is turned about the x axis by 180 degrees, what is then the induced electromotive force? Explain both in terms of the magnetic flux and of the induced electric field. — *(a) $e(t) = -2S\,dB(t)/dt$ in both cases. (b) $e(t) = 2S\,dB(t)/dt$ in both cases. (c) $e(t) = -2S\,dB(t)/dt$ in the first case, zero in the second.*

Q14.33. A solid conducting body is placed near a loop with time-varying current. Are any forces acting on free charges inside the body? Explain. — *(a) No, because there is only a magnetic field inside the body. (b) Yes, because there is an induced electric field inside the body. (c) Yes, because inside the body there is both a magnetic and an electric field.*

Q14.34. Is there a magnetic force between the body and the loop from the preceding example? Explain. — *(a) Yes, there is an electric force. (b) Yes, there is a magnetic force on the induced currents. (c) No, because there is no magnetic field inside the body.*

S **Q14.35.** A planar insulated loop with time-varying current is placed on the surface of a plane conducting sheet. What happens in the sheet? Is the power required to drive the current in the loop different when it resides on the sheet than when it is isolated in space? Explain. — *Hint: there is an induced electric field in the sheet due to the loop current, but also an induced electric field along the loop due to the currents induced in the sheet.*

Answer. A current will be induced in the sheet. It is accompanied by both losses and change in the loop inductance, and so the power required to drive the current in the loop is different when the sheet is present.

Q14.36. Of two closed conducting loops, C_1 and C_2, C_1 is connected to a generator of time-varying voltage. Is there a current in C_2? Explain. — *Hint: think in terms of the induced electric field. Can you think of an example in which there is no induced current in C_2?*

Q14.37. An elastic metal circular ring carrying a steady current I is periodically deformed to a flat ellipse, and then released to retain its original circular shape. Is an electromotive force induced in the loop? Explain. — *(a) No, because the current is not time-varying. (b) No, because there is no time variation of flux through the ring. (c) Yes, because the segments of the loop are moving the magnetic field of the current in the loop.*

Q14.38. In the introduction to this Chapter, it was mentioned that the flux through a super-conducting loop cannot be changed. Does this mean that the flux through a superconducting loop cannot be changed by *any* means? — *(a) It cannot be changed by any means. (b) It cannot be changed by the induced electric field. (c) It can be changed by the induced electric field which varies in time very slowly.*

Q14.39. Why does the shape of voltmeter leads influence the time-varying voltage it measures? Why does this influence increase with frequency? — *Hint: have in mind the induced electric field and Eq. (14.3).*

PROBLEMS

S **P14.1.** Starting from Eq. (14.3), prove that the lines of the induced electric field vector of a circular current loop with a time-varying current are circles centered at the loop axis. — *Hint: refer to Fig. P14.1 and have in mind the symmetry of the structure.*

Solution. Let the loop be in the plane of the paper (Fig. P14.1). Consider an arbitrary point P in space (not necessarily in the plane of the loop). Let $d\boldsymbol{E}_{\text{ind}}$ and $d\boldsymbol{E}'_{\text{ind}}$ be the induced electric field vectors at P due to two symmetrical current elements, $i\,d\boldsymbol{l}$ and $i\,d\boldsymbol{l}'$, shown in the figure. In accordance to the expression in Eq. (14.3), these vectors are such that the vector $d\boldsymbol{E}_{\text{ind}} + d\boldsymbol{E}'_{\text{ind}}$ is tangential to the circular contour C centered at the loop axis and indicated in dashed line in the figure. The circular contour C is thus a line of the vector $\boldsymbol{E}_{\text{ind}}$.

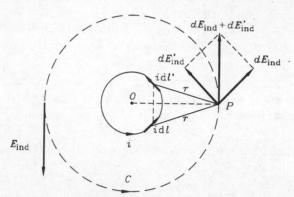

Fig. P14.1. A circular loop with a time-varying current.

P14.2. Assume that you know the induced electric field $\boldsymbol{E}_{\text{ind}}(t)$ along a circular line C of radius a in problem P14.1. A wire loop of radius a coincides with C. Evaluate the total electromotive force induced in the loop. Prove that this is, actually, the voltage of the Thévenin generator equivalent to the distributed infinitesimal generators around the loop. — *(a) $e(t) = E_{\text{ind}}(t)\,2\pi a$. (b) $e(t) = -E_{\text{ind}}(t)\,2\pi a$. (c) $e(t) = E_{\text{ind}}(t)\,\pi a$.*

S **P14.3.** Two coaxial solenoids shown in Fig. P14.3 are connected in series. A current $i(t) = 1.5\sin 1000t\,\text{A}$, where time is in seconds, flows through the solenoids. The dimensions are $a = 1\,\text{cm}$, $b = 2\,\text{cm}$, and $L = 50\,\text{cm}$. The number of turns in both is the same, $N_1 = N_2 = 1000$. Find the approximate induced electric field at points A_1 (surface of the inner solenoid), A_2 (halfway between the two solenoids), and A_3 (right outside of the outer solenoid). — *Hint: the points A_1, A_2 and A_3 are far from the solenoid ends, so the solenoids can be considered to be infinitely long. Therefore, for $N_1 = N_2$, $B = \mu_0 N_1 i / L$ ($a < r < b$), and $B = 0$ ($r < a$, $r > b$). Noting that the lines of the electric field are circles, use Faraday's law to find $E_{\text{ind}}(r)$.*

Solution. The points A_1, A_2 and A_3 are far from the solenoid ends, so we can approximately find the required fields by assuming that the solenoids are infinitely long. Therefore, since $N_1 = N_2$, the magnetic flux density is given by

$$B = \frac{\mu_0 N_1 i}{L} \quad (a < r < b), \qquad B = 0 \quad (r < a, \ r > b),$$

where r is the distance from the axis of the solenoids (what is the reference direction of $\boldsymbol{B}$?).

Faraday's law tells us that

$$E_{\text{ind}}\,2\pi r = -\frac{d\Phi}{dt},$$

Φ being the magnetic flux through the surface spanned over the circular contour of radius r (what is the reference direction of E_{ind}?). Hence we obtain

$$E_{\text{ind}} = 0 \quad (r < a), \qquad E_{\text{ind}} = -\frac{r^2 - a^2}{2r}\frac{dB}{dt} \quad (a < r < b), \qquad E_{\text{ind}} = -\frac{b^2 - a^2}{2r}\frac{dB}{dt} \quad (r > b).$$

It is left to the reader to find numerical values of E_{ind} at points A_1, A_2 and A_3.

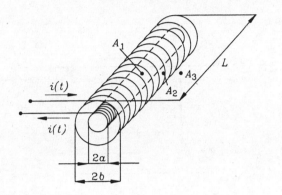

Fig. P14.3. Two coaxial solenoids.

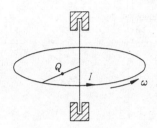

Fig. P14.4. Rotating current loop and charge.

P14.4. A circular loop of radius a, with a current I, rotates about the axis normal to its plane with an angular frequency ω. A small charge Q is fastened to a loop radius and rotates with the loop, as in Fig. P14.4. Is there a force on the charge? If it exists, determine its direction. — *(a) There is no force on the charge. (b) There is a force, radial. (c) There is a force, parallel to the axis of rotation.*

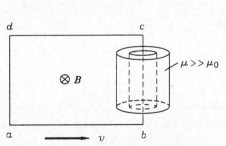

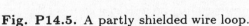

Fig. P14.5. A partly shielded wire loop.

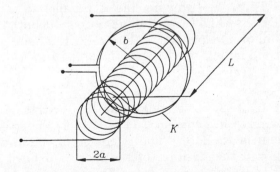

Fig. P14.6. A solenoid and a coil.

P14.5. A side of a rectangular wire loop is partly shielded from the magnetic field normal to the loop plane with a hollow ferromagnetic cylinder, as in Fig. P14.5. Therefore the sides ad and bc are situated in a different magnetic field. If the loop, together with the cylinder, moves in the indicated direction with a velocity v, will there be a current in the loop? If the answer is yes, this could serve for measuring the velocity with respect to the earth's magnetic field. — *Hint: use superposition. (a) There will be no current in the loop. (b) There will be a current in the loop. (c) Current in the loop exists, but its magnitude is too small to be measured.*

P14.6. A current $i(t) = I_m \sin(2\pi ft) = 2.5 \sin 314t$ A is flowing through the solenoid in Fig. P14.6, where frequency is in hertz and time is in seconds. The solenoid has $N_1 = 50$ turns of wire, and the coil K shown in the figure has $N_2 = 3$ turns. Calculate the emf induced in the coil, as well as the amplitude of the induced electric field along the coil turns. The dimensions indicated in the figure are: $a = 0.5$ cm, $b = 1$ cm, and $L = 10$ cm. Plot the induced emf as a function of N_1, N_2, and f. — (a) $e = 0.116 \cos 314t\,mV$, $E_{\text{ind m}} = 0.616\,mV/m$. (b) $e = -0.116 \cos 314t\,mV$, $E_{\text{ind m}} = 0.313\,mV/m$. (c) $e = -0.116 \cos 314t\,mV$, $E_{\text{ind m}} = 0.616\,mV/m$.

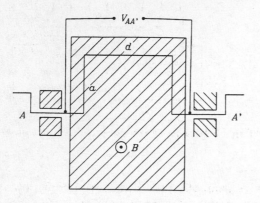

Fig. P14.7. A rotating conductor.   **Fig. P14.8.** A conductor and two-wire line.

P14.7. If the conductor AA' in Fig. P14.7 is rotating at a constant angular velocity and makes n turns per second, find the voltage $V_{AA'}$ as a function of time. Assume that at $t = 0$ the conductor was in the position shown in the figure. — (a) $V_{AA'}(t) = 2\pi n Bad \cos(2\pi nt)$. (b) $V_{AA'}(t) = 2\pi n Bad \sin(2\pi nt)$. (c) $V_{AA'}(t) = -2\pi n Bad \sin(2\pi nt)$.

P14.8. A two-wire line is parallel to a long straight conductor with a dc current I (Fig. P14.8). The two-wire line is open at both ends, and a conductive bar is sliding along it with a uniform velocity v, as shown in the figure. Find the potential difference between the two line conductors. — (a) $V_{A'A}(t) = \mu_0 Iv \ln[(d+a)/a]/(2\pi)$. (b) $V_{A'A}(t) = \mu_0 Iv \ln[(d+a)/a]/\pi$. (c) $V_{A'A}(t) = \mu_0 Iv \ln[(d+2a)/a]/(2\pi)$.

S **P14.9.** A rectangular wire loop with sides of length a and b is moving away from a straight wire with a current I (Fig. P14.9). The velocity of the loop, v, is constant. Find the induced emf in the loop. The reference direction of the loop is shown in the figure. Assume that at $t = 0$ the position of the loop is defined by $x = a$. — *For the reference direction in Fig. P14.9, (a)* $e = \mu_0 Iabv/[2\pi(a + vt)(2a + vt)]$. *(b)* $e = -\mu_0 Iabv/[2\pi(a + vt)(2a + vt)]$. *(c)* $e = -\mu_0 Iabv/[\pi(a + vt)(2a + vt)]$.

Solution. The magnetic flux through the surface S of the rectangle, oriented towards the reader, is given by

$$\Phi = -\int_S B\,dS = -\int_x^{x+a} \frac{\mu_0 I}{2\pi r} b\,dr = -\frac{\mu_0 Ib}{2\pi} \ln\frac{x+a}{x},$$

where $x = a + vt$. The induced emf in the loop, for the reference direction in Fig. P14.9, is

$$e = -\frac{d\Phi}{dt} = -\frac{\mu_0 I a b v}{2\pi(a + vt)(2a + vt)}.$$

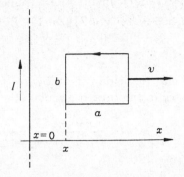

Fig. P14.9. A moving frame in magnetic field.

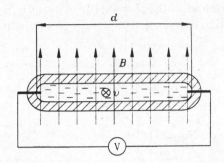

Fig. P14.11. Measurement of fluid velocity.

P14.10. The current flowing through the straight wire from the preceding problem is now $i(t)$ (a function of time). Find the induced emf in the loop, which is moving away from the wire as in the preceding problem. What happens to your expression for the emf when (1) the frame stops moving, or (2) when $i(t)$ becomes a dc current, I? — *Hint: there are two components of the emf, one due to the motion of the frame, and the other to the magnetic field varying in time.*

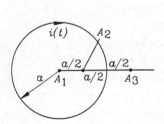

Fig. P14.12. Cross section of a solenoid.

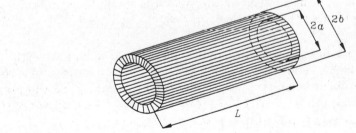

Fig. P14.13. A tubular coil.

P14.11. A liquid with a small but finite conductivity is flowing through a flat insulating pipe with an unknown velocity v. The velocity of the fluid is roughly uniform over the cross section of the pipe. To measure the fluid velocity, the pipe is in a magnetic field with a flux density vector $\mathbf{B}$ normal to the pipe, as shown in Fig. P14.11. Two small electrodes are in contact with the fluid at the two ends of the pipe. A voltmeter with large input impedance shows a voltage V when connected to the electrodes. Find the velocity of the fluid. — *(a)* $v = V/(2Bd)$. *(b)* $v = 2V/(Bd)$. *(c)* $v = V/(Bd)$.

P14.12. Shown in Fig. P14.12 is the cross section of a very long solenoid of radius $a = 1$ cm, with $N' = 2000$ turns/m. In the time interval $0 \leq t \leq 1$ s, a current $i(t) = 50t$ A flows through the solenoid. Determine the acceleration of an electron at points A_1, A_2, and A_3 indicated in the figure. (Note: the acceleration, $\mathbf{a}$, is found from the relation $\mathbf{F} = m\mathbf{a}$, where $\mathbf{F}$ is the force on the electron.) — (a) $a_1 = 0$, $a_2 = 3.5 \cdot 10^7$ m/s^2 and $a_3 = 5.3 \cdot 10^7$ m/s^2. (b) $a_1 = 0$, $a_2 = 5.5 \cdot 10^7$ m/s^2 and $a_3 = 7.3 \cdot 10^7$ m/s^2. (c) $a_1 = 0$, $a_2 = 2.5 \cdot 10^7$ m/s^2 and $a_3 = 4.3 \cdot 10^7$ m/s^2.

S **P14.13.** Determine approximately the induced electric field strength inside the tubular coil sketched in Fig. P14.13. The current intensity in the coil is $I = 0.02 \cos 10^6 t$ A, the number of turns is $N = 100$, and the coil dimensions are $a = 1$ cm, $b = 1.5$ cm, and $L = 10$ cm. — (a) $E_{\text{ind}} = 0.162 \sin 10^6 t$ V/m. (b) $E_{\text{ind}} = 1.12 \sin 10^6 t$ V/m. (c) $E_{\text{ind}} = -6.31 \sin 10^6 t$ V/m.

Solution. Starting from Eq. (14.3), we conclude that the induced electric field in the tube (for $r < a$) is axial and approximately uniform. The magnetic flux density vector in the tube is circular (the lines of B are circles centered at the tube axis), of intensity

$$B = \frac{\mu_0 N i}{2 \pi r}$$

(r is the distance from the tube axis). By neglecting the induced electric field outside the tube (this can be done since $L \gg b$), we can write

$$E_{\text{ind}} L = -\frac{d\Phi}{dt}, \qquad \Phi = \frac{\mu_0 N i L}{2\pi} \ln \frac{b}{a},$$

so that $E_{\text{ind}} = 0.162 \sin 10^6 t$ V/m.

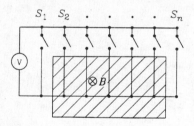

Fig. P14.14. A test of electromagnetic induction.

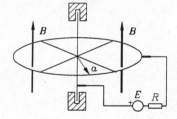

Fig. P14.15. An idealized electric motor.

P14.14. Sketched in Fig. P14.14 is an experimental setup for the analysis of electromagnetic induction. By closing sequentially the switches $S_1, \ldots, S_n$, it is possible to change the magnetic flux through the closed contour shown from zero to a maximal value. Will the voltmeter indicate an emf induced in the circuit? Explain. — *Hint: note that there is neither a motion of the contour in the magnetic field, nor a time variation of the magnetic field in which the contour is situated.*

P14.15. Find the angular velocity of the rotor of an idealized electric motor shown in Fig. P14.15 for the case when no load is connected to it. Does the value of R influence the angular velocity? The rotor is in the form of a metal wheel with four spokes, situated in a uniform magnetic field of magnetic flux density B, as shown in the figure. What is the direction of rotation of the rotor? — (a) $\omega = E/(Ba^2)$, clockwise. (b) $\omega = 2E/(Ba^2)$, clockwise. (c) $\omega = E/(Ba^2)$, counterclockwise.

P14.16. A circular loop of radius a rotates with an angular velocity ω about the axis lying in its plane and containing the center of the loop. It is situated in a uniform magnetic field of flux density $B(t)$ normal to the axis of rotation. Determine the induced emf in the loop. At $t = 0$, the position of the loop is such that B is normal to its plane. — (a) $e = 2B\pi a^2 \omega \sin \omega t$. (b) $e = -B\pi a^2 \omega \sin \omega t$. (c) $e = B\pi a^2 \omega \sin \omega t$.

P14.17. A winding of $N = 1000$ turns of wire with sinusoidal current of amplitude $I_m = 200\,\text{mA}$ is wound on a thin toroidal ferromagnetic core of mean radius $a = 10\,\text{cm}$. Fig. P14.17a shows the idealized hysteresis loop of the core corresponding to the sinusoidal magnetization of the core to saturation in both directions. Also wound on the toroid are several turns of wire over the first winding. Plot the emf induced in the second winding during one period of the sinusoidal current in the first winding. — *Compare your results with those in Fig. P14.17b.*

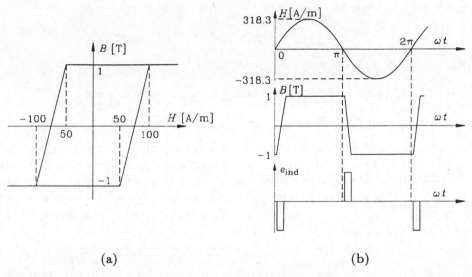

(a)   (b)

Fig. P14.17. (a) An idealized hysteresis loop. (b) Time dependence of H, B, and e_{ind}.

S **P14.18.** Shown in Fig. P14.18 is a rectangular loop encircling a very long solenoid of radius R and with N' turns of wire per unit length. The amplitude of current in the winding is I_m, and its angular frequency is ω. Determine the emf induced in the entire rectangular loop, as well as in its sides a and b separately. — *Hint: note that the induced electric field is normal to the loop diagonals, and determine the flux through the four triangles obtained by dividing the loop by its diagonals.*

Solution. The magnetic flux through the loop is

$$\Phi = \mu_0 N' I_m \cos \omega t.$$

Consider the triangle with a side a and a vertex coinciding with the center of the loop (there are two such triangles in Fig. P14.18), and also the triangle with a side b. The fluxes through these triangles are

$$\Phi_a = \frac{\pi - \alpha}{2\pi}\,\Phi, \qquad \Phi_b = \frac{\alpha}{2\pi}\,\Phi,$$

where $\alpha/2 = \tan^{-1}(b/a)$ (see the figure).

There is no emf induced in the sides of triangles indicated in dashed line in the figure (vector $\boldsymbol{E}_{\text{ind}}$ is normal to these sides). So, the induced electromotive forces in the entire loop, in side a, and in side b, are respectively given by

$$e = -\frac{\mathrm{d}\Phi}{\mathrm{d}t}, \quad e_a = -\frac{\mathrm{d}\Phi_a}{\mathrm{d}t}, \quad e_b = -\frac{\mathrm{d}\Phi_b}{\mathrm{d}t}.$$

The same results can be obtained as line integrals of the induced electric field, but it is not so simple.

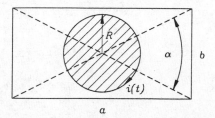

Fig. P14.18. A rectangular loop encircling a solenoid.

P14.19. The cross section of a thick coil with a large number, N, of turns of thin wire, is shown in Fig. P14.19. The coil is situated in a time-varying magnetic field of flux density $\boldsymbol{B}(t)$, in the indicated direction. Determine the emf induced in the coil. — *Hint: note that different layers of turns have different fluxes through them. The induced emf in the coil is $e = -\mathrm{d}\Phi/\mathrm{d}t$, where: (a) $\Phi = \pi B(t)N(a^2 + ab + b^2)\cos\alpha$. (b) $\Phi = -(1/6)\pi B(t)N(a^2 + ab + b^2)\cos\alpha$. (c) $\Phi = -(1/3)\pi B(t)N(a^2 + ab + b^2)\cos\alpha$.*

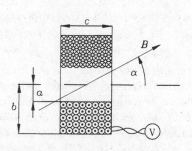

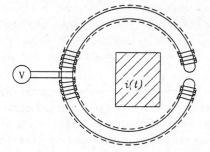

Fig. P14.19. A coil of rectangular cross section.   **Fig. P14.20.** A conductor encircled by a coil.

S   **P14.20.** The conductor whose cross section is shown shaded in Fig. P14.20 carries a sinusoidal current of amplitude I_{m} and angular frequency ω. The conductor is encircled by a flexible thin rubber strip of cross-sectional area S, densely wound along its length with N' turns of wire per unit length. The measured amplitude of the voltage between the terminals of the strip winding is V_{m}. Determine I_{m}. — (a) $I_{\text{m}} = V_{\text{m}}/(\mu_0 N' S)$. (b) $I_{\text{m}} = 2V_{\text{m}}/(\mu_0 N' S)$. (c) $I_{\text{m}} = V_{\text{m}}/(2\mu_0 N' S)$.

Solution. There are $\mathrm{d}N = N'\mathrm{d}l$ turns of wire on a length $\mathrm{d}l$ of the strip. The magnetic flux through a single turn is $\Phi_0 = \boldsymbol{B} \cdot \boldsymbol{S}$, and that through $\mathrm{d}N$ turns is

$$\mathrm{d}\Phi = \Phi_0 \mathrm{d}N = N'\mathrm{d}l \cdot BS,$$

since S and $\mathrm{d}l$ are colinear. The total flux through all the turns of the flexible solenoid is thus

$$\Phi = \oint_C \mathrm{d}\Phi = N'S \oint_C B \cdot \mathrm{d}l = \mu_0 N'S\,i(t),$$

according to the Ampère law applied to the contour C along the strip. The induced emf in the winding is $e = -\mathrm{d}\Phi/\mathrm{d}t$, so that, finally, the expression for the amplitude of $i(t)$ reads $I_{\mathrm{m}} = V_{\mathrm{m}}/(\mu_0 N'S\omega)$.

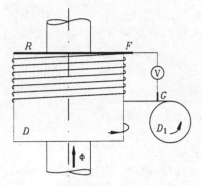

Fig. **P14.21.** A test of electromagnetic induction.

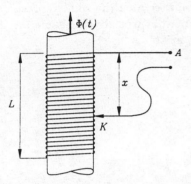

Fig. **P14.22.** A coil with a sliding contact.

P14.21. Wire is being wound from a drum D_1 onto a drum D, at a rate of N' turns per unit time (Fig. P14.21). The end of the wire on drum D is fastened to the ring R, which has a sliding contact F. A voltmeter is connected between the contact F and another contact G. Through the drum D there is a constant flux Φ, as indicated. What is the electromotive force measured by the voltmeter? — *Hint: recall that the emf can be induced in only two ways: by moving sections of the contour in the magnetic field, or by varying the magnetic field in time.*

S **P14.22.** A cylindrical coil is tightly wound around a ferromagnetic core with time-varying magnetic flux $\Phi(t) = \Phi_{\mathrm{m}} \cos \omega t$, as shown in Fig. P14.22. The length of the coil is L, and the number of turns in the coil is N. If a sliding contact K moves along the coil according to the law $x = L(1 + \cos \omega_1 t)/2$, what is the time dependence of the electromotive force between contacts A and K? Plot your result. — *(a)* $e = N\omega\Phi_{\mathrm{m}} \sin \omega t\,(1 + \cos \omega_1 t)$. *(b)* $e = (N\omega\Phi_{\mathrm{m}}/2) \sin \omega t\,(1 + \cos \omega_1 t)$. *(c)* $e = (3N\omega\Phi_{\mathrm{m}}/2) \sin \omega t\,(1 + \sin \omega_1 t)$.

Solution. According to Faraday's law of electromagnetic induction, an electromotive force is induced in each turn of the coil, and its value is

$$e_1 = -\frac{\mathrm{d}\Phi}{\mathrm{d}t} = \omega\Phi_{\mathrm{m}} \sin \omega t.$$

Since between the contacts A and K there are at the same instant

$$n = \frac{Nx}{L} = \frac{N}{2}(1 + \cos \omega_1 t)$$

turns, the electromotive force induced in all of them is

$$e = ne_1 = \frac{N\omega\Phi_m}{2}\sin\omega t\,(1 + \cos\omega_1 t).$$

Note that in this case Faraday's law cannot be written in the form

$$e = -\frac{d\Phi_{total}}{dt} = -\frac{d(n\Phi)}{dt} = -n\frac{d\Phi}{dt} - \Phi\frac{dn}{dt}.$$

The second term on the right represents the emf that would be induced as the result of an increase of the number of turns. By the same arguments as in the preceding problem, it is clear that such an emf does not exist.

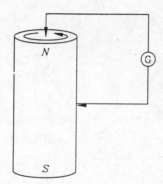

Fig. P14.23. A rotating magnet with sliding contacts.

P14.23. A cylindrical conducting magnet of circular cross section rotates about its axis with a uniform angular velocity. A galvanometer G is connected to the equator of the magnet and to the center of one of its bases by means of sliding contacts, as shown in Fig. P14.23. If such an experiment is performed, the galvanometer indicates a certain current through the circuit (Faraday, 1832). Where is the electromotive force induced: in the stationary conductors connecting the sliding contacts with the galvanometer, or in the magnet itself? — *Hint: note that the magnetic field is time-constant in spite of the magnet rotation, and the material of the magnet rotates in this field.*

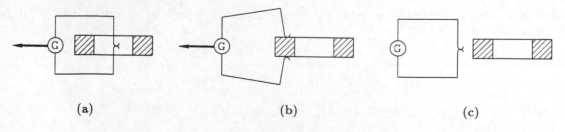

Fig. P14.24. (a) A moving elastic loop encircles a magnetized toroid. (b) The loop is moving, but closed by conducting toroid body. (c) The loop does not move and does not encircle the toroid.

P14.24. A ferromagnetic toroid with no air gap is magnetized so that no magnetic field exists outside it. The toroid is encircled by an elastic metal loop, as in Fig. P14.24a. The loop is now taken out of the toroid in such a way that during the process, the loop is always electrically closed through the conducting material of the toroid, as in Figs. P14.24b and c. The magnetic flux through the contour was obviously changed from a value Φ, the flux through the toroid, to zero. A formal application of Faraday's law leads to the conclusion that a certain charge will flow through the circuit during the process, but in this case it is not possible to detect any current (Herring, 1908). Explain the negative result of the experiment. — *Hint: recall that the emf can be induced in only two ways: by moving sections of the contour in the magnetic field, or by varying the magnetic field in time.*

P14.25. Discuss the possibility of constructing a generator of electromotive force constant in time, operating on the basis of electromagnetic induction. — *Hint: solve the equation* $\frac{d\Phi}{dt} = -E = $ constant, *which gives the flux necessary to obtain an emf constant in time.*

S **P14.26.** In a straight copper wire of radius $a = 1\,\mathrm{mm}$ there is a sinusoidal current $i(t) = 1\cos\omega t\,\mathrm{A}$. A voltmeter is connected between points 1 and 2, with leads of the shape shown in Fig. P14.26. If $b = 50\,\mathrm{cm}$ and $c = 20\,\mathrm{cm}$, evaluate the voltage measured by the voltmeter for (1) $\omega = 314\,\mathrm{rad/s}$, (2) $\omega = 10^4\,\mathrm{rad/s}$, and (3) $\omega = 10^6\,\mathrm{rad/s}$. Assume that the resistance of the copper conductor per unit length, R', is approximately that for a dc current (which actually is *not* the case, due to the so-called skin effect), and evaluate for the three cases the difference between the potential difference $V_1 - V_2 = R'bi(t)$ and the voltage induced in the leads of the voltmeter. — *The rms value of the potential difference $V_1 - V_2 = 1.98\,mV$. The difference between this potential difference and the voltage indicated by the voltmeter for the three specified frequencies is: (a) (1) $217.4\,\mu V$, (2) $5.75\,mV$, and (3) $5.75\,V$. (b) (a) (1) $117.4\,\mu V$, (2) $3.75\,mV$, and (3) $3.75\,V$. (c) (a) (1) $87.4\,\mu V$, (2) $2.75\,mV$, and (3) $2.75\,V$.*

Solution. The voltage between the terminals of the voltmeter is

$$v_{\text{voltmeter}} = (V_1 - V_2) - e = R'bi - e,$$

where $R' = 1/(\sigma_{\mathrm{Cu}}\pi a^2)$ ($\sigma_{\mathrm{Cu}} = 5.7 \cdot 10^7\,\mathrm{S/m}$), and e is the induced emf in the rectangular contour containing the voltmeter and the wire segment between points 1 and 2. This emf is given by (see solution to problem P14.10)

$$e = \frac{\mu_0 b}{2\pi}\frac{di}{dt}\ln\frac{c+a}{a}.$$

The rms value of the potential difference $V_1 - V_2$ amounts to $1.98\,\mathrm{mV}$, and does not depend on frequency. The difference between this potential difference and the voltage indicated by the voltmeter for the three specified frequencies is: (1) $117.4\,\mu\mathrm{V}$, (2) $3.75\,\mathrm{mV}$, and (3) $3.75\,\mathrm{V}$. This difference represents an error in measuring the potential difference using the voltmeter with such leads. We see that in case (2) the relative error is as large as 189%, and that in case (3) such a measurement is meaningless.

P14.27. A circular metal loop of radius R, conductivity σ and cross-sectional area S encircles a long solenoid with a time-varying current $i(t)$ (Fig. P14.27). The solenoid has N' turns of wire per unit length, and its radius is r. Determine the current in the loop, and the voltage between points A and B of the loop along paths a, b, c, and d. Neglect the induced electric field in the loop due to the loop current itself. Determine also the voltage between points A and

C, along the path AcC, and along the path $AbBC$. — *Let the loop be oriented clockwise, and let the emf induced in it be* e. (a) $V_{AaB} = V_{AbB} = -V_{AcB} = -V_{AdB} = e/3$, $V_{AcC} = -3e/4$, $V_{AbBC} = e/4$. (b) $V_{AaB} = V_{AbB} = -V_{AcB} = -V_{AdB} = -e/2$, $V_{AcC} = e/4$, $V_{AbBC} = -e/4$. (c) $V_{AaB} = V_{AbB} = -V_{AcB} = -V_{AdB} = e/2$, $V_{AcC} = -e/4$, $V_{AbBC} = 3e/4$.

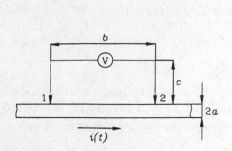

Fig. P14.26. Measurement of ac voltage.

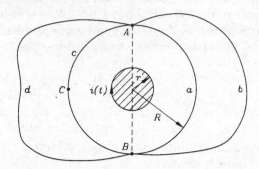

Fig. P14.27. A solenoid encircled by a circular loop.

P14.28. Repeat problem P14.27 assuming that the two halves of the loop have different conductivities, σ_1 and σ_2, and they meet at points A and B. — *Let the conductivity of the semicircle* AaB *be* σ_1, *and that of* AcB σ_2. (a) $V_{AaB} = V_{AbB} = e\sigma_2/(\sigma_1 + \sigma_2)$, $V_{AcB} = V_{AdB} = -e\sigma_1/(\sigma_1 + \sigma_2)$, $V_{AcC} = V_{AcB}/2$, $V_{AbBC} = V_{AaB}/2$. (b) $V_{AaB} = V_{AbB} = e\sigma_2/(2\sigma_1 + \sigma_2)$, $V_{AcB} = V_{AdB} = -e\sigma_1/(2\sigma_1 + \sigma_2)$, $V_{AcC} = V_{AcB}/2$, $V_{AbBC} = V_{AaB}/2$. (c) $V_{AaB} = V_{AbB} = e\sigma_2/(\sigma_1 + 2\sigma_2)$, $V_{AcB} = V_{AdB} = -e\sigma_1/(\sigma_1 + 2\sigma_2)$, $V_{AcC} = V_{AcB}/2$, $V_{AbBC} = V_{AaB}/2$.

15. Inductance

• A time-varying current in one wire loop induces an emf in another loop. This is called *mutual induction*, or *magnetic coupling*. The electromagnetic parameter that enables a simple evaluation of this emf in linear media is the *mutual inductance*. A single wire loop with time-varying current creates a time-varying induced electric field along itself, resulting in an induced emf in the loop, an effect known as *self-induction*. In linear media, this emf is evaluated in terms of the electromagnetic parameter known as the inductance, or *self-inductance*, of the loop.

• The emf induced in a loop C_2 by a current $i_1(t)$ in a nearby loop C_1 is given by

$$e_{12}(t) = \oint_{C_2} \boldsymbol{E}_{1\text{ind}}(t) \cdot \mathrm{d}\boldsymbol{l}_2 \qquad \text{(V)}, \qquad (15.1)$$

where $\boldsymbol{E}_{1\text{ind}}(t)$ is the induced electric field of current $i_1(t)$ along loop C_2. In linear media the magnetic flux of current $i_1(t)$ through C_2 is proportional to $i_1(t)$,

$$\Phi_{12}(t) = L_{12} i_1(t) \qquad \text{(Wb)}. \qquad (15.2)$$

• It can be proved that $L_{21} = L_{12}$, so that

$$L_{12} = \frac{\Phi_{12}(t)}{i_1(t)} = L_{21} = \frac{\Phi_{21}(t)}{i_2(t)} \qquad (\text{henry} - \text{H}), \qquad (15.3)$$

which is the definition of *mutual inductance* (also denoted by M). Mutual induction for dc currents does not exist, but Eq. (15.3) is valid in that case also, and is used frequently for the determination of L_{12}. By combining the Faraday law and Eq. (15.2), mutual inductance can also be defined in terms of the induced emf,

$$e_{12}(t) = -\frac{d\Phi_{12}(t)}{dt} = -L_{12}\frac{di_1(t)}{dt} \qquad (\text{V}). \qquad (15.4)$$

• For a single wire loop in a linear medium, the induced emf in the loop by the current $i(t)$ in the loop is given by

$$e(t) = -\frac{d\Phi_{\text{self}}(t)}{dt} = -L\frac{di(t)}{dt} \qquad (\text{V}), \qquad (15.5)$$

where L is the *self-inductance* of the loop,

$$L = \frac{\Phi_{\text{self}}(t)}{i(t)} \qquad (\text{H}). \qquad (15.6)$$

• For two coupled loops, the self-inductances and the mutual inductance satisfy the inequality

$$L_{11}L_{22} \geq L_{12}^2, \qquad (15.7)$$

or

$$L_{12} = k\sqrt{L_{11}L_{22}}, \quad -1 \leq k \leq 1. \qquad (15.8)$$

The coefficient k is called the *coupling coefficient*.

QUESTIONS

S **Q15.1.** What does the expression in Eq. (15.1) for the emf induced in a wire loop actually represent? — *(a) The emf of a generator concentrated at a point of the loop. (b) The emf distributed along the loop. (c) The Thévenin equivalent to all the elemental electromotive forces induced around the loop.*

Answer. The Thévenin equivalent to all the elemental electromotive forces induced around the loop.

Q15.2. Why does mutual (and self) inductance have no practical meaning in the dc case? — *(a) Magnetic coupling in the dc case exists, but is of no interest. (b) Magnetic coupling in the dc case does not exist at all. (c) Magnetic coupling in the dc case is too small to be of practical interest.*

Q15.3. Explain why mutual inductance for a toroidal coil and a wire loop encircling it (e.g., see Fig. P15.1) does not depend on the shape of the wire loop. — *(a) The magnetic flux due to the current in the toroidal coil through all such loops is the same. (b) The magnetic flux due to the current in the loop through the toroidal coil evidently does not depend on the shape of the loop. (c) The lines of the induced electric field of current in the toroidal coil are closed lines.*

Q15.4. Explain in terms of the induced electric field why the emf induced in a coil encircling a toroidal coil and consisting of several loops (e.g., see Fig. P15.1) is proportional to the number of turns of the loop. — *Hint: follow the integration path in Eq. (15.1).*

Q15.5. Can mutual inductance be negative as well as positive? Explain by considering reference directions of the loops. — *(a) It can be only positive. (b) It can have both signs. (c) It can be only negative.*

S **Q15.6.** Mutual inductance of two simple loops is L_{12}. We replace the two loops by two very thin coils of the same shapes, with N_1 and N_2 turns of very thin wire. What is the mutual inductance between the coils? Explain in terms of the induced electric field. — *(a)* $(N_1/N_2)L_{12}$. *(b)* $\sqrt{N_1 N_2}L_{12}$. *(c)* $N_1 N_2 L_{12}$.

Answer. When evaluating the induced electromotive force in one loop by the current in the other, we need to adopt a direction around that other loop for the integration of the induced electric field. This electromotive force can, therefore, be both positive and negative — it is positive if it acts in the adopted direction, and negative otherwise. Consequently, the mutual inductance can be both positive and negative.

Q15.7. A two-wire line crosses another two-wire line at a distance d. The two lines are normal. Prove that the mutual inductance is zero, starting from the induced electric field. — *Hint: use the expression for the induced electric field in Eq. (14.3).*

S **Q15.8.** The self-inductance of a toroidal coil is proportional to the *square* of the number of turns of the coil. Explain this in terms of the induced electric field and induced voltage in the coil due to the current in the coil. — *Hint: follow the integration path in the evaluation of the induced emf.*

Answer. The induced electric field is proportional to the number of turns. The induced voltage in the coil, obtained as an integral of the induced electric field along all the turns, is therefore proportional to the square of the number of turns of the coil.

Q15.9. A thin coil is made of N turns of very thin wire pressed tightly together. If the self-inductance of a single turn of wire is L, what do you expect is the self-inductance of the coil? Explain in terms of the induced electric field. *(a)* NL. *(b)* $N^2 L$. *(c)* $\sqrt{N}L$.

Q15.10. Explain in your own words what the meaning of self-inductance of a coaxial cable is. *Hint: note that the lines of $\mathbf{B}$ are circles, and that the flux between the inner and outer conductor is that through a strip spanned between the two conductors.*

Q15.11. Is it physically sound to speak about the mutual inductance between two wire segments belonging either to two loops or to a single loop? Explain. — *Hint: recall the definition of the induced electric field and the emf induced in a wire element.*

Q15.12. Is it physically sound to speak about the self-inductance of a segment of a closed loop? Explain. *Hint: the same as for the preceding question.*

Q15.13. To obtain a resistive wire with the smallest self-inductance possible, the wire is sharply bent in the middle and the two mutually insulated halves are pressed tightly together, as shown in Fig. Q15.13. Explain why the self-inductance is minimal in terms of the induced electric field and in terms of the magnetic flux through the loop. — *Hint: have in mind Eq. (14.3) and note that the flux increases as the area increases (why?).*

S **Q15.14.** Can self-inductance be negative as well as positive? Explain in terms of the flux. — *(a) Yes, it depends on the reference directions. (b) No, reference directions are fixed. (c) Yes, since reference directions are arbitrary.*

Answer. The self inductance of a loop is defined as the flux due to the current in the loop, divided by that current. In that, the reference directions of the flux and current are always connected by the right-hand rule. Therefore the self inductance is always positive.

Q15.15. The self-inductance of two identical loops is L. What is approximately the mutual inductance between them if they are pressed together? Explain in terms of the induced electric field and in terms of the magnetic flux. — *(a) $|M| \simeq L/2$. (b) $|M| \simeq L$. (c) $|M| \simeq 2L$.*

Q15.16. Two coils are connected in series. Does the total (equivalent) inductance of the connection depend on their mutual position? Explain. — *(a) It does not. (b) Only if they have a small number of turns. (b) It does.*

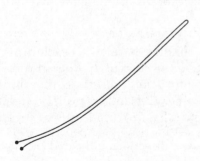

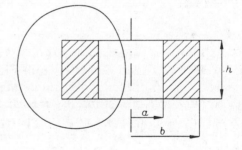

Fig. Q15.13. A loop with small self-inductance.

Fig. P15.1. A toroidal coil and wire loop.

S **Q15.17.** Pressed onto a thin conducing loop is an identical thin *superconducting* loop. What is the self-inductance of the conducting loop? Explain. — *Hint: recall the Lentz law, which is exact for a superconducting loop.*

Answer. Zero. A current is induced in the superconducting loop such that it cancels the flux due to current in the other loop. So, the flux will also be zero in the other loop, and thus the self-inductance of the other loop is zero.

Q15.18. A loop is connected to a source of voltage $v(t)$. As a consequence, a current $i(t)$ exists in the loop. Another conducting loop with no source is brought near the first loop. Will the current in the first loop be changed? Explain. — *(a) It will not be changed, because there is no current in the other loop. (b) It will be changed, due to the current in the other loop. (c) It will not be changed, because there is no source in the other loop.*

Q15.19. Answer question Q15.18 assuming that the source in the first loop is a dc source. Explain. — *The answers are the same as for the preceding question.*

Q15.20. A thin, flat loop of self-inductance L is placed over a flat surface of very high permeability. What is the new self-inductance of the loop? — *Hint: replace the surface by the image of the loop.*

PROBLEMS

P15.1. Find the mutual inductance between an arbitrary loop and the toroidal coil in Fig. P15.1. There are N turns around the torus, and the permeability of the core is μ. — *Assume the coil and the loop are oriented in the same way. (a) $L_{12} = (\mu N h/\pi)\ln(b/a)$. (b) $L_{12} = (2\mu N h/\pi)\ln(b/a)$. (c) $L_{12} = (\mu N h/2\pi)\ln(b/a)$.*

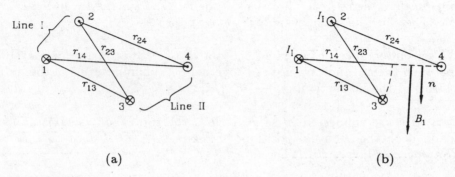

(a)   (b)

Fig. P15.2. (a) Two parallel two-wire lines, and (b) the surface for determining the magnetic flux.

S **P15.2.** Find the mutual inductance of two two-wire lines running parallel to each other. The cross section of the lines is shown in Fig. P15.2a. — *Hint: let a current I_I flow in line I. Use the surface in Fig. 15.2b to evaluate the magnetic flux per unit length of line II due to current I_I in conductor 1, and similarly for conductor 2. (a) $L'_{I,II} = (\mu_0/2\pi)\ln[(r_{14}r_{23})/(r_{13}r_{24})]$. (b) $L'_{I,II} = (\mu_0/2\pi)\ln[(r_{24}r_{23})/(r_{13}r_{14})]$. (c) $L'_{I,II} = (\mu_0/2\pi)\ln[(r_{13}r_{23})/(r_{14}r_{24})]$.*

Solution. Suppose that a current I_I flows through line I, in the directions indicated in the figure. Let us first evaluate the magnetic flux per unit length of line II due to current I_I in conductor 1. The simplest surface for determining the flux is that shown in Fig. 15.2b in dashed line. So we can write

$$(\Phi'_{I,II})_{\text{due to current in conductor 1}} = \int_{r_{13}}^{r_{14}} B_1(r)\, dr = \frac{\mu_0 I_I}{2\pi}\int_{r_{13}}^{r_{14}}\frac{dr}{r} = \frac{\mu_0 I_I}{2\pi}\ln\frac{r_{14}}{r_{13}}.$$

By analogy, the magnetic flux per unit length of line II due to current I_1 in conductor 2 is

$$(\Phi'_{I,II})_{\text{due to current in conductor 2}} = -\frac{\mu_0 I_I}{2\pi}\ln\frac{r_{24}}{r_{23}}.$$

Hence the total flux per unit length of line II is

$$\Phi'_{I,II} = (\Phi'_{I,II})_{\text{due to current in conductor 1}} + (\Phi'_{I,II})_{\text{due to current in conductor 2}} = \frac{\mu_0 I_I}{2\pi}\ln\frac{r_{14}r_{23}}{r_{13}r_{24}}.$$

The mutual inductance per unit length of the two lines is hence

$$L'_{I,II} = \frac{\Phi'_{I,II}}{I_I} = \frac{\mu_0}{2\pi} \ln \frac{r_{14} r_{23}}{r_{13} r_{24}}.$$

What is the order of magnitude of this inductance per one meter? What happens if $r_{14} r_{23} <$ $r_{13} r_{24}$, or if $r_{14} r_{23} = r_{13} r_{24}$? Explain.

P15.3. A cable-car track runs parallel to a two-wire phone line, as in Fig. P15.3. The cable-car power line and track form a two-wire line. The amplitude of the sinusoidal current through the cable-car wire is I_m and its angular frequency is ω. All conductors are very thin compared to the distances between them. Find the amplitude of the induced emf in the section of the phone line b long. — *Hint: use the expression for mutual inductance from the preceding problem.*

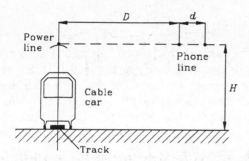

Fig. P15.3. Cable car parallel to phone line.

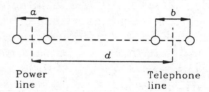

Fig. P15.4. Parallel power and phone lines.

P15.4. Parallel to a two-wire symmetrical power line along a distance h is a two-wire telephone line, as shown in Fig. P15.4. (1) Find the mutual inductance between the two lines. (2) Find the amplitude of the emf induced in the telephone line when there is a sinusoidal current of amplitude I_m and frequency f in the power line. As a numerical example, assume the following: $f = 100$ Hz, $I_m = 100$ A, $h = 50$ m, $d = 10$ m, $a = 50$ cm, and $b = 25$ cm. *Hint: use the results of problem P15.2.* *(a)* $L_{ab} = 18.5 \, nH$, $(e_{ind})_m = 0.985 \, mV$. *(b)* $L_{ab} = 32.5 \, nH$, $(e_{ind})_m = 1.785 \, mV$. *(c)* $L_{ab} = 12.5 \, nH$, $(e_{ind})_m = 0.785 \, mV$.

S **P15.5.** Two coaxial thin circular loops of radii a and b are in the same plane. Assuming that $a \gg b$ and that the medium is air, determine approximately the mutual inductance of the loops. As a numerical example, evaluate the mutual inductance if $a = 10$ cm, and $b = 1$ cm. — *(a)* $L_{12} = \pm 2.97 \, nH$. *(b)* $L_{12} = \pm 1.97 \, nH$. *(c)* $L_{12} = \pm 3.97 \, nH$.

Solution. Assume a current I_1 in the larger loop. The magnetic flux through the smaller loop can be approximately evaluated as $\Phi_{12} = B_1 \pi b^2$, where $B_1 = \mu_0 I_1/(2a)$ is the magnetic flux density of the larger loop in its center (see problem P12.7). Thus, the mutual inductance of the loops if oriented in the same direction is $L_{12} = \Phi_2/I_1 = \mu_0 \pi b^2/(2a) = 1.97$ nH.

P15.6. Two coaxial thin circular loops of radii a and b are in air a distance d $(d \gg a, b)$ apart. Determine approximately the mutual inductance of the loops. As a numerical example, evaluate the mutual inductance if $a = b = 1$ cm and $d = 10$ cm. — *(a)* $L_{12} = 19.7 \, pH$. *(b)* $L_{12} = 29.7 \, pH$. *(c)* $L_{12} = 39.7 \, pH$.

P15.7. Inside a very long solenoid wound with N' turns per unit length is a small flat loop of surface area S. The plane of the loop makes an angle θ with the solenoid axis. Determine and plot the mutual inductance between the solenoid and the loop as a function of θ. The medium is air. — (a) $L_{12} = \pm\mu_0 N' S \cos\theta$. (b) $L_{12} = \pm 2\mu_0 N' S \sin\theta$. (c) $L_{12} = \pm\mu_0 N' S \sin\theta$.

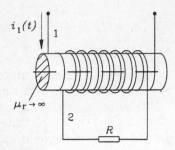

Fig. P15.8. Two coupled circuits.

P15.8. Assume that within a certain time interval the current in circuit 1 in Fig. P15.8 grows linearly, $i_1(t) = I_0 + It/t_1$. Will there be any current in circuit 2 during this time? If yes, what is the direction and magnitude of the current? The number of turns of the two coils is the same. — *Adopt the reference direction of current in the resistor, $i_2(t)$, to be from left to the right and note that $\mu_r \to \infty$ (a perfect transformer).* (a) $i_2(t) = 5\, i_1(t)/6$. (b) $i_2(t) = i_1(t)$. (c) $i_2(t) = 6\, i_1(t)/5$.

P15.9. Three coupled closed circuits have self-inductances equal to L_1, L_2, and L_3, resistances R_1, R_2, and R_3, and mutual inductances L_{12}, L_{13}, and L_{23}. Write the equations for the currents in all three circuits if a voltage $v_1(t)$ is connected to circuit 1 only. Then write the equations for the case when three sources, of voltages $v_1(t)$, $v_2(t)$, and $v_3(t)$ are connected to circuits 1, 2, and 3, respectively. — *Hint: write circuit-theory equations, noting that in all three circuits there are two additional emf's due to the currents in the other two circuits.*

S **P15.10.** A coaxial cable has conductors of radii a and b. The inner conductor is coated with a layer of ferrite d thick ($d < b - a$) and of permeability μ. The rest of the cable is air-filled. Find the external self-inductance per unit length of the cable. What should your expression reduce to when (1) $d = 0$, and (2) when $d = b - a$? — (a) $L' = (1/2\pi)\{\mu \ln[(a + 2d)/a] + \mu_0 \ln[b/(a+d)]\}$. (b) $L' = (1/\pi)\{\mu \ln[(a+d)/a] + \mu_0 \ln[b/(a+d)]\}$. (c) $L' = (1/2\pi)\{\mu \ln[(a+d)/a] + \mu_0 \ln[b/(a+d)]\}$.

Solution. Assume a current I in the cable. The flux through the flat surface of length l and width $b - a$ between the cable conductors is given by

$$\Phi = \frac{Il}{2\pi}\left(\mu \int_a^{a+d} \frac{dr}{r} + \mu_0 \int_{a+d}^b \frac{dr}{r}\right) = \frac{Il}{2\pi}\left(\mu \ln\frac{a+d}{a} + \mu_0 \ln\frac{b}{a+d}\right).$$

The external self-inductance per unit length of the cable is thus

$$L' = \frac{\Phi}{Il} = \frac{1}{2\pi}\left(\mu \ln\frac{a+d}{a} + \mu_0 \ln\frac{b}{a+d}\right).$$

See also Example 15.5.

When (1) $d = 0$ and (2) $d = b - a$, the expression for L' reduces to that of (1) an air-filled, and (2) a ferrite-filled coaxial cable. (Convince yourself that this is true.)

P15.11. The conductor radii of a two-wire line are a and the distance between them is d ($d \gg a$). Both conductors are coated with a thin layer of ferrite b thick ($b \ll d$) and of permeability μ. The ferrite is an insulator. Calculate the external self-inductance per unit length of the line. — (a) $L' = (1/\pi)\{\mu \ln[(a + b)/a] + \mu_0 \ln[d/(a + b)]\}$. (b) $L' = (1/2\pi)\{\mu \ln[(a + b)/a] + \mu_0 \ln[d/(a + b)]\}$. (c) $L' = (1/2\pi)\{\mu \ln[(2a + b)/a] + \mu_0 \ln[d/(2a + b)]\}$.

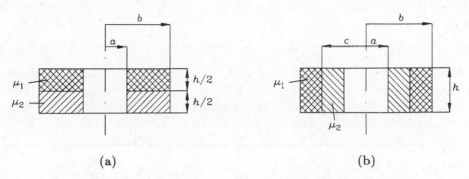

(a)   (b)

Fig. P15.12. Two toroidal coils with inhomogeneous core.

P15.12. The core of a toroidal coil of N turns consists of two materials, of respective permeabilities μ_1 and μ_2, as in each part of Fig. P15.12. Find the self-inductance of the toroidal coil and the mutual inductance between the coil and the loop positioned as in Fig. P15.1 if (1) the ferrite layers are of equal thicknesses, $h/2$, in Fig. P15.12a, and (2) the ferrite layers are of equal height h and the radius of the surface between them is c ($a < c < b$), in Fig. P15.12b. — *Hint: note that in both cases in the core $H = NI/(2\pi r)$, where I is the current in the coil, and r is the radial distance from the torus axis. Note also that, in both cases, $L_{12} = L/N$ (why?).* (a) $L^{(1)} = \{[(\mu_1 + \mu_2)N^2 h]/(2\pi)\} \ln(b/a)$, $L^{(2)} = (N^2 h/\pi)[\mu_1 \ln(c/a) + \mu_2 \ln(b/c)]$. (b) $L^{(1)} = \{[(\mu_1 + \mu_2)N^2 h]/(4\pi)\} \ln(b/a)$, $L^{(2)} = (N^2 h/2\pi)[\mu_1 \ln(c/a) + \mu_2 \ln(b/c)]$. (c) $L^{(1)} = \{[(\mu_1 + \mu_2)N^2 h]/(\pi)\} \ln(b/a)$, $L^{(2)} = (N^2 h/\pi)[\mu_1 \ln(c/a) + \mu_2 \ln(b/c)]$.

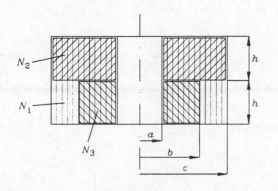

Fig. P15.13. Three toroidal coils.

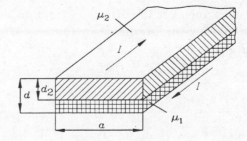

Fig. P15.14. A strip line with a two-layer dielectric.

P15.13. Three toroidal coils are wound in such a way that the coils 2 and 3 are inside coil 1, as in the cross section shown in Fig. P15.13. The medium is air. Find the self-inductances L_1, L_2, and L_3, and mutual inductances L_{12}, L_{13}, and L_{23}. What are the different values of inductance that can be obtained by connecting the three windings in series in different ways? — *We give the results for self-inductances only.* (a) $L_1 = (\mu_0 N_1^2 h/2\pi)\ln(c/a)$, $L_2 = (\mu_0 N_2^2 h/\pi)\ln(c/a)$, $L_3 = (\mu_0 N_3^2 h/\pi)\ln(b/a)$. (b) $L_1 = (\mu_0 N_1^2 h/4\pi)\ln(c/a)$, $L_2 = (\mu_0 N_2^2 h/4\pi)\ln(c/a)$, $L_3 = (\mu_0 N_3^2 h/4\pi)\ln(b/a)$. (c) $L_1 = (\mu_0 N_1^2 h/\pi)\ln(c/a)$, $L_2 = (\mu_0 N_2^2 h/2\pi)\ln(c/a)$, and $L_3 = (\mu_0 N_3^2 h/2\pi)\ln(b/a)$.

P15.14. The width of the strips of a long, straight strip line is a and their distance is d (Fig. P15.14 for $d_2 = 0$). Between the strips is a ferrite of permeability μ. Neglecting edge effects, find the inductance of the line per unit length. — *The correct solution is given in the answers to the next problem, for $d_2 = 0$ and $\mu_1 = \mu$.*

S **P15.15.** The width of the strips of a strip line is a and their distance is d. Between the strips there are two ferrite layers of permeabilities μ_1 and μ_2, and the latter is d_2 thick, as in Fig. P15.14. Neglecting edge effects, find the inductance of the line per unit length. — *Hint: note that between the strips $H = I/a$.* (a) $L' = [2\mu_1(d - d_2)/a] + \mu_2 d_2/a$. (b) $L' = [\mu_1(d - d_2)/a] + 2\mu_2 d_2/a$. (c) $L' = [\mu_1(d - d_2)/a] + \mu_2 d_2/a$.

Solution. See problem P13.13. The magnetic field intensity vector between the strips is parallel to them and lies in the transversal cross section of the line. Its magnitude is equal to $H = I/a$. The magnetic flux through the surface spanned between the strips, which we adopt to be normal to $\mathbf{H}$, and of length l, is given by

$$\Phi = \mu_1 H(d - d_2)l + \mu_2 H d_2 l,$$

so that the inductance per unit length of the line is

$$L' = \frac{\Phi}{Il} = \frac{\mu_1(d - d_2) + \mu_2 d_2}{a}.$$

P15.16. A long thin solenoid of length b and cross-sectional area S is situated in air and has N tightly wound turns of thin wire. Neglecting edge effects, determine the inductance of the solenoid. — (a) $L = 2\mu_0 N^2 S/b$. (b) $L = \mu_0 N^2 S/(4\pi b)$. (c) $L = \mu_0 N^2 S/b$.

P15.17. A thin toroidal core of permeability μ, mean radius R and cross-sectional area S is densely wound with two coils of thin wire with N_1 and N_2 turns, respectively. The windings are wound one over the other. Determine the self- and mutual inductances of the coils, and the coefficient of coupling between them. — (a) $L_1 = \mu N_1^2 S/(2\pi R)$, $L_2 = \mu N_2^2 S/(2\pi R)$, $L_{12} = \mu N_1 N_2 S/(2\pi R)$, $k = 1$. (b) $L_1 = \mu N_1^2 S/(4\pi R)$, $L_2 = \mu N_2^2 S/(4\pi R)$, $L_{12} = \mu N_1 N_2 S/(4\pi R)$, $k = 1$. (c) $L_1 = \mu N_1^2 S/R$, $L_2 = \mu N_2^2 S/R$, $L_{12} = \mu N_1 N_2 S/R$, $k = 1$.

P15.18. A thin toroidal ferromagnetic core, of mean radius R and cross-sectional area S is densely wound with N turns of thin wire. A current $i(t) = I_0 + I_m \cos \omega t$, where I_0 and I_m are constants, and $I_0 \gg I_m$, is flowing through the coil. Which permeability would you adopt in approximately determining the coil self-inductance? Assuming that this permeability is μ, determine the self-inductance of the coil. Does it depend on I_0? — (a) Differential *permeability.* $L = \mu N^2 S/(2\pi R)$. *(b) Normal permeability.* $L = \mu N^2 S/(\pi R)$. *(c) Initial permeability.* $L = \mu N^2 S/(4\pi R)$.

S **P15.19.** A thin solenoid is made of a large number of turns of very thin wire tightly wound in several layers. The radius of the innermost layer is a, of the outermost layer b, and the solenoid length is d ($d \gg a, b$). The total number of turns is N, and the solenoid core is made out of cardboard. Neglecting edge effects, determine approximately the solenoid self-inductance. Note that the magnetic flux through the turns differs from one layer to the next. Plot this flux as a function of radius, assuming the layers of wire are very thin. — *(a)* $L = \mu_0 \pi N^2 (3a^2 + 2ab + b^2)/(3d)$. *(b)* $L = \mu_0 \pi N^2 (3a^2 + 6ab + b^2)/d$. *(c)* $L = \mu_0 \pi N^2 (3a^2 + 2ab + b^2)/(6d)$.

Solution. There is a total of $dN = N dr/(b - a)$ turns in a layer of radius r and thickness dr. Such a layer can be considered as a very long solenoid, so it produces a uniform magnetic field inside itself. The magnetic flux density is given by

$$dB = \frac{\mu_0 \, dNI}{d}.$$

Outside the layer, $dB = 0$. By integrating the above expression, we get the magnetic flux density due to all the turns to be

$$B(r) = \frac{\mu_0 NI}{d} \quad (r < a), \qquad B(r) = \frac{\mu_0 NI(b - r)}{(b - a)d} \quad (a < r < b), \qquad B(r) = 0 \quad (r > b).$$

The magnetic flux through a single turn, of radius r, is

$$\Phi_0 = \frac{\mu_0 \pi NI(3br^2 - 2r^3 - a^3)}{3(b - a)d}.$$

Finally, the total flux we obtain as

$$\Phi = \int_{r=a}^{b} \Phi_0 \, dN,$$

so that the self-inductance of the solenoid is

$$L = \frac{\Phi}{I} = \frac{\mu_0 \pi N^2 (3a^2 + 2ab + b^2)}{6d}.$$

Check the result by considering the case with $(b - a) \to 0$.

P15.20. Repeat problem P15.19 for a thin toroidal core. Assume that the mean toroid radius is R, the total number of turns N, the radius of the innermost layer a, and that of the outermost layer b, with $R \gg a, b$. — *Hint: the solution is practically the same as that of problem P15.19; just set $d = 2\pi R$.*

P15.21. The current intensity in a circuit of self-inductance L and negligible resistance was kept constant during a period of time at a level I_0. Then during a short time interval Δt, the current was linearly reduced to zero. Determine the emf induced in the circuit during this time interval. Does this have any connection with a spark you have probably seen inside a switch you turned off in the dark? Explain. — *Adopt the reference direction of the circuit to be the same as that of the current. (a) $e = -2L\,I_0/\Delta t$. (b) $e = -L\,I_0/\Delta t$. (c) $e = -L\,I_0/(2\Delta t)$.*

16. Energy and Forces in the Magnetic Field

• Almost any device that makes use of electric or magnetic forces can be made in an "electric version" and in a "magnetic version". Realizable magnetic forces are much greater than electric forces, and therefore devices based on magnetic forces are used much more often.

• Assume that n rigid and stationary contours, with currents $i_1(t), i_2(t), \ldots, i_n(t)$, and resistances $R_1, R_2, \ldots, R_n$, are connected to generators of emf's $e_1(t), e_2(t), \ldots, e_n(t)$, and that the *total* fluxes through the contours are $\Phi_1(t), \Phi_2(t), \ldots, \Phi_n(t)$. The elemental work done by the generators in all the contours in a time dt satisfies the equation

$$\mathrm{d}A_\mathrm{g} = \mathrm{d}A_\mathrm{J} + \mathrm{d}A_\mathrm{m}. \tag{16.1}$$

The first term on the right are Joule's losses, and the second is the work done in changing the magnetic field.

• The work of generators in individual contours can be expressed as

$$(\mathrm{d}A_\mathrm{g})_k = R_k i_k^2(t)\mathrm{d}t + i_k(t)\mathrm{d}\Phi_k(t), \quad k = 1, 2, \ldots, n, \tag{16.2}$$

so that the work of all the generators is

$$\mathrm{d}A_\mathrm{g} = \sum_{k=1}^{n} R_k i_k^2(t)\mathrm{d}t + \sum_{k=1}^{n} i_k(t)\mathrm{d}\Phi_k(t). \tag{16.3}$$

The second term on the right-hand side represents the energy used to change the magnetic field,

$$\mathrm{d}A_\mathrm{m} = \mathrm{d}A_\mathrm{g} - \mathrm{d}A_\mathrm{J} = \sum_{k=1}^{n} i_k(t)\mathrm{d}\Phi_k(t). \tag{16.4}$$

This is the law of conservation of energy for the n current contours. $\mathrm{d}A_\mathrm{m}$ is the work necessary to change the fluxes through the n contours by $\mathrm{d}\Phi_1, \mathrm{d}\Phi_2, \ldots, \mathrm{d}\Phi_n$.

• The total work A_m needed to establish dc currents $I_1, I_2, \ldots, I_n$ in the contours, for which the fluxes through the contours are $\Phi_1, \Phi_2, \ldots, \Phi_n$, is hence

$$(A_m)_{\text{in establishing currents}} = \sum_{k=1}^{n} \int_0^{\Phi_k} i_k(t) \mathrm{d}\Phi_k(t) \qquad \text{(J)}, \qquad (16.5)$$

which for linear media becomes the energy stored in the magnetic field,

$$W_m = \frac{1}{2} \sum_{k=1}^{n} I_k \Phi_k \qquad \text{(J)}. \qquad (16.6)$$

• In terms of the self and mutual inductances of the contours and currents in them this has the form

$$W_m = \frac{1}{2} \sum_{j=1}^{n} \sum_{k=1}^{n} L_{jk} I_j I_k \qquad \text{(J)}, \qquad (16.7)$$

which for a single contour becomes

$$W_m = \frac{1}{2} I \Phi = \frac{1}{2} L I^2 \qquad \text{(J)}. \qquad (16.8)$$

• The density of work that needs to be done to change the magnetic flux density at a point from B_1 to B_2 is given by

$$\frac{\mathrm{d}A_m}{\mathrm{d}v} = \int_{B_1}^{B_2} H(t) \mathrm{d}B(t) \qquad \text{(J/m}^3\text{)}. \qquad (16.9)$$

For linear media, this yields the density of energy at a point of the magnetic field,

$$\frac{\mathrm{d}W_m}{\mathrm{d}v} = \frac{1}{2} \frac{B^2}{\mu} = \frac{1}{2} \mu H^2 = \frac{1}{2} B H \qquad \text{(J/m}^3\text{)}. \qquad (16.10)$$

The energy contained in the magnetic field *in a linear medium* is thus

$$W_m = \int_v \frac{1}{2} \mu H^2 \mathrm{d}v \qquad \text{(J)}. \qquad (16.11)$$

• If a sinusoidal magnetic field exists in a piece of ferromagnetic material, there are *hysteresis losses* in the material. They are proportional to the *hysteresis loop area* and to *frequency*. If the field is *uniform*, the losses are also proportional to the volume of the ferromagnetic material.

• The part of inductance associated with the energy contained inside a conductor is known as the *internal inductance*, and is defined by

$$L_{\text{internal}} = \frac{2(W_m)_{\text{inside conductor}}}{i^2} \qquad \text{(H)}. \tag{16.12}$$

The *external inductance* is associated with the energy outside the conductor,

$$L_{\text{external}} = \frac{2(W_m)_{\text{outside conductor}}}{i^2} \qquad \text{(H)}. \tag{16.13}$$

For a long straight wire of circular cross section the internal inductance per unit length is given by

$$L'_{\text{internal}} = \frac{\mu}{8\pi} \qquad \text{(H/m)}. \tag{16.14}$$

• If the distribution of currents in a magnetically *homogeneous* medium is known, the Biot-Savart law can be used for determining the magnetic flux density. Combined with the relation $d\boldsymbol{F_m} = I d\boldsymbol{l} \times \boldsymbol{B}$, the magnetic force can then be found on any part of the current distribution. In some important cases the magnetic force can be calculated as a derivative of the magnetic energy. This can be done under two assumptions: (1) the fluxes through all the contours are kept constant, or (2) the currents in all the contours are kept constant:

$$F_x = - \left(\frac{dW_m}{dx} \right)_{\Phi=\text{constant}}, \tag{16.15}$$

$$F_x = + \left(\frac{dW_m}{dx} \right)_{I=\text{constant}}. \tag{16.16}$$

Using these formulas, the *magnetic pressure* on boundary surfaces between two media is obtained as

$$p = \frac{1}{2}(\mu_2 - \mu_1) \left(H_{\text{tang.}}^2 + \frac{B_{\text{norm.}}^2}{\mu_1\mu_2} \right) \qquad \text{(reference direction into medium 1)}. \tag{16.17}$$

The maximal obtainable magnetic pressure is on the order of 10^4 times greater than the maximal obtainable electric pressure.

QUESTIONS

Q16.1. What does Eq. (16.1) actually represent? — *(a) The law of conservation of energy. (b) An equation for evaluating the power of Joule's losses. (c) An equation for evaluating the power of generators.*

S **Q16.2.** Explain why the expression $dA_g = e(t)\, i(t)\, dt$ is the work done by a generator. — *Hint: recall the definition of the emf of a generator, of the potential difference, and of the current intensity.*

Answer. The generator does the work against the electric forces of charges on its electrodes. The work of electric forces is obtained as the potential difference between the electrodes, $V(t)$, times the charge $dq = i\,dt$ $(dq > 0)$ transferred by the electric field from the positive to the negative electrode. Since the electromotive force has the same value as $V(t)$, and the generator moves the charges in the opposite direction, $e(t)\,i(t)\,dt$ is the work done by the generator.

Q16.3. For a simple circuit of resistance R, with an emf $e(t)$, $e(t) = Ri(t) + d\Phi(t)/dt$. Explain the physical meaning of the last term. — *Hint: recall Faraday's law.*

S   **Q16.4.** Why does the energy of a system of current loops not depend how the currents in the loops attained their final values? — *(a) Because Joule's losses are the same. (b) Because the induced electromotive forces along the loops are always the same. (c) Because otherwise the law of conservation of energy would be violated.*

Answer. This would violate the law of conservation of energy. We could then always make the currents attain their final value in a way using less energy and then make them go to zero in a way that requires more energy to build up the currents, and therefore returns more energy when the currents are reduced to zero. In this way we would make a surplus of energy with no source of energy.

Q16.5. Is Eq. (16.7) valid for nonlinear magnetic media? Explain. — *(a) Yes, because the derivation does not depend on the medium. (b) No, because for nonlinear media the flux is not proportional to the current. (c) No, because for nonlinear media the starting equation, Eq. (16.1), is not valid.*

Q16.6. The current in a thin loop 1 is increased from zero to a constant value I. A thin resistive loop 2 has no generators in it, but is in the magnetic field of the current in loop 1. Both loops are made of a linear magnetic material. Are the power $p_g(t)$ of the generator in loop 1 and the final value W_m of the energy *stored* in the system affected by the presence of loop 2? — *(a) $p_g(t)$ is affected, W_m not. (b) $P_g(t)$ is not affected, W_m is. (c) Neither is affected.*

Q16.7. Repeat question Q16.6 with loop 2 open-circuited. — *The answers are the same as for the preceding questions.*

Q16.8. A body of a linear magnetic material is placed in the vicinity of loop 1 of question Q16.6. Is some energy associated with the magnetization of the body? — *(a) None. (b) Yes, some. (c) Depends on the size of the body.*

Q16.9. Eq.(16.6) was derived by assuming that the currents were increased inside *stationary* conductors. Using the law of conservation of energy as an argument, prove that this expression must be valid for the magnetic energy of the system considered, irrespective of the process by which the current system is obtained. — *Hint: note that, if this were true, different processes would require different energies to obtain the* same *system.*

Q16.10. Using a sound physical argument, explain why the work in Eq. (16.5) done by the generators in establishing a given *time-constant* magnetic field is a function of the process by which the system of currents is established when ferromagnetic materials are present in the field. — *Hint: recall hysteresis losses.*

S   **Q16.11.** Will the magnetic energy of a system of fixed quasi-filamentary dc current loops be changed if a closed conducting loop with *no* current is introduced into the system? Explain.

— *(a) It will not be changed. (b) It will be changed. (c) It will be changed only if the loop is superconductive.*

Answer. Only if the loop is superconductive, because then a current will be induced (and remain) in it, which would change the magnetic field, and its energy. If the loop has resistance, however small, a current will be induced in the loop while it is introduced in the field, but it will soon become zero once the loop is in its final position.

Q16.12. Is it possible to determine theoretically the self-inductances and mutual inductances in a system of current loops by starting from Eq. (16.7) if W_m is known? Explain. — *(a) Yes, because the process is reciprocal. (b) No, because this is just one equation. (c) Yes, if the number of loops is less than three.*

Q16.13. Two equal thin loops of self-inductance L are pressed onto each other, so that $|L_{12}| \simeq L$. If the currents in the loops are I_1 and I_2, what is the magnetic energy of the system? Answer the question if the two currents are (1) in the same direction and (2) in opposite directions. — *(a) $W_{m1} = 2LI^2 = W_{m2}$. (b) $W_{m1} = 2LI^2$, $W_{m1} = -2LI^2$. (c) $W_{m1} = 2LI^2$, $W_{m1} = 0$.*

S   **Q16.14.** How would you make an electric version of a generator of sinusoidal emf? — *Hint: e.g., start from a rectangular metal plate cut into two mutually insulated halves. Let it rotate in a uniform electric field, the two plate halves being connected by sliding contacts to a resistor.*

Answer. Imagine the page you are reading to be cut in half, parallel to the printed lines, with the two halves insulated by a thin dielectric rod representing the axis of rotation. Let the two plates be connected by sliding contacts to a resistor. If this structure rotates in a uniform electric field, the plates will be charged according to the sinusoidal time variation, and there will be a current of this time variation in the resistor. Efficiency of such a generator, however, is extremely small, which the reader can readily prove himself/herself.

Q16.15. How would you make an electric version of a generator of "rectified" sinusoidal emf? — *Hint: having in mind the answer to the preceding question, recall the commutator in generators based on electromagnetic induction.*

Q16.16. Imagine somebody came to you with a piece of a ferromagnetic material he developed and stated that the working point moves along the hysteresis loop in the clockwise direction. Would you believe him? Explain. — *(a) Yes, if we are certain that he is a reliable person. (b) Yes, because the motion of the working point can be either way. (c) No, because such a piece of material would produce energy, instead of dissipating it.*

Q16.17. Is it possible to *derive* Eq. (16.9) from Eq. (16.10)? Explain. — *Hint: recall which of the two is more general.*

Q16.18. If a hysteresis loop was obtained by a sinusoidally varying $H(t)$, will the hysteresis losses be exactly equal to the area of this loop if $H(t)$ varies as a triangular function of time (that is, varies *linearly* from $-H_m$ to H_m, then back to $-H_m$, and so on)? Explain. — *(a) They will be the same, provided the areas of the triangle and the sinusoid are the same. (b) They do depend on the time variation of the field, and will not be the same. (c) They will be the same, because the shape of the hysteresis loop does not depend on $H(t)$, as long as it is periodic and of the same amplitude.*

Q16.19. If the frequency of the alternating current producing a magnetic field is f (cycles per second), what is the power per unit volume necessary to maintain the field in a piece of ferromagnetic substance? *(a) It equals f^2 times the integral around the hysteresis loop of $H(t)\mathrm{d}B(t)$. (b) f times that integral. (c) $\sqrt{f}$ times that integral.*

S **Q16.20.** According to the expression in Eq. (16.10), the volume density of magnetic energy is always greater in a vacuum than in a paramagnetic or idealized linear ferromagnetic material for the same B. Using a sound physical argument, explain this result. — *Hint: recall what is done when such materials are magnetized.*

Answer. A paramagnetic or ferromagnetic material is magnetized so that it enhances the magnetic field. Less energy will be needed to obtain the same B inside such a material, than in a vacuum.

Q16.21. The magnetization curve of a real ferromagnetic material is approximated by a nonlinear, but single-valued, function $B(H)$ (not by a hysteresis loop). Is it possible to speak about the energy density of the magnetic field inside the material? If you think it is, what is the energy density equal to? — *(a) No, it is not possible, because the material is not linear. (b) No, it is not possible, because such an approximation is not possible to make. (c) Yes, it is equal to the integral of $H(t)\mathrm{d}B(t)$ from $B = 0$ to the value of B at the point considered.*

S **Q16.22.** A thin toroidal ferromagnetic core is magnetized to saturation and the current in the excitation coil is reduced to zero, so that the operating point in the $H - B$ plane is $H = 0$, $B = B_\mathrm{r}$. Is it possible to speak in that case about the energy of the magnetic field stored in the core? Is it possible to speak about the energy used to create the field? Explain. — *Hint: recall that when we say "energy stored in the core", we mean how much energy we would get if the field goes to zero.*

Answer. When we say "energy stored in the core", we mean how much energy we get if the field goes to zero. In this case we cannot make the field go to zero except by introducing more energy in the core. So we cannot define any magnetic energy stored in the core. The energy used to create this field is defined, however, by the integral of $H(t)\mathrm{d}B(t)$ from $B = 0$ to $B = B_\mathrm{r}$ along the magnetization curve, multiplied by the volume of the core. We know that one part of this energy was lost to hysteresis losses.

Q16.23. The first part of the magnetization curve can be approximated as $B(H) = CH^2$, where C is a constant. How can you evaluate in that case the energy density necessary for the magnetization of the material? Is that also the energy density of the magnetic field? — *Hint: use Eq. (16.9).*

Q16.24. Is it possible for the initial magnetization curve to be partly decreasing in B as H increases? What would that mean? — *(a) Yes, it depends on the material. (b) No, because this means a flow of energy from the material back to the sources during the magnetization process. (b) Yes, for any material, with appropriate time variation of current in the coil.*

Q16.25. Evaluate approximately the density of hysteresis losses per cycle for the hysteresis loop in Fig. Q16.25 if $H_\mathrm{m} = 200\,\mathrm{A/m}$ and $B_\mathrm{m} = 0.5\,\mathrm{T}$. — *(a) $30\,J/m^3$. (b) $80\,J/m^3$. (c) $150\,J/m^3$.*

Q16.26. Why are hysteresis losses linearly proportional to frequency? — *Hint: use Eq. (16.9) as an argument.*

S **Q16.27.** Is the volume density of hysteresis losses in a thick toroidal ferromagnetic core with a coil carrying a sinusoidal current the same at all points of the core? Is the answer the same if the current intensity is such that at all points of the core saturation is attained, and if it is not? — *Hint: recall that H is not of the same value in the entire core.*

Answer. It is not the same, since the hysteresis loops have different sizes for different distances from the toroid axis. If the saturation is attained at all points, this argument does not hold any more, and the density of hysteresis losses is then the same at all points of the core.

Q16.28. Why can the self-inductance of a thick conductor not be defined naturally in terms of the induced emf or flux through the conductor? — *(a) There is no possibility to define a unique closed loop. (b) The closed loop must not be inside a conductor. (c) The axis of a thick loop is not defined.*

Q16.29. Why is it very difficult to obtain the internal inductance using the flux definition of self-inductance? To answer the question, consider two wires, one thin and the other thick, with the same current I flowing through them. — *Hint: discuss the possibility of defining a closed contour for defining the total flux in such cases.*

Q16.30. Direct current due to a lightning stroke on a three-phase line propagates along the three conductors. Will the force repel or attract the conductors? — *(a) There will be no force on the conductors. (b) The force will be repulsive. (c) The force will be attractive.*

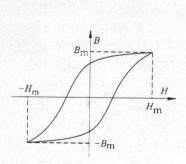

Fig. Q16.25. A hysteresis loop.

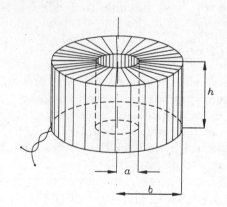

Fig. P16.6. A thick toroidal coil.

Q16.31. If a lightning stroke hits a transformer, in some cases it may be noticed that the transformer "swells". Explain why. — *(a) It is heated up and explodes. (b) The magnetic forces always tend to increase the size of a current loop. (c) It is due to chemical changes in the transformer insulation.*

PROBLEMS

S **P16.1.** Write the explicit expression for the magnetic energy of three current loops with currents I_1, I_2, and I_3. Assume that the self-inductances and mutual inductances of the loops are known. — *Hint: use Eq. (16.7) with n = 3. Note that the double sum in that case results in a total of nine terms.*

Solution. The energy of such a system is

$$W_m = \frac{1}{2} \sum_{k=1}^{3} \sum_{j=1}^{3} L_{kj} I_k I_j$$

$$= \frac{1}{2} L_{11} I_1^2 + \frac{1}{2} L_{22} I_2^2 + \frac{1}{2} L_{33} I_3^2 + L_{12} I_1 I_2 + L_{13} I_1 I_3 + L_{23} I_2 I_3.$$

(This is Eq. (16.11) for $n = 3$.)

P16.2. Find the magnetic energy per unit length in the dielectric of a coaxial cable of inner conductor radius a and outer conductor radius b, carrying a current I. The permeability of the dielectric is μ_0. Show that $W'_m = L'_{external} I^2/2$. — (a) $W'_m = (\mu_0 I^2/2\pi) \ln(b/a)$. (b) $W'_m = (\mu_0 I^2/\pi) \ln(b/a)$. (c) $W'_m = (\mu_0 I^2/4\pi) \ln(b/a)$.

P16.3. Find the total inductance per unit length of a coaxial cable that has an inner conductor of radius a and an outer conductor with inner radius b and outer radius c. The permeability of the conductors and the dielectric is μ_0, and current is distributed uniformly over the cross sections of the two conductors. — *Hint: add the two internal inductances to the external inductance.*

P16.4. A thin ferromagnetic toroidal core is made of a material that can be characterized approximately by a constant permeability $\mu = 4000\mu_0$. The mean radius of the core is $R=10$ cm and the core cross-sectional area is $S=1$ cm^2. A current of $I=0.1$ A is flowing through $N=500$ turns wound around the core. Find the energy spent on magnetizing the core. Is this equal to the energy contained in the magnetic field in the core? — (a) $W_m = 1\,mJ$. (b) $W_m = 2.1\,mJ$. (c) $W_m = 3.2\,mJ$.

P16.5. In the toroidal core of the preceding problem, a small part of length $l_0=2$ mm is cut out so that now there is a small air gap in the core. The current in the coil is kept constant while the piece is being cut out. Find the energy contained in the magnetic field in this case. — (a) $38\,\mu J$. (b) $56\,\mu J$. (c) $73\,\mu J$.

P16.6. Show that the same expression for the self-inductance of the toroidal coil in Fig. P16.6 is obtained from the expression $\Phi = LI$, and the expression $2W_m = LI^2$. — *Hint: the first expression requires the determination of the magnetic flux through all the coil turns, and the second expression requires the evaluation of the magnetic energy in the toroid core using Eq. (16.11).*

S **P16.7.** On a thin ferromagnetic toroidal core of cross-sectional area $S = 1$ cm^2, $N = 1000$ turns of thin wire are tightly wound. The mean radius of the core is $R = 16$ cm. It may be assumed that the magnetic field is uniform over the cross section of the toroid. The idealized initial magnetization curve is shown in Fig. P16.7. Determine the work A_m done in establishing the magnetic field inside the toroid if the intensity of the current through the coil is $I = 2$ A. — (a) $0.021\,J$. (b) $0.050\,J$. (c) $0.075\,J$. *Repeat the problem if (1) $I = 1\,A$, and (2) $I = 0.5\,A$.*

Solution. The magnetic field intensity in the core, $H = NI/(2\pi R) \simeq 2000$ A/m, is larger than 1000 A/m, so the density of work done in establishing the field is

$$\frac{dA_m}{dv} = \int_0^{1\,T} H dB = \frac{1}{2} 1000\,\frac{A}{m} \cdot 1\,T = 500\,\frac{J}{m^3}$$

(the density of work is proportional to the area of the triangle between the magnetization curve and the B axis in Fig. P16.7). The total work is

$$A_m = \frac{dA_m}{dv}\,2\pi RS = 0.05\,J.$$

Repeat the problem if (1) $I = 1\,A$, and (2) $I = 0.5\,A$.

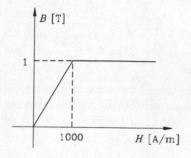

Fig. P16.7. An idealized magnetization curve.

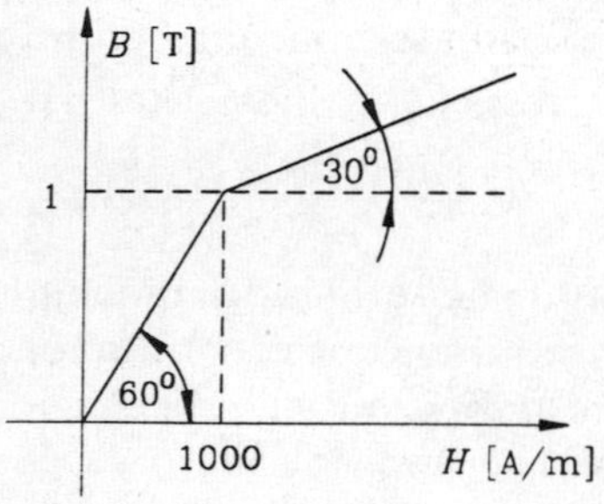

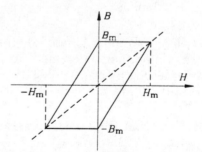

Fig. P16.8. An idealized magnetization curve.

P16.8. Repeat problem P16.7 if the idealized initial magnetization curve is as shown in Fig. P16.8. — *(a) 0.031 J. (b) 0.07 J. (c) 0.10 J.*

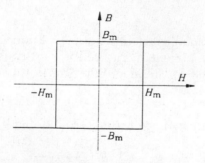

Fig. P16.12. An idealized hysteresis loop.

Fig. P16.13. An idealized hysteresis loop.

P16.9. On the toroidal core shown in Fig. P16.6, $N = 650$ turns of thin wire are tightly wound. The intensity of the time-constant current in the coil is $I = 2\,A$ and the idealized initial magnetization curve of the core is as shown in Fig. P16.7. Determine the work done in establishing the magnetic field in the core if $a = 5\,cm$, $b = 15\,cm$, and $h = 10\,cm$. — *(a) 5.41 J. (b) 8.23 J. (c) 3.14 J.*

P16.10. Repeat the preceding problem for intensities of current through the coil of (1) 0.5 A, and (2) 1 A. — *The result in case 2: (a) 2.54 J. (b) 6.38 J. (c) 8.22 J.*

P16.11. The initial magnetization curve of a ferromagnetic material can be approximated by $B(H) = B_0 H/(H_0 + H)$, where B_0 and H_0 are constants. Determine the work done per unit volume in magnetizing this material from zero to a magnetic field intensity H. — *Hint: first determine* dB, *and then the work done per unit volume using Eq. (16.9).*

S **P16.12.** The idealized hysteresis loops of the ferromagnetic core in Fig. P16.6 are as in Fig. P16.12. Determine the power of hysteresis losses in the core if it is wound with N turns of wire with sinusoidal current of amplitude I_m and frequency f. Assume that saturation is not reached at any point, and neglect the field of eddy currents. — *Hint: the energy lost per unit volume of the core in one cycle is numerically equal to the area of the hysteresis loop at the considered point of the core. Note that, in the present case, the hysteresis loops are different for different points.* $P_{\text{hysteresis losses}} = f W_{\text{hysteresis losses}}, \text{ where } W_{\text{hysteresis losses}} =$

$$(a) \ \frac{\mu_{\text{normal}} N^2 I_m^2 h}{2\pi} \ln \frac{b}{a}, \quad (b) \ \frac{\mu_{\text{normal}} N^2 I_m^2 h}{4\pi} \ln \frac{b}{a}, \quad (c) \ \frac{2\mu_{\text{normal}} N^2 I_m^2 h}{\pi} \ln \frac{b}{a},$$

and μ_{normal} *is the normal permeability,* $\mu_{\text{normal}} = B_m/H_m$.

Solution. The energy lost per unit volume of the core in one cycle is numerically equal to the area of the hysteresis loop at the considered point of the core. In the present case, the hysteresis loops are different for different points, since the amplitude of the magnetic field intensity is

$$H_m(r) = \frac{N I_m}{2\pi r}.$$

Consequently, we have to determine the elemental losses in thin circular rings of radius r and thickness dr ($a < r < b$), the total losses being given as their sum (integral). Note that (Fig. P16.12)

$$\frac{B_m(r)}{H_m(r)} = \frac{B_m}{H_m} = \text{const} = \mu_{\text{normal}},$$

where μ_{normal} is the normal permeability of the core.

Losses in one such ring in one cycle are numerically equal to the area of the loop in Fig. P16.12 multiplied by the volume of the ring:

$$dW_{\text{hysteresis losses}} = [4B_m(r)H_m(r)](2\pi rhdr) = 8\mu_{\text{normal}} H_m^2(r)\pi rhdr.$$

Losses in the entire core per cycle are

$$W_{\text{hysteresis losses}} = \int_a^b \frac{2\mu_{\text{normal}} N^2 I_m^2 h}{\pi r} \, dr = \frac{2\mu_{\text{normal}} N^2 I_m^2 h}{\pi} \ln \frac{b}{a}.$$

Finally, noting that f is the number of cycles per unit time, the power of the hysteresis losses is given by

$$P_{\text{hysteresis losses}} = f W_{\text{hysteresis losses}}.$$

P16.13. Repeat problem P16.12 for idealized hysteresis loops shown in Fig. P16.13, assuming B_m/H_m for all the loops is the same and that saturation is not reached at any point. Neglect the field of eddy currents. — *Hint: the losses are one half of those in the preceding problem (why?).*

P16.14. A ferromagnetic core of a solenoid is made of thin, mutually insulated sheets. To estimate the eddy current and hysteresis losses, the *total* power losses were measured at two frequencies, f_1 and f_2, for the same amplitude of the magnetic flux density. The total power losses were found to be P_1 and P_2, respectively. Determine the power of hysteresis losses and of eddy current losses at both frequencies. *Hint: note that $P_{1/2} = P_{\text{hysteresis losses }1/2} +$ $P_{\text{eddy current losses }1/2} = C_1 f_{1/2} + C_2 f_{1/2}^2$, and solve the two equations obtained at the two frequencies for the constants C_1 and C_2.*

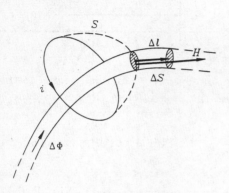

Fig. P16.15 Tube of flux of a current loop.

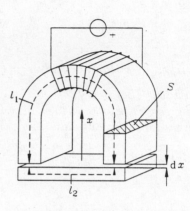

Fig. P16.17. An electromagnet.

*P16.15. Prove that Eq. (16.9) is valid for any magnetic field, not necessarily uniform. — *Hint: consider an arbitrary current contour, divide the field into elemental tubes of magnetic flux, as in Fig. P16.15, and use Eq. (16.5).*

S **P16.16.** Two coaxial solenoids of radii a and b, lengths l_1 and l_2, and number of turns N_1 and N_2 have the same current I flowing through them. Find the axial force that the solenoids exert on each other if the thinner solenoid is pulled by a length x $(x < l_1, l_2)$ into the other solenoid. Neglect edge effects and assume that the medium is air. *Hint: assume that $b < a$, that the turns of both solenoids (and the currents in them) are oriented in the same direction, and find the mutual inductance of the solenoids as a function of x. Then use Eq. (16.16). (a) $F_x = \mu_0 N_1^2 N_2^2 \pi b^2 I^2/(l_1 l_2)$. (b) $F_x = \mu_0 N_1 N_2 \pi b^2 I^2/(l_1 l_2)$. (c) $F_x = 4\mu_0 N_1 N_2 \pi b^2 I^2/(l_1 l_2)$.*

Solution. Assume that $b < a$, and that the turns of both solenoids (and the currents in them) are oriented in the same direction. The mutual inductance of the solenoids is then

$$L_{12} = \frac{\mu_0 N_1 N_2 \pi b^2 x}{l_1 l_2}.$$

The total magnetic energy of the system is

$$W_m = \left(\frac{1}{2}L_1 + \frac{1}{2}L_2 + L_{12}\right) I^2.$$

Note that only the inductance L_{12} depends on x. The force on the inner solenoid is given by Eq. (16.21),

$$F_x = \left.\frac{dW_m}{dx}\right|_{I=\text{constant}} = \frac{dL_{12}}{dx}I^2 = \frac{\mu_0 N_1 N_2 \pi b^2 I^2}{l_1 l_2}.$$

Since $F_x > 0$, we conclude that the thicker solenoid tends to pull in the thinner one, i.e., the centers of the two solenoids tend to be at the same point.

Does the result remain the same when we change the direction of current in one of the solenoids?

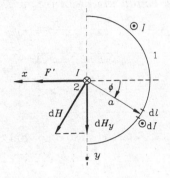

Fig. **P16.18.** A two-conductor line.

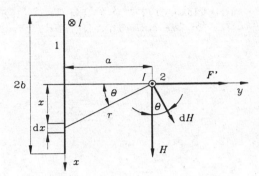

Fig. **P16.19** A two-conductor line.

P16.17. An electromagnet and the weight it is supposed to lift are shown in Fig. P16.17. The dimensions are $S = 100\,\text{cm}^2$, $l_1 = 50\,\text{cm}$, $l_2 = 20\,\text{cm}$. Find the current through the winding of the electromagnet and the number of turns in the winding so that it can lift a load that is $W = 300$ kiloponds (a kp is 9.81 N) heavy. The electromagnet is made of a material whose magnetization curve can be approximated by $B(H) = 2H/(400 + H)$, where B is in T and H is in A/m. — *Note: the solution is not unique. (a) For $N = 200$, $I = 2.431\,A$. (b) For $N = 100$, $I = 1.223\,A$. (c) For $N = 50$, $I = 10.38\,A$.*

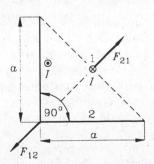

Fig. **P16.20.** A two-conductor line.

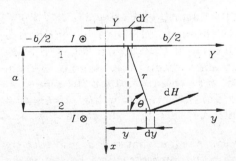

Fig. **P16.21.** Cross section of a stripline.

P16.18. One of the conductors of a two-conductor line is in the form of one half of a thin circular cylinder. The other conductor is a thin wire running along the axis of the first (Fig. P16.18). If a current I flows through the two conductors in opposite directions, determine the force per unit length on the conductors. — *(a) $F'_{\text{on conductor 2}} = \mu_0 I^2/(\pi^2 a)$. (b) $F'_{\text{on conductor 2}} = \mu_0 I^2/(4\pi^2 a)$. (c) $F'_{\text{on conductor 2}} = 2\mu_0 I^2/(\pi^2 a)$.*

P16.19. A thin conductor 2 runs parallel to a thin metal strip 1 (Fig. P16.19). Both a and b are much larger than the thickness of the strip. Determine the force per unit length on the two conductors for a current I flowing through them in opposite directions. — (a) $F' = [(\mu_0 I^2)/(\pi b)]\cdot \tan^{-1}(b/a)$. (b) $F' = [(\mu_0 I^2)/(2\pi b)] \ln(b/a)$. (c) $F' = [(\mu_0 I^2)/(2\pi b)] \tan^{-1}(b/a)$.

P16.20. Determine the force per unit length on the conductors of the line with a cross section as shown in Fig. P16.20. The current in the conductors is I and the medium is air. — (a) $F'_{12} = (\mu_0 I^2 \sqrt{2})/(4\pi b)$. (b) $F'_{12} = (\mu_0 I^2 \sqrt{2})/(4b)$. (c) $F'_{12} = (\mu_0 I^2 \pi \sqrt{2})/(2b)$.

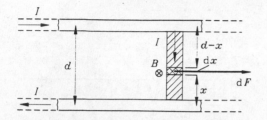

Fig. P16.22. Short-circuited two-wire line.

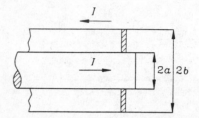

Fig. P16.23. Short-circuited coaxial cable.

*P16.21. Determine the force per unit length on the conductors of the stripline with a cross section as shown in Fig. P16.21. The current in the conductors is I, in opposite directions. — *The solution is relatively complex, and we give only the correct answer:*

$$F' = F'_x = \frac{\mu_0 I^2}{2\pi b^2} \left(2b \tan^{-1} \frac{b}{a} - a \ln \frac{a^2 + b^2}{a^2} \right).$$

P16.22. A thin two-wire line has conductors of circular cross section of radius a and the distance between their axes d and is short-circuited by a straight conducting bar, as shown in Fig. P16.22. If a current I flows through the line, what is the force on the bar? Assume that the section of the line to the left of the bar is very long, and that the medium is air. — (a) $F \simeq [(\mu_0 I^2)/(4\pi)] \ln(d/a)$. (b) $F \simeq [(\mu_0 I^2)/(4\pi)] \ln(2d/a)$. (c) $F \simeq [(\mu_0 I^2)/(2\pi)] \ln(d/a)$.

S **P16.23.** A long air-filled coaxial cable is short-circuited at its end by a thin conducting plate, as shown in Fig. P16.23. Determine the force on the end plate corresponding to a current of intensity I through the cable. — *The force is axial.* (a) $F_x = [(2\mu_0 I^2)/\pi] \ln(b/a)$. (b) $F_x = [(\mu_0 I^2)/\pi] \ln(b/a)$. (c) $F_x = [(\mu_0 I^2)/(4\pi)] \ln(b/a)$.

Solution. The force on the plate can be obtained by integrating the elemental forces acting on the plate elements. Here we solve the problem in a different way, starting from energy considerations.

Let us introduce the z axis along the cable axis, directed towards the plate. Imagine that a magnetic force on the plate, which we need to find, has moved the plate by dz along the z axis, and that the current I in the cable has been kept constant during such an experiment. It is easy to conclude that during the experiment the magnetic energy contained in the cable has been increased by

$$dW_m = \frac{1}{2} L' dz\, I^2,$$

where L' is the external self inductance per unit length of the coaxial cable,

$$L' = \frac{\mu_0}{2\pi} \ln \frac{b}{a}.$$

The force on the plate is hence

$$F_x = \left.\frac{dW_\mathrm{m}}{dx}\right|_{I=\text{constant}} = \frac{1}{2}L'I^2 = \frac{\mu_0 I^2}{4\pi} \ln \frac{b}{a}.$$

P16.24. Determine approximately the force between two parallel coaxial circular loops with currents I_1 and I_2. The radii of the loops are a and b, respectively, with $a \gg b$, and the distance between their centers z. — *The force is axial. If the currents in the loops are in the same direction, it is attractive.*

$$(a)\ F_z = \frac{3\mu_0 \pi a^2 b^2 I_1 I_2 z}{2(z^2 + a^2)^{3/2}}, \quad (b)\ F_z = \frac{\mu_0 \pi a^2 b^2 I_1 I_2 z}{(z^2 + a^2)^{3/2}}, \quad (c)\ F_z = \frac{\mu_0 \pi a^2 b^2 I_1 I_2 z}{2(z^2 + a^2)^{3/2}}.$$

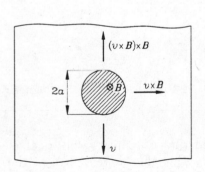

Fig. P16.25. A strip moving in magnetic field.

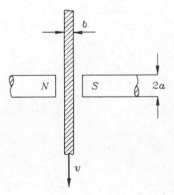

Fig. P16.26. A plate falling in magnetic field.

S **P16.25.** A metal strip of conductivity σ and of small thickness b moves with a uniform velocity v between the round poles of a permanent magnet. The radius of the poles of the magnet is a and the width of the strip is much larger than a (Fig. P16.25). The flux density $\boldsymbol{B}$ is very nearly constant over the circle shown hatched and practically zero outside it. Assuming that the induced current density in that circle is given by $\boldsymbol{J} = \sigma \boldsymbol{v} \times \boldsymbol{B}/2$, determine the force on the strip. — *(a) $F = \pi\sigma a^2 bvB^2/4$. (b) $F = \pi\sigma a^2 bvB^2/2$. (c) $F = \pi\sigma a^2 bvB^2/3$.*

Solution. The magnetic force on the current element $\boldsymbol{J}dv$ is given by $d\boldsymbol{F} = \boldsymbol{J}dv \times \boldsymbol{B}$, so that the magnetic force per unit volume acting on the currents in the hatched circle is

$$\frac{d\boldsymbol{F}}{dv} = \boldsymbol{J} \times \boldsymbol{B} = \frac{\sigma}{2}(\boldsymbol{v} \times \boldsymbol{B}) \times \boldsymbol{B}.$$

The vector dF/dv is constant in the circle. It is directed in the direction opposite to that of v, that is, it opposes the motion of the strip. The total force is in the same direction, of intensity

$$F = \frac{dF}{dv}\,\pi a^2 b = \frac{1}{2}\pi\sigma a^2 bvB^2.$$

P16.26. A thin metal plate is falling between the poles of a permanent magnet (Fig. P16.26) under the action of the gravitational field. The pole radius is $a = 2\,\mathrm{cm}$ and the flux density in the gap is $B = 1\,\mathrm{T}$. Determine approximately the velocity of the plate, if its thickness is $b = 0.5\,\mathrm{mm}$, its surface area $S = 100\,\mathrm{cm}^2$, its conductivity $\sigma = 36\cdot 10^6\,\mathrm{S/m}$ (aluminum), and its mass density $\rho_\mathrm{m} = 2.7\,\mathrm{g/cm}^3$ (aluminum). — *(a) $v \simeq 2.34\,cm/s$. (b) $v \simeq 3.58\,cm/s$. (c) $v \simeq 1.17\,cm/s$.*

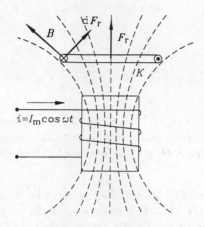

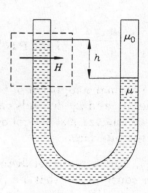

Fig. P16.27. A ring in a magnetic field.   **Fig. P16.28.** A U-shaped tube in a magnetic field.

P16.27. A metal ring K of negligible resistance is placed above a short cylindrical electromagnet, as shown in Fig. P16.27. Determine qualitatively the time dependence of the total force on the ring if the current in the electromagnet coil is of the form $i(t) = I_\mathrm{m}\cos\omega t$. — *With the reference direction of the force upward, (a) $F_\mathrm{r} = F_\mathrm{rm}\cos\omega t$. (b) $F_\mathrm{r} = F_\mathrm{rm}\cos^3\omega t$. (c) $F_\mathrm{r} = F_\mathrm{rm}\cos^2\omega t$.*

P16.28. A U-shaped glass tube is filled with a paramagnetic liquid of unknown magnetic susceptibility χ_m. A part of the tube inside the dashed square in Fig. P16.28 is exposed to a uniform magnetic field of intensity H. Under the influence of magnetic forces, a difference h of the levels of the liquid in the two sections of the tube is observed. Given that the mass density of the liquid is ρ_m and that the medium above the liquid is air, determine χ_m. — *Hint: use Eq. (16.17). (a) $\chi_\mathrm{m} = \rho_\mathrm{m}gh/(2\mu_0 H^2)$. (b) $\chi_\mathrm{m} = 2\rho_\mathrm{m}gh/(\mu_0 H^2)$. (c) $\chi_\mathrm{m} = 2\rho_\mathrm{m}gh/(\pi\mu_0 H^2)$. ($g \simeq 9.81\,m/s^2$).*

P16.29. Plot the "scale calibration curve" $F_\mathrm{tot}(I)$ for the ammeter sketched in Fig. P16.29. $F_\mathrm{tot}(I)$ is the total force acting on the iron nail for a given current I in the coil. Given are $a = 1\,\mathrm{mm}$, $l = 5\,\mathrm{cm}$, $N' = 10\,\mathrm{turns/cm}$ (you need to look up the relative permeability for iron and its mass density). — *Hint: the downward magnetic force, $F_\mathrm{tot}(I)$, can be evaluated* *by means of Eq. (16.16) once the inductance of the coil with the nail a distance x inside it is known. (a) $F \simeq \mu S N'^2 I^2$. (b) $F \simeq \frac{1}{2} \mu S N'^2 I^2$. (c) $F \simeq 2 \mu S N'^2 I^2$.*

S **P16.30.** Derive the general expression for pressure of magnetic forces, Eq. (16.17). — *Hint: use boundary conditions combined with Eqs. (16.15) and (16.16) for the force on a patch ΔS of the boundary, assuming it to move by Δx into medium 1 under the pressure at the interface.*

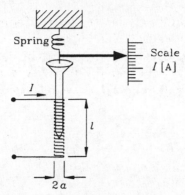

Fig. P16.29. Sketch of a simple ammeter.

Solution. The normal component of force (per unit area) acting on the interface between two media of permeabilities μ_1 and μ_2 depends on the direction of $\boldsymbol{B}$ and $\boldsymbol{H}$ in both media. All other cases can be deduced if we analyze just the following two: (1) when $\boldsymbol{B}$ and $\boldsymbol{H}$ are normal, and (2) when they are parallel to the interface.

In case (1), during a virtual displacement dx of the interface, $\boldsymbol{B}$, and hence also the magnetic flux Φ, remain constant, so that Eq. (16.20) applies. Assume that the reference direction of the displacement is into the medium of permeability μ_1. According to the boundary condition in Eq. (13.18), the change in energy located in the magnetic field is

$$dW_{\mathrm{m}} = \frac{1}{2} \left(\frac{1}{\mu_2} - \frac{1}{\mu_1} \right) B_{\mathrm{normal}}^2 \Delta S dx,$$

where ΔS is the area of a patch at the interface. So the pressure on the surface for a reference direction into medium of permeability μ_1 is

$$p^{(1)} = -\frac{1}{\Delta S} \frac{dW_{\mathrm{m}}}{dx} = \frac{1}{2} \left(\frac{1}{\mu_1} - \frac{1}{\mu_2} \right) B_{\mathrm{normal}}^2.$$

In case (2), $\boldsymbol{H}$ (i.e. the currents) remains constant, and Eq. (16.21) applies. The boundary condition in Eq. (13.17) tells us that

$$dW_{\mathrm{m}} = \frac{1}{2} (\mu_2 - \mu_1) H_{\mathrm{tang}}^2 \Delta S dx,$$

from which

$$p^{(2)} = \frac{1}{\Delta S} \frac{dW_{\mathrm{m}}}{dx} = \frac{1}{2} (\mu_2 - \mu_1) H_{\mathrm{tang}}^2.$$

Noting that

$$\mu H^2 = \mu H_{\text{tang}}^2 + \frac{B_{\text{normal}}^2}{\mu},$$

from the expressions for $p^{(1)}$ and $p^{(2)}$ above it may be deduced that the pressure on the boundary surface in the general case is given by the expression in Eq. (16.17).

17. Some Examples and Applications of Time-Invariant and Slowly Time-Varying Magnetic Field

• The largest permanent magnet around us is the earth itself. Its magnetic field is similar to the field of a giant current loop with an axis declined 11° with rе pect to the earth's axis of rotation. The planet's North Pole is approximately the *south* magnetic pole. The average magnitude of the earth's magnetic flux density is about $50\,\mu\text{T}$ (at the poles it is about $60\,\mu\text{T}$). The region in which the earth's magnetic field can be detected is called the *magnetosphere*.

• Charged particles always move when found in both electric and magnetic fields. In many instances one of the fields is of much less influence than the other, but in some cases the effects of both fields on a moving particle are of the same order of magnitude. The motion of charged particles in electric and magnetic fields may be in a vacuum (or very rarefied gas), in gases, and in solid or liquid conductors.

• The force on a charge Q moving in an electric and a magnetic field with a velocity v is the Lorentz force,

$$\boldsymbol{F} = Q\boldsymbol{E} + Q\boldsymbol{v} \times \boldsymbol{B}. \tag{17.1}$$

If a charge of mass m moves in a vacuum, then this force at any instant must be equal in magnitude and opposite in direction to the force of inertia,

$$m\frac{\mathrm{d}\boldsymbol{v}}{\mathrm{d}t} = Q\boldsymbol{E} + Q\boldsymbol{v} \times \boldsymbol{B}. \tag{17.2}$$

E and B in general are functions of space coordinates and of time. If a charge moves in a material (a gas, a liquid or a solid), it collides with the particles in the material, and Eq. (17.2) is not valid for average (drift) velocity.

• If a charge Q $(Q > 0)$ moves in a uniform, x-directed electric field of intensity E, with arbitrary initial velocity $\boldsymbol{v}_0 = \boldsymbol{v}_{0x} + \boldsymbol{v}_{0y}$, after integration Eq. (17.2) yields

$$x(t) = \frac{QE}{2m}t^2 + v_{0x}t + x_0, \tag{17.3}$$

$$y(t) = v_{0y}t + y_0, \tag{17.4}$$

where x_0 and y_0 are the initial x- and y-coordinates of the charge. Consequently, the charge will move along a parabola in the x-y plane. This is used, for example, for deflecting an electron beam between the plates of a charged parallel-plate capacitor.

- Since the magnetic force on the charge is $\boldsymbol{F}_\mathrm{m} = Q\boldsymbol{v} \times \boldsymbol{B}$, it is *always perpendicular to the direction of motion*. Therefore, a magnetic field cannot change the magnitude of the charge velocity (i.e., the charge kinetic energy), but only the direction of its motion. For example, in a uniform magnetic field of flux density B normal to the direction of motion of a particle of charge Q and mass m, the particle describes a circle of radius

$$R = \frac{mv}{QB}. \tag{17.5}$$

It makes a full circle during a time

$$T = \frac{2\pi R}{v} = \frac{2\pi m}{QB}. \tag{17.6}$$

The frequency of rotation of the particle,

$$f = \frac{1}{T} = \frac{QB}{2\pi m}, \tag{17.7}$$

is known as the *cyclotron frequency*, because this effect is used in *cyclotrons*, devices for accelerating charged particles.

- If a strip of width d with current of density J is situated in a uniform magnetic field of flux density B normal to the strip, there is a voltage between the strip edges. The effect is known as the *Hall effect*, and the voltage as the *Hall voltage*, which is given by

$$|V_{12}| = \frac{Jd}{NQ}B, \tag{17.8}$$

where N is the number of free charge carriers, of charge Q, per unit volume. So, if we know the coefficient Jd/NQ, by measuring V_{12} we can measure $\boldsymbol{B}$. Such a measuring strip is called a *Hall element*.

- The hard (and floppy) disk in every computer is a magnetic memory. We write to the disk by magnetizing a small piece of the disk surface, and we read from the disk by inducing a voltage in a small loop ("magnetic head") which is moving in close proximity to the magnetized disk surface element. The magnetic head is a magnetic circuit.

- A *transformer* is a magnetic circuit with (usually) two windings, "primary" (e.g., with N_1 turns), and "secondary" (with N_2 turns), on a common ferromagnetic core. If v_1 and v_2 are the primary and secondary voltages, and i_1 and i_2 the respective currents, the basic equations of an *ideal transformer* (a transformer with the core of infinitely large permeability) are

$$\frac{v_1}{v_2} = \frac{N_1}{N_2}, \tag{17.9}$$

$$\frac{i_1}{i_2} = \frac{N_2}{N_1}.$$ (17.10)

If the secondary winding of an ideal transformer is connected to a resistor of resistance R_2, the resistance seen from the primary terminals is

$$R_1 = R_2 \left(\frac{N_1}{N_2}\right)^2.$$ (17.11)

• *Electric motors* are devices which are used to transform continuously the electric energy into mechanical energy. Of several types of electric motors, two types, the *synchronous* and *asynchronous* motors, use rotating magnetic field. This is a magnetic field that keeps its magnitude constant in time, but the vector B rotates about an axis normal to itself with a constant angular velocity ω. It can be obtained by two perpendicular stationary coils with equal currents shifted 90° in phase, or by three stationary coils at 120° angle with symmetrical three-phase currents. In synchronous motors, the rotating part (the *rotor*) is a permanent magnet, which rotates in synchronism with the field. In asynchronous motors, the rotating part consists of short-circuited wire loops. The rotating field induces currents in the loops, and there is a moment on these currents, but the rotor in this case need not rotate in synchronism with the field.

• Voltages induced in our body when we are close to power lines are much smaller than the normal electric impulses flowing through our nerve cells, and are on the order of pV.

QUESTIONS

Q17.1. Where is the earth's south magnetic pole? — *(a) Close to the earth South Pole. (b) Close to the earth North Pole. (c) Exactly at the earth North Pole.*

Q17.2. What is the order of magnitude of the earth's magnetic flux density? — *(a) 0.05 mT. (b) 0.05 μT. (c) 0.05 T.*

S **Q17.3.** Approximately how fast would you need to spin around your axis in the magnetic field of the earth to induce 1 mV around the contour of your body? — *(a) About three revolutions in a second. (b) About one third of a revolution in a second. (c) About 11 revolutions in a second.*

Answer. From the answer to the preceding question, and assuming the area of the equivalent loop to be $S = 1\,\text{m}^2$, we have the equation $e(t) = -d\phi(t)/dt = -\omega BS \cos \omega t$. So the amplitude of this induced electromotive force, or voltage, would be 1 mV for an angular velocity of the body of $\omega = 0.001/(5 \cdot 10^{-5}) = 2\,\text{rad/s}$. This corresponds to about three revolutions in a second.

Q17.4. Turn your computer monitor sideways or upside down while it is on (preferably with some brightly colored pattern on it). Do you notice changes on the screen? If yes, what and why? — *(a) Nothing happens. (b) Since the electrons are heavy, the gravitational field will change the pattern somewhat. (c) Since the electrons are moving in the magnetic field of the earth, the pattern will be somewhat changed.*

Q17.5. What do you expect to happen if a magnet is placed close to a monitor? If you have a small magnet, perform the experiment (note that the effect might remain after you remove the magnet, but it is not permanent). Explain. — *(a) The pattern on the screen is shifted and distorted, due to the action of the magnetic field on the electron beam. (b) Nothing happens. (c) The pattern on the screen is shifted and distorted because the magnet is heavy, and acts by the gravitational force on the electron beam.*

S **Q17.6.** Explain how the Hall effect can be used to measure the magnetic flux density. — *Hint: note that the Hall voltage will exist for any angle between the strip and the B-lines. So the experiment needs to be repeated to obtain the maximal Hall voltage (why?).*

Answer. We put the Hall element into the magnetic field, and find its orientation corresponding to the maximal Hall voltage. From this voltage we obtain the magnitude of B, and the vector $\boldsymbol{B}$ is normal to the Hall element.

Q17.7. Explain how the Hall effect can be used to determine whether a semiconductor is p- or n- doped. — *Hint: note that the Hall voltage sign depends on the sign of free charge carriers.*

Q17.8. What magnetic material properties are chosen for the tracks and heads in a hard disk? — *(a) Tracks have low B_{remanent}, head has high B_{remanent}. (b) Tracks have high B_{remanent}, head has low B_{remanent}. (c) Both tracks and head have high B_{remanent}.*

Q17.9. Sketch and explain the time-domain waveform of the induced emf (or current) in the magnetic head coil in "read" mode as it passes over a piece of information recorded on a computer disk as "110". (Assume that a "1" is a small magnet along the track with a S-to-N orientation inside the magnet in the direction of motion of the head, and a "0" is a small magnet magnetized in the opposite direction.) — *Hint: check if the sketch of the time-domain waveform of the induced emf as given in Fig. Q17.9 is what you expect.*

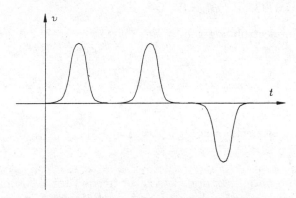

Fig. Q17.9. Sketch of induced emf.

S **Q17.10.** Write Ampère's law for an ideal transformer, and derive the voltage, current and impedance (resistance) transformation ratio. The number of turns in the primary and secondary are N_1 and N_2. — *Correct results are given in Eqs. (17.9)-(17.11).*

Answer. $N_1 i_1 = N_2 i_2$, so that $i_1/i_2 = N_2/N_1$. For primary and secondary voltages, $v_1/v_2 = N_1/N_2$. *The impedance is defined only in sinusoidal steady state and complex notation.* If the secondary is closed by an impedance Z_2, the input impedance of the transformer is $Z_1 = V_1/I_1 = (N_1/N_2)^2 V_2/I_2 = (N_1/N_2)^2 Z_2$.

Q17.11. What are the loss mechanisms in a real transformer, and how does each of the contributors to loss depend on frequency? — *(a) There are no losses in a real transformer. (b) The losses exist only in the transformer windings. (c) The losses exist in the transformer windings, and in the transformer core (hysteresis plus eddy-current losses).*

S **Q17.12.** Explain how a synchronous motor works. — *(a) The rotor is a short-circuited wire loop, in which current is induced, and the moment of magnetic forces acts on these currents. (b) The rotor is a permanent magnet, and the rotating magnetic field makes it rotate. (c) The current in the stator and rotor windings are in synchronism.*

Answer. It uses a rotating magnetic field to rotate a rotor in the form of a permanent magnet, with the same angular velocity as the magnetic field (i.e., in synchronism with it).

Q17.13. How is an asynchronous motor different from the synchronous type? — *Hint: use the answers to the preceding question.*

Q17.14. Describe the two mechanisms by which ac currents can affect our body. Use formulas in your description. — *Hint: recall that our body is conducting, and consider the influence of the quasistatic and induced components of the ac electric field produced by ac currents.*

PROBLEMS

P17.1. What is the minimum magnitude of a magnetic flux density vector that will produce the same magnetic force on an electron moving at $100\,\text{m/s}$ as that a $10\,\text{kV/cm}$ electric field produces? — *(a) $B = 1\,T$. (b) $B = 10\,T$. (c) $B = 10\,kT$.*

S **P17.2.** Calculate the velocity of an electron in a 10 kV CRT. The electric field is used to accelerate the electrons, and the magnetic field to deflect them. — *(a) $v = 3.93 \cdot 10^6\,m/s$. (b) $v = 5.93 \cdot 10^7\,m/s$. (c) $v = 1.93 \cdot 10^7\,m/s$.*

Solution. The velocity of an electron is given by Eq. (11.4),

$$v = \sqrt{\frac{2eV}{m_e}} = 5.93 \cdot 10^7 \text{ m/s},$$

where e is the absolute value of electron charge ($e = 1.602 \cdot 10^{-19}\,\text{C}$), and m_e is the electron mass ($m_e = 9.108 \cdot 10^{-31}\,\text{kg}$).

S **P17.3.** How large is the magnetic flux density vector needed for a 20 cm deflection in the CRT in problem P17.2, if the length of the tube is 25 cm? — *Refer to Fig. P17.3. (a) $B = 1.32\,mT$. (b) $B = 0.32\,mT$. (c) $B = 0.18\,T$.*

Solution. According to Fig. P17.3,

$$(R - x)^2 + d^2 = R^2,$$

from which the radius of curvature is $R = (x^2 + d^2)/(2x) = 25.62\,\text{cm}$. Since, at the same time,

$$R = \frac{m_e v}{eB}$$

(see Example 17.3), the required magnetic flux density is $B = m_e v/(eR) = 1.32\,\text{mT}$.

P17.4. A thin conductive ribbon is placed perpendicularly to the field lines of a uniform B-field. When the current is flowing in the direction shown in Fig. P17.4, there is a measured negative voltage V_{12} between the two edges of the ribbon. Are the free charges in the conductive ribbon positive or negative? — (a) If $V_{12} < 0$ and $B > 0$, $Q < 0$. (b) If $V_{12} < 0$ and $B > 0$, $Q > 0$. (c) If $V_{12} > 0$ and $B > 0$, $Q > 0$.

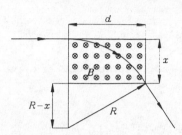

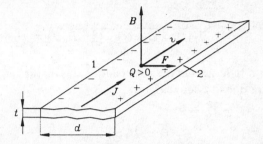

Fig. P17.3. A stream of electrons in a CRT.   **Fig. P17.4.** A current strip in a magnetic field.

P17.5. What is the voltage V_{12} equal to in problem P17.4 if $B = 0.8\,\text{T}$, $t = 0.5\,\text{mm}$, $I = 0.8\,\text{A}$ and the concentration of free carriers in the ribbon is $N = 8 \cdot 10^{28}\,\text{m}^{-3}$? — (a) $V_{12} \simeq -1 \cdot 10^{-7}$ V. (b) $V_{12} \simeq -5 \cdot 10^{-7}$ V. (c) $V_{12} \simeq -8 \cdot 10^{-7}$ V.

P17.6. The magnetic head in the form of a toroidal coil wound on a thin ferromagnetic toroidal core with a narrow air gap is in write mode. Calculate the magnitude of the current I in the winding that would be needed to produce a $B_0 = 1\,\mu\text{T}$ field in the gap. There are $N = 5$ turns on the core, the core can approximately be considered as linear, of relative permeability of $\mu_r = 1000$, the gap is $L_0 = 20\,\mu\text{m}$ wide, the cross-sectional area of the core is $S = 10^{-9}\,\text{m}^2$, and the mean radius of the core is $r = 0.1\,\text{mm}$. Assume that the fringing field in the gap makes the gap cross-sectional area effectively 10% larger than that of the core. — (a) $I = 3.76\,pA$. (b) $I = 5.08\,pA$. (c) $I = 6.34\,pA$.

P17.7. The head and the tracks in magnetic hard disks are made of different magnetic materials. Sketch and explain the preferred hysteresis curves for the two materials, indicating the differences. Which has higher loss in an ac regime? — (a) The hysteresis loop for the head material should be wide, and that for the disk material should be narrow. (b) Opposite to the preceding answer. (c) Both materials should have narrow hysteresis loops.

P17.8. A CRT needs $10\,\text{kV}$ to produce an electric field for electron acceleration. Design a wall-plug transformer to convert from 110 V in the U.S. and Canada or 220 V in Europe and Asia. Assume you have a core made of a magnetic material that has a very high permeability. — Assume the number of turns in the primary winding to be ten. The number of the turns in the secondary winding in the two cases is then: (a) 110 and 220. (b) 400 and 200. (c) 450 and 900.

S **P17.9.** Assume that for obtaining a rotating magnetic field you use three coils. The axes of the coils are at 60 degrees, not 90 degrees with respect to each other. What is the relative phasing of three sinusoidal currents (of equal amplitude) in the coils that will give a rotating magnetic field? Plot the current waveforms as a function of time.

Solution. Assume that the three coils are in the xy plane. Let the first coil be along the x axis, producing a magnetic flux density vector $B_1(t)$ in the x direction. Let the second coil produce a magnetic flux density vector $B_2(t)$ at an angle of 120 degrees with the x axis, and the third coil produce a vector $B_3(t)$ at an angle of 240 degrees (or -120 degrees) with the x axis. Assume that the three vectors are of the same amplitude and have initial phases 0 degrees, -120 degrees and -240 degrees, respectively, i.e., that

$$B_1(t) = B_m \cos \omega t, \quad B_2(t) = B_m \cos(\omega t - 2\pi/3), \quad B_1(t) = B_m \cos(\omega t - 4\pi/3).$$

It is necessary to consider the instantaneous values of the x- and y-components of the magnetic flux density separately:

$$B_x(t) = B_1(t) - B_2(t) \cos \pi/3 - B_3(t) \cos \pi/3 = B_1(t) - \frac{1}{2}[B_2(t) + B_3(t)],$$

$$B_y(t) = [B_2(t) - B_3(t)] \cos \pi/6.$$

The sum $B_2(t) + B_3(t)$ and the difference $B_2(t) - B_3(t)$ are obtained by representing the cosines in the expessions for $B_2(t)$ and $B_3(t)$ as $\cos(a - b) = \cos a \cos b + \sin a \sin b$. The final result is

$$B_x(t) = \frac{3}{2} B_m \cos \omega t.$$

$$B_y(t) = \frac{3}{2} B_m \sin \omega t.$$

These are the same expressions as for the two B-field components in a two-phase system, and the B-vector rotates with an angular velocity ω, as in the two-phase case.

Part 4: Transmission Lines

18. Transmission Lines

- Transmission lines consist most frequently of two parallel conductors (such as a twin lead or coaxial cable), but may have more than two (e.g., a three-phase transmission line). Strictly, transmission lines are electromagnetic structures, but they can also be analyzed by circuit-theory methods. Any small length Δz of a line is represented as a combination of circuit elements. For "lossless" lines (lines with negligible losses), this representation includes a series inductor of inductance $L'\Delta z$ and a shunt capacitor of capacitance $C'\Delta z$. If losses cannot be neglected, a series resistor of resistance $R'\Delta z$ and a shunt capacitor of conductance $G'\Delta z$ are added in the representation. R', L', G', and C' are the line resistance, inductance, conductance, and capacitance per meter.

- For "lossless" lines, the fundamental differential equations governing the voltage and current along the line (assumed along the z axis) are

$$\frac{\partial v(t,z)}{\partial z} = -L'\frac{\partial i(t,z)}{\partial t}, \qquad \frac{\partial i(t,z)}{\partial z} = -C'\frac{\partial v(t,z)}{\partial t}. \qquad (18.1)$$

These equations are known as the *transmission-line equations*, or *telegraphers' equations* (for lossless lines).

- Combining these equations, the second-order differential equations are obtained with only one unknown, voltage or current,

$$\frac{\partial^2 v(t,z)}{\partial t^2} - \frac{1}{L'C'}\frac{\partial^2 v(t,z)}{\partial z^2} = 0, \qquad \frac{\partial^2 i(t,z)}{\partial t^2} - \frac{1}{L'C'}\frac{\partial^2 i(t,z)}{\partial z^2} = 0. \qquad (18.2)$$

These are the (one-dimensional) *wave equations*. Their solutions are the forward ($+z$-directed) and backward ($-z$-directed) *voltage and current waves*, of the form (for the voltage)

$$v(t,z) = V_+ f(t - z/c) + V_- g(t + z/c) \qquad \text{(V)}, \qquad (18.3)$$

where V_+ and V_- are constants, f and g are *arbitrary* functions, and

$$c = \frac{1}{\sqrt{L'C'}} \qquad \text{(m/s)} \qquad (18.4)$$

is the velocity of wave propagation along a line. The wave with the argument $(t - z/c)$ propagates in the $+z$-direction, and that with the argument $(t + z/c)$ in the opposite direction. The first is the *forward travelling wave*, or the *incident wave*, and the second is the *backward travelling wave*, or the *reflected wave*. The velocity of propagation along lines with air dielectric equals the velocity of light in a vacuum ($c_0 \simeq 3 \cdot 10^8$ m/s), and along a typical dielectric-filled coaxial cable about 2/3 of that velocity.

• In the frequency domain, the lossless-transmission-line equations in complex form are

$$\frac{\mathrm{d}V(z)}{\mathrm{d}z} = -\mathrm{j}\omega L' I(z), \qquad \frac{\mathrm{d}I(z)}{\mathrm{d}z} = -\mathrm{j}\omega C' V(z), \qquad (18.5)$$

and the corresponding wave equation for the voltage is

$$\frac{\mathrm{d}^2 V(z)}{\mathrm{d}z^2} = -\omega^2 L' C' V(z). \qquad (18.6)$$

The solution of this wave equation is

$$V(z) = V_+ \mathrm{e}^{-\mathrm{j}\beta z} + V_- \mathrm{e}^{+\mathrm{j}\beta z} \qquad \text{(V)}, \qquad (18.7)$$

where the first term is the forward wave, the second is the backward wave, and

$$\beta = \omega\sqrt{L'C'} = \frac{\omega}{c} \qquad \text{(1/m)} \qquad (18.8)$$

is the *phase constant*. The *wavelength* of the sine wave along the line is

$$\lambda = \frac{2\pi}{\beta} = \frac{2\pi}{(\omega/c)} = \frac{c}{f} \qquad \text{(m)}. \qquad (18.9)$$

• Analogous expressions are valid for the current wave in time and frequency domain,

$$i(t,z) = \frac{V_+}{Z_0} f(t - z/c) - \frac{V_-}{Z_0} g(t + z/c) \qquad \text{(A)}, \qquad (18.10)$$

$$I(z) = \frac{V_+}{Z_0} \mathrm{e}^{-\mathrm{j}\beta z} - \frac{V_-}{Z_0} \mathrm{e}^{+\mathrm{j}\beta z} \qquad \text{(A)}, \qquad (18.11)$$

where

$$Z_0 = \sqrt{\frac{L'}{C'}} \qquad (\Omega) \qquad (18.12)$$

is known as the *characteristic impedance* of the (lossless) line. The value of the characteristic impedance depends on the physical parameters of the line (geometry, size, properties of the dielectric, etc.).

- If *only a forward wave* exists along a line, then

$$\frac{v(t,z)}{i(t,z)} = Z_0.$$
(18.13)

If *only a backward wave* propagates along the line,

$$\frac{v(t,z)}{i(t,z)} = -Z_0.$$
(18.14)

Consequently, we can eliminate the reflected wave on a line of any length by terminating the line in its characteristic impedance. If we do this, we say that the line is *matched*. The generators driving transmission lines are most often matched to the lines.

- If a lossless line of finite length ζ is terminated in an impedance Z_L, it is convenient to shift the coordinate origin, $z = 0$, from the generator to the load. Defining the *reflection coefficient* as

$$\rho = \frac{V_-}{V_+} \qquad \text{(dimensionless)},$$
(18.15)

the total phasor voltage and current along the line can be written in the form

$$V(z) = V_+ e^{-j\beta z}(1 + \rho e^{2j\beta z}) \qquad \text{(V)},$$
(18.16)

$$I(z) = \frac{V_+}{Z_0} e^{-j\beta z}(1 - \rho e^{2j\beta z}) \qquad \text{(A)}.$$
(18.17)

The impedance at any point along the line is $Z(z) = V(z)/I(z)$.

- The *transmission coefficient* is defined by

$$\tau = \frac{V_{\text{load}}}{V_+} = \frac{V_+ + V_-}{V_+} = 1 + \rho \qquad \text{(dimensionless)}.$$
(18.18)

The time-average power delivered to the load can be expressed as

$$P_{\text{load av}} = \text{Re}\{V(0)I(0)^*\} = \frac{|V_+|^2}{Z_0}(1 - |\rho|^2) \qquad \text{(W)}.$$
(18.19)

- Adopting the ζ axis with the origin at the load and directed towards the generator, the impedance of the line terminated in a load Z_L, at any point ζ and looking towards the load, is

$$Z(\zeta) = Z_0 \frac{1 + \rho e^{-j2\beta\zeta}}{1 - \rho e^{-j2\beta\zeta}} \qquad (\Omega).$$
(18.20)

In particular, for $\zeta = 0$ we have that $Z(\zeta) = Z_L$, so that

$$Z_L = Z_0 \frac{1+\rho}{1-\rho} \qquad (\Omega), \qquad (18.21)$$

from which

$$\rho = \frac{Z_L - Z_0}{Z_L + Z_0} \qquad \text{(dimensionless)}, \qquad (18.22)$$

$$\tau = \frac{2Z_L}{Z_L + Z_0} \qquad \text{(dimensionless)}. \qquad (18.23)$$

Eq. (18.20) can also be written in the form

$$Z(\zeta) = Z_0 \frac{Z_L \cos \beta\zeta + jZ_0 \sin \beta\zeta}{Z_0 \cos \beta\zeta + jZ_L \sin \beta\zeta} = Z_0 \frac{Z_L + jZ_0 \tan \beta\zeta}{Z_0 + jZ_L \tan \beta\zeta} \qquad (\Omega). \qquad (18.24)$$

• The *voltage standing-wave ratio* (VSWR) is defined as the ratio of the maximal to minimal voltage along the line,

$$\text{VSWR} = \frac{V(z)_{\text{max}}}{V(z)_{\text{min}}} = \frac{1 + |\rho|}{1 - |\rho|} \qquad \text{(dimensionless)}. \qquad (18.25)$$

For a matched load VSWR $= 1$, and for open and for short circuit VSWR $= \infty$.

• The phasor transmission-line equations for *lossy transmission lines* are

$$\frac{dV(z)}{dz} = -(R' + j\omega L')I(z), \qquad \text{and} \qquad \frac{dI(z)}{dz} = -(G' + j\omega C')V(z), \qquad (18.26)$$

and the characteristic impedance of such lines is

$$Z_0 = \sqrt{\frac{R' + j\omega L'}{G' + j\omega C'}}. \qquad (18.27)$$

Instead of the phase coefficient, β, we now have the *propagation coefficient*,

$$\gamma = \alpha + j\beta = \sqrt{(R' + j\omega L')(G' + j\omega C')}, \qquad (18.28)$$

where α as the *attenuation constant (coefficient)*, and β the *phase constant (coefficient)*. This means that, for lossy lines and a forward wave, instead of $e^{-j\beta z}$ we have $e^{-(\alpha + j\beta)z} =$

$e^{-\alpha z} e^{-j\beta z}$, i.e., the amplitudes of the voltage and current waves are *exponentially attenuated* as they propagate. For transmission lines with small losses ($R' << \omega L'$ and $G' << \omega C'$),

$$\alpha \simeq \frac{1}{2}\left(R'\sqrt{\frac{C'}{L'}} + G'\sqrt{\frac{L'}{C'}} \right) \quad , \quad \beta \simeq \omega\sqrt{L'C'}. \tag{18.29}$$

- The attenuation of a forward voltage wave over a distance d is expressed in *nepers* (Np), as

$$\text{Attenuation in nepers} = \alpha d \quad (\alpha : \text{ nepers per meter } - \text{ Np/m}). \tag{18.30}$$

More frequently, the attenuation is expressed in *decibels* (dB),

$$\text{Attenuation in decibels} = (20 \log e)\, \alpha d \quad (\alpha : \text{ decibels per meter } - \text{ dB/m}). \tag{18.31}$$

Since $20 \log e = 8.686$, we have that $1\,\text{Np} = 8.686\,\text{dB}$.

- For detecting faults in cables, an instrument known as the *time domain reflectometer* (TDR) is used. The principle is very simple: the instrument sends a voltage step, and waits for the reflected signal. If there is a fault in the cable, it will be equivalent to some rapid change in cable properties, and part of the voltage step wave will reflect off the discontinuity. The analysis requires the analysis of transmission-lines in time domain.

- The graphical solution of lossless-line problems is possible using a *Smith chart*. The Smith chart enables us to make a direct determination of the complex reflection coefficient $\rho(0)$ at the load, corresponding to a given load impedance Z_L and the characteristic impedance of the line. Conversely, if $\rho(0)$ is determined experimentally, we can read directly from the chart the load impedance Z_L, if Z_0 is known. The Smith chart is used in the calculation of impedance matching, transmission-line circuit design, amplifier design, and in many other applications. The chart is based on the expression in Eq. (18.21), modified to evaluate the *normalized load impedance*, Z_L/Z_0:

$$z = Z_L/Z_0 = \frac{1+\rho}{1-\rho} \qquad \text{(dimensionless)}. \tag{18.32}$$

QUESTIONS

S **Q18.1.** Why it is not practically possible to obtain a coaxial cable of characteristic impedance $Z_0 = 500\,\Omega$? Can you have a two-wire line of this characteristic impedance? — *Hint: note that for a coaxial line $Z_0 = (\sqrt{\mu/\epsilon}/2\pi)\ln(b/a)$, and for an air two-wire line $Z_0 \simeq 120\ln(d/a)$, and calculate the needed ratios b/a and d/a, respectively.*

Answer. Recall that $Z_0 = \sqrt{L'/C'} = 1/(cC')$, where $c = 1/\sqrt{\epsilon\mu}$ is the velocity of wave propagation along the line. Since for a coaxial cable $C' = 2\pi\epsilon/\ln(b/a)$, the characteristic impedance for the cable is

$$Z_0 = \frac{1}{2\pi}\sqrt{\frac{\mu}{\epsilon}}\ln\frac{b}{a}.$$

For an air-filled coaxial cable this becomes $Z_0 \simeq 60\ln(b/a)$, which implies a large, practically unobtainable, ratio b/a for $Z_0 = 500\,\Omega$.

For an air thin two-wire line, $Z_0 \simeq 120\ln(d/a)$, where a is the wire radius, and d the distance between the wire axes. Evidently, the two-wire line can have a characteristic impedance on the order of several hundred ohms.

Q18.2. Assume that a transmission line is made of two parallel, highly resistive wires. Can this line be analyzed using fundamental transmission-line equations? Explain. — *(a) Only if the wire resistance is not too high. (b) Yes, in all cases. (c) No, because theory does not allow highly resistive wires.*

Q18.3. A coaxial cable is filled with water. Does it represent a transmission line? Explain. — *(a) Only if water is distilled (nonconductive). (b) Yes, for any conductivity. (c) No, for any conductivity.*

Q18.4. Two wires several wavelengths long serve as a connection between a generator and a receiver. The distance between the wires is small but is not constant, varying as a smooth function of the coordinate along the line. Can you use the transmission-line equations for the analysis of this line? Explain. — *(a) No, the distance between the wires must be constant. (b) Yes, the equations remain exactly the same. (c) Yes, except that C', L', R', and G' are functions of z.*

Q18.5. Explain how you can obtain (1) a forward wave only; (2) a backward wave only along a transmission line. — *Hint: think where the generator in the two cases must be, and how the line must be terminated.*

S **Q18.6.** Describe at least three ways of obtaining simultaneously a forward and a backward sinusoidal wave of the same amplitude along a transmission line. — *One answer only is correct, find another two ways. (a) Open-circuited lossy line, matched generator. (b) Lossless line terminated in the line characteristic impedance. (c) Short-circuited lossless line, matched generator.*

Answer. The line must be lossless, and needs to be driven at its beginning with a sinusoidal generator of impedance equal to the line characteristic impedance. It then can be open- or short-circuited at its end, or a purely reactive load (a capacitor or a coil) can be connected at the end, and the reflected (backward) wave will be of the same magnitude as the incident (forward) wave.

Q18.7. Why you can replace an infinitely long end of a transmission line with a resistor of resistance equal to the line characteristic impedance? — *Because the input impedance of the removed infinite part is (a) Z_0, (b) $Z_0/2$, (c) zero.*

Q18.8. Can a voltage (or a current) wave along a transmission line be described by the expression of the form $u(xy)$, where $u(xy)$ is a function of the product of the arguments $x = (t - z/c)$ and $y = (t + z/c)$? Explain. — *(a) No, because xy does not represent a traveling wave. (b) Yes, because xy does represent a traveling wave. (c) Yes, because both x and y represent waves.*

Q18.9. Can we adopt the negative instead of positive value of the square root in Eq. (18.4) for the velocity of wave propagation along transmission lines? Explain. — *(a) Yes, nothing else needs to be changed. (b) Yes, but the expressions for the forward and backward waves need to be exchanged. (c) No, because the velocity cannot be negative.*

S **Q18.10.** Why must the exponent of the forward voltage and current waves in Eqs. (18.7) and (18.11) be negative? Why must those of the backward waves be positive? — *Hint: note that, in complex notation, the factor* $e^{-j\beta z}$ *contains the phase of the wave along the line.*

Answer. In complex notation, the term multiplying j in the expression $e^{-j\beta z}$, i.e. $-\beta z$, is the phase of the wave at location z. If z increases as the wave progresses, the minus sign means that the phase of the wave at points further away along the z axis lags behind that at points for smaller z. This must be so for a forward wave, since at a fixed z the phase is getting larger, and so the phase of the wave further away is always smaller. The opposite argument can be used to explain why for a backward wave this exponent must be positive.

Q18.11. Is the wavelength along an air line greater or less than that in the same line filled with a dielectric? What is the answer if the dielectric has relative permeability greater than one? Explain. — *(a) It is greater. (b) It is less. (c) It is the same.*

Q18.12. What are the SI units for the following quantities: (1) the attenuation constant (α), (2) the phase constant (β), (3) the reflection coefficient (ρ), and (4) the voltage standing-wave ratio (VSWR)? — *(a) Neper, radian, volt, dimensionless. (b) Neper per meter, radian per meter, dimensionless, dimensionless. (c) Decibel times meter, radian times meter, volt per meter, volt per meter.*

Q18.13. What is the magnitude of the reflection coefficient, $|\rho|$, and of the VSWR, for which one half of the power of the incident wave is transferred to the load? — *(a) $|\rho| = 0.707$, VSWR = 5.83. (b) $|\rho| = 0.5$, VSWR = 3. (c) $|\rho| = 0.63$, VSWR = 4.4.*

S **Q18.14.** Why is the voltage at the termination Z of a transmission line with characteristic impedance Z_0 equal to $V = 2V_+ Z/(Z + Z_0)$? — *Hint: note that the generator is matched to the line, and that the forward wave sees the line as infinite (it "does not know" how long it is). Note also that the equivalent Thévenin generator from the load terminals towards the generator has an electromotive force $V = 2V_+$ (prove!) and internal impedance Z_0.*

Answer. The generator connected to the line is always matched to it, i.e., its internal impedance is Z_0. Let the generator electromotive force be V. Then the forward voltage between the line conductors at the generator terminals is $V_+ = V/2$ (the voltage across the input impedance Z_0 of the line seen by the forward wave). The equivalent Thévenin generator from the load terminals towards the generator has an electromotive force $V = 2V_+$ and internal impedance Z_0. So the current in the load is $I_{\text{load}} = 2V_+/(Z_0 + Z)$, and the voltage across the load is this current times the load impedance, Z.

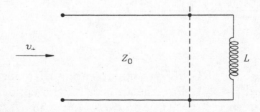

Fig. Q18.18. Calculating the step response of an inductor.

Q18.15. What are the input impedances to lossless lines of lengths $\lambda/4$ and $\lambda/2$ if they are (1) open-circuited or (2) short-circuited? — *Hint: use Eqs. (18.24) and (18.9).*

Q18.16. Can a resistive load of *any* resistance R be matched in practice to a transmission line of characteristic impedance Z_0? Explain. — *Hint: let the $\lambda/4$ matching line section have a characteristic impedance Z_0'. Then Z_0' is defined by the relation $Z_0' = \sqrt{RZ_0}$. Note that characteristic impedances of transmission lines are available in only limited range (what is this range approximately?).*

Q18.17. The characteristic impedance of a lossy line in Eq. (18.27) is real if $R' = 0$ and $G' = 0$. Can it be real for some other relation between R', L', G', and C'? — *(a) No. (b) Yes, if $G' \to \infty$. (c) Yes, if $R'/L' = G'/C'$.*

Q18.18. Why could we not use simple transmission-line analysis when calculating the step response of an inductor, as in Fig. Q18.18? — *Hint: note that the transmission-line analysis with complex currents and voltages is valid for a single frequency.*

S **Q18.19.** If you had a break in the dielectric of a cable causing a large shunt conductance, what do you expect to see reflected if you excite the cable with a short pulse (practical delta function)? — *As if at the break position there is (a) a short circuit, (b) an open circuit, (c) a resistor of resistance Z_0.*

Answer. Such a fault in the cable is approximately the same as a short circuit. Therefore the reflected pulse will be of approximately the same magnitude as that of the incident pulse.

Q18.20. If you had a break in the outer conductor of a cable, causing a large series resistance, what do you expect to see reflected if you excite the cable with a short pulse (imperfect delta function)? — *The answers are the same as for the preceding question.*

Q18.21. What do the reflected waves off a series inductor and shunt capacitor in the middle of a transmission line look like for a short pulse excitation, assuming that $\omega L \gg Z_0$ and $\omega C \gg 1/Z_0$? — *The answers are the same as for question Q18.19.*

Q18.22. Using the Smith chart, determine the complex reflection coefficient on a 60-Ω line if it is terminated by (1) 80 Ω, (2) $(30 - j40)\,\Omega$, or (3) $(40 + j90)\,\Omega$. — *The normalized impedances are $(0.5 - j0.67)$, $(0.67 + j1.5)$, and $(1.33 + j0)$. The complex reflection coefficients are $\rho = 0.69$, $\theta_\rho = 60°$; $\rho = 0.19$, $\theta_\rho = 0$, $\rho = 0.57$, $\theta_\rho = -99°$. Find matching values for cases (1), (2), and (3).*

Q18.23. Using the Smith chart, determine the terminating impedance of a 70-Ω line, if it was found experimentally that the complex voltage reflection coefficient is (1) 0.8, (2) $0.2e^{-j\pi/4}$, or (3) $0.5e^{j\pi/3}$. — *The load impedances are $(89.6 - j27.3)\,\Omega$, $(406 + j0)\,\Omega$, and $(29.4 + j114.8)\,\Omega$. Find matching values for cases (1), (2), and (3).*

PROBLEMS

P18.1. Given a high-frequency RG-55/U coaxial cable with $a = 0.5\,\text{mm}$, $b = 2.95\,\text{mm}$, $\epsilon_r = 2.25$ (polyethylene), and $\mu_r = 1$, find the values for the capacitance and inductance per unit length of the cable. — *(a) $C' = 90.52\,pF/m$, $L' = 155\,nH/m$. (b) $C' = 70.52\,pF/m$, $L' = 355\,nH/m$. (c) $C' = 75.52\,pF/m$, $L' = 455\,nH/m$.*

S **P18.2.** Assume that the coaxial cable from problem P18.1 is not lossless, but that the losses are small, resulting in an attenuation constant in *decibels per meter* at $10\,\mathrm{GHz}$ of $\alpha = 0.5\,\mathrm{dB/m}$. Assuming the dielectric in the cable to be perfect, find the resistance per unit length that causes the losses in the conductors. — *(a)* $R' = 4.17\,\Omega/m$. *(b)* $R' = 6.17\,\Omega/m$. *(c)* $R' = 8.17\,\Omega/m$.

Solution. Since the dielectric is assumed to be perfect, the shunt conductance per unit length, G', is zero. So, according to Eqs. (18.36) and (18.38), the resistance per unit length is

$$R' = 2\sqrt{\frac{L'}{C'}}\,\frac{\alpha_{\mathrm{dB/m}}}{8.686} = 8.17\,\Omega/m.$$

P18.3. The distance d between wires of a lossless two-wire line is a smooth, slowly varying function of the coordinate z along the line so that the line capacitance and inductance per unit length, L', and C', are also smooth functions of z, $L' = L'(z)$ and $C' = C'(z)$. Derive the transmission-line equations for such a *nonuniform* transmission line. Check if these equations become the transmission-line equations (18.1) for $L'(z)$ and $C'(z)$ constant. — *The circuit approximately equivalent to a nonuniform transmission line is sketched in Fig. P18.3. The correct result is*

$$\frac{\partial v(t,z)}{\partial z} = -L'(z)\,\frac{\partial i(t,z)}{\partial t} \qquad \text{and} \qquad \frac{\partial i(t,z)}{\partial z} = -C'(z)\,\frac{\partial v(t,z)}{\partial t}.$$

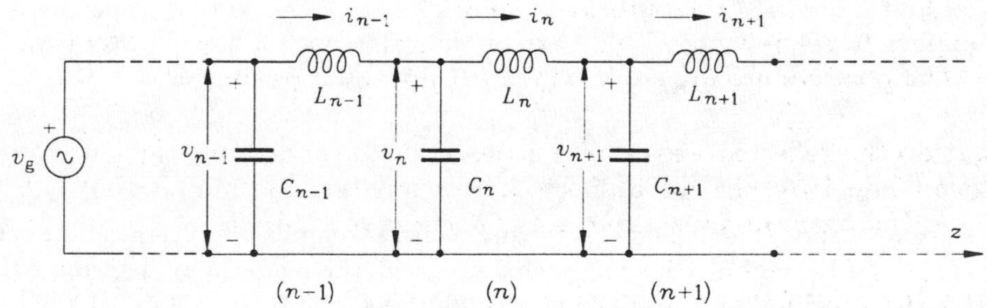

Fig. P18.3. The circuit approximately equivalent to a nonuniform transmission line.

S **P18.4.** Using circuit theory, analyze approximately a matched, lossless, air-filled coaxial transmission line of length $l = \lambda$ and conductor radii $a = 1\,\mathrm{mm}$ and $b = 3\,\mathrm{mm}$ as a connection of n cells, as in Fig. P18.4, for $n = 1, 2, \ldots, 20$. Such a circuit-theory approximation to transmission lines is known as an *artificial transmission line*. Note that an artificial transmission line can be analyzed as a simple ladder network. Assume the artificial line to be terminated in the actual characteristic impedance, and compare current in series concentrated inductive elements and voltage across parallel concentrated capacitive elements with exact results. Solve the problem so that you can vary L', C', and n. — *Hint: Fig. P18.4 shows the artificial line approximating the actual coaxial line, of length l, inductance per unit length L', and capacitance per unit length C'. The actual line characteristic impedance is $Z_0 = \sqrt{L'/C'}$. The parameters of the concentrated elements in the artificial line are $L = L'l/n$ and $C = C'l/n$. The ladder networks are most easily solved if the solution procedure starts from the load.*

Solution. Fig. P18.4 shows the artificial line approximating the actual coaxial line, of length l, inductance per unit length L', and capacitance per unit length C'. The actual line characteristic impedance is $Z_0 = \sqrt{L'/C'}$. The parameters of the concentrated elements in the artificial line are $L = L'l/n$ and $C = C'l/n$. Let the voltage of the generator be V_g. We know, however, that the ladder networks are most easily solved if the solution procedure starts *from the load*. So, referring to Fig. P18.4, we have

$$I_1 = \frac{V_0}{Z_0}, \qquad V_1 = V_0 + j\omega L\, I_1, \qquad I_2 = I_1 + j\omega C\, V_1, \qquad V_2 = V_1 + j\omega L\, I_2, \qquad \ldots,$$

$$I_n = I_{n-1} + j\omega C\, V_{n-1}, \qquad V_n = V_{n-1} + j\omega L\, I_n.$$

We can assume now that the load voltage is $V_0 = 1\,\text{V}$, and solve for I_1, V_1, I_2, $\ldots$, V_n. The generator voltage producing 1 V across the load is hence $V_g' = V_n$. Due to linearity of the circuit,

$$\frac{V_0}{V_g} = \frac{1\,\text{V}}{V_g'},$$

from which the load voltage produced by the actual generator is $V_0 = (V_g/V_n)\,\text{V}$.

The line is *matched*, so that there is no backward wave. The exact expressions for the voltage and current along the line are therefore

$$V(\zeta) = V_0\, e^{j\beta\zeta} = V_g\, e^{j\beta(\zeta-l)}, \qquad I(\zeta) = \frac{V_0}{Z_0}\, e^{j\beta\zeta} = \frac{V_g}{Z_0}\, e^{j\beta(\zeta-l)},$$

where $\beta = \omega\sqrt{L'C'}$ is the phase coefficient of the line.

Based on the above approximate and exact expressions, a computer program can be easily written, with inputs l, L', C', ω, V_g, and n. By using such a program, the reader may determine the error due to the approximation of the actual line by an artificial line, given by

$$\delta_{V_k} = \frac{V_k - V\,(\zeta = kl/n)}{V\,(\zeta = kl/n)}, \qquad \delta_{I_k} = \frac{I_k - I\,(\zeta = kl/n)}{I\,(\zeta = kl/n)}, \qquad k = 1, 2, \ldots, n.$$

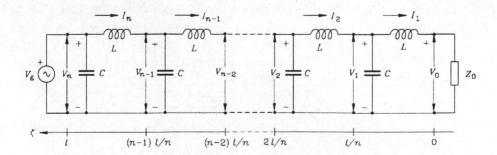

Fig. P18.4. An artificial transmission line can be analyzed as a ladder network.

P18.5. Noting that $c = 1/\sqrt{\epsilon\mu}$ for all transmission lines, prove that the inductance per unit length and the characteristic impedance of a lossless transmission line can be expressed in terms of c and C'. — (a) $L' = 2/(c^2 C')$, $Z_0 = 2/(cC')$. (b) $L' = c^2 C'$, $Z_0 = cC'$. (c) $L' = 1/(c^2 C')$, $Z_0 = 1/(cC')$.

S **P18.6.** Express $V(z)$ in Eq. (18.7) and $I(z)$ in Eq. (18.11) for lossless lines in terms of the sending-end voltage $V(0)$ and sending-end current $I(0)$. — *Hint: note that* $V(0) = V_+ + V_-$ *and* $I(0) = V_+/Z_0 - V_-/Z_0$.

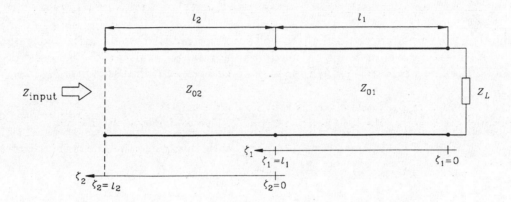

Fig. P18.8. Two transmission lines connected in series.

Solution. Since

$$V(0) = V_+ + V_- \quad \text{and} \quad I(0) = \frac{V_+}{Z_0} - \frac{V_-}{Z_0},$$

we find that

$$V_+ = \frac{1}{2}\left[V(0) + Z_0 I(0)\right] \quad \text{and} \quad V_- = \frac{1}{2}\left[V(0) - Z_0 I(0)\right].$$

Consequently, $V(z)$ can be expressed as

$$V(z) = \frac{1}{2}\left\{\left[V(0) + Z_0 I(0)\right] e^{-j\beta z} + \left[V(0) - Z_0 I(0)\right] e^{+j\beta z}\right\} = V(0)\cos\beta z - j Z_0 I(0)\sin\beta z.$$

Similarly, the expression for $I(z)$ becomes

$$I(z) = I(0)\cos\beta z - j\frac{V(0)}{Z_0}\sin\beta z.$$

P18.7. Prove that it is possible to obtain the characteristic impedance of any lossless line by measuring the input impedance of a section of the line when it is open-circuited, and when it is short-circuited. — *Hint: use Eq. (18.24) and note that for an open-circuited line $Z_L \to \infty$, and for a short-circuited line $Z_L = 0$.*

P18.8. A lossless line of characteristic impedance Z_{01} and length l_1 is terminated in an impedance Z_L. The line serves as a load for another lossless line of characteristic impedance Z_{02} and length l_2. The dielectric in both lines is air and the angular frequency of the current is ω. Determine general expressions for the input impedance of the second line, the reflection coefficient in both lines, and the voltage standing-wave ratio in both lines. — *Hint: refer to Fig. P18.8 and use Eqs. (18.24), (18.22) and (18.25).*

S **P18.9.** A short and then an open load are connected to a 50-Ω transmission line at $z = 0$. Make a plot of the impedance, normalized voltage ("normalized" means that you divide the voltage by its maximal value to get a maximum normalized voltage of 1), and normalized current along the line up to $z = -3\lambda/2$ for the two cases. — *Hint: use Eqns. (18.24), (18.16) and (18.17).*

Solution. See problem P18.6. Let us introduce a new coordinate, $\theta = -\beta z$, representing the distance from the load *in electrical units*. [Note that the range of θ corresponding to the specified range of z, $0 \geq z \geq -3\lambda/2$, is $0 \leq \theta \leq 3\pi$, since $\theta = -(2\pi/\lambda)z$.] With this, when the line is short-circuited, we have

$$Z(\theta) = j Z_0 \tan\theta, \qquad V(\theta) = j Z_0 I(0) \sin\theta, \qquad I(\theta) = I(0) \cos\theta \qquad \text{(short load)},$$

while in the case of an open load,

$$Z(\theta) = -j Z_0 \cot\theta, \qquad V(\theta) = V(0) \cos\theta, \qquad I(\theta) = j \frac{V(0)}{Z_0} \sin\theta \qquad \text{(open load)}.$$

It is left to the reader to plot these functions in the interval $0 \leq \theta \leq 3\pi$.

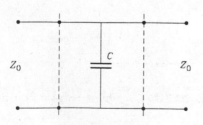

Fig. P18.10. A shunt capacitor in a line.

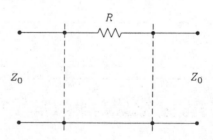

Fig. P18.11. A series resistor in a line.

S **P18.10.** A *lumped* capacitor is inserted into a transmission-line section, as shown in Fig. P18.10. Find the reflection coefficient for a wave incident from the left. Assume the line is terminated to the right so that there is no reflection off the end of the line. Find a simplified expression that applies when C is small. The characteristic impedance of the line is Z_0. — *For C small: (a) $\rho \simeq -j\omega C Z_0/2$. (b) $\rho \simeq -j\omega C Z_0$. (c) $\rho \simeq -j2\omega C Z_0$.*

Solution. The equivalent impedance "seen" by a wave which is incident from the left is that of a parallel connection of Z_0 and C, i.e.,

$$Z_e = \frac{Z_0/(1/j\omega C)}{Z_0 + 1/(j\omega C)} = \frac{Z_0}{1 + j\,\omega C Z_0}.$$

The corresponding reflection coefficient is

$$\rho = \frac{Z_e - Z_0}{Z_e + Z_0} = \frac{-j\omega C Z_0}{2 + j\omega C Z_0} = -\frac{\omega^2 C^2 Z_0^2}{4 + \omega^2 C^2 Z_0^2} - j\,\frac{2\omega C Z_0}{4 + \omega^2 C^2 Z_0^2}.$$

If C is small, the above expression can be simplified to read

$$\rho \simeq -j\,\frac{\omega C Z_0}{2} \qquad \left(C \ll \frac{2}{\omega Z_0}\right),$$

i.e., ρ becomes a small negative imaginary number.

P18.11. Repeat problem P18.10 assuming that a *lumped* resistor is inserted into a transmission-line section as shown in Fig. P18.11. Find a simplified expression that applies when R is small. — *For R small: (a) $\rho \simeq R/2Z_0$. (b) $\rho \simeq R/Z_0$. (c) $\rho \simeq 2R/Z_0$.*

P18.12. Repeat problem P18.10 assuming that a *lumped* inductor is inserted into a transmission-line section as shown in Fig. P18.12. Find a simplified expression that applies when L is small. — *For L small: (a) $\rho \simeq j\omega L/2Z_0$. (b) $\rho \simeq j\omega L/Z_0$. (c) $\rho \simeq j2\omega L/Z_0$.*

P18.13. Repeat problem P18.10 assuming that a *lumped* resistor is inserted into a transmission-line section as shown in Fig. P18.13. Find a simplified expression that applies when R is large. — *For R large: (a) $\rho \simeq -Z_0/2R$. (b) $\rho \simeq -Z_0/R$. (c) $\rho \simeq -2Z_0/R$.*

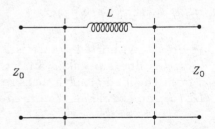

Fig. P18.12. A series coil in a line.

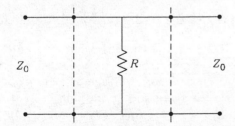

Fig. P18.13. A shunt resistor in a line.

P18.14. A 50-Ω transmission line needs to be connected to a 100-Ω load. The setup is used at 1 GHz. What would you connect between the line and the load to have no reflected voltage on the line? — *Hint: use a quarter-wave transformer and calculate its length if the dielectric in the quarter-wave transformer is polyethylene ($\epsilon_r = 2.25$ and $\mu_r = 1$). In that case the length of the transformer line section is about: (a) $d = 3\,cm$. (b) $d = 4\,cm$. (c) $d = 5\,cm$.*

P18.15. In problem P18.14, the load is a 100-Ω resistor, but the leads are long and represent a 2 nH inductor in series with the resistor. How would you get rid of the reflected voltage on the line in this case? — *Hint: note that a short short-circuited segment of any line behaves as an inductor, and that a short open-circuited segment behaves as a capacitor. There are several possibilities for solving the problem.*

P18.16. Find the transmission coefficient for the transmission line in Fig. P18.10. — *Hint: use the result of Problem P18.10 and recall that $1 + \rho = \tau$.*

P18.17. Find the reflection and transmission coefficients for the transmission line in Fig. P18.17. Because the reflection coefficient is defined by voltage, the power is given by its square. What are the reflected and transmitted power equal to? Does the power balance make sense? — *Hint: note that the equivalent impedance looking to the right from points 1 and 2 is that of a parallel connection of Z_0 and $2Z_0$. (a) $\rho = -0.2$, $\tau = 0.8$. (b) $\rho = 0.2$, $\tau = -0.8$. (c) $\rho = 0.2$, $\tau = 0.8$.*

P18.18. Derive the normalized input impedance (i.e., the impedance divided by Z_0) for a section that is $n\lambda/8$ long and is shorted at the other end, for $n = 1, 2, 3, 4$, and 5. Plot the impedance as a function of electrical line length (length measured in wavelengths along the line). — *Hint: use Eq. (18.24).*

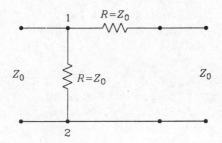

Fig. P18.17. Two resistors in a line.

P18.19. Repeat problem P18.18 for an open-ended line. — *Hint: use Eq. (18.24).*

P18.20. Find the total current and voltage at the beginning of a $\lambda/4$ shorted transmission line of characteristic impedance Z_0. What circuit element does this line look like? Plot the total current and voltage as a function of electrical line length from the load (length measured in wavelengths along the line). — *Hint: use Eqs. (18.22), (18.16), and (18.17). You need to replace z in the last two equations by $-\zeta$ (why?).*

P18.21. Repeat the previous problem for an open-ended line. — *Hint: the same as in the previous problem.*

P18.22. Find the reflection and transmission coefficients for an ideal $n : 1$ transformer, as in Fig. P18.22, where n is the voltage transformation ratio. — *(a) $\rho = (n^2 Z_0 - Z_1)/(n^2 Z_0 + Z_1)$. (b) $\rho = (Z_0 - Z_1)/(Z_0 + Z_1)$. (c) $\rho = (nZ_0 - Z_1)/(nZ_0 + Z_1)$.*

P18.23. Find the input impedance for the circuit in Fig. P18.23. — *Hint: note that the open-circuited stub is a reactance connected in parallel to the input terminals of the 75-Ω line. Use Eq. (18.24) to find the input impedance (or admittance) of the 75-Ω line. (a) $Z = (56.25 - j\,18.75)\,\Omega$. (b) $Z = (33.05 + j\,15.25)\,\Omega$. (c) $Z = (86.44 - j\,21.06)\,\Omega$.*

P18.24. A coaxial transmission line with a characteristic impedance of $150\,\Omega$ is $2\,\mathrm{cm}$ long and is terminated in a load impedance of $Z = 75 + \mathrm{j}150\,\Omega$. The dielectric in the line has a relative permittivity of $\epsilon_r = 2.56$. Find the input impedance and VSWR on the line at $f = 3\,\mathrm{GHz}$. — *Hint: it is first necessary to find the electrical length of the line. Once this is known, use Eqs. (18.24) and (18.25). (a) $Z = (62.3 - \mathrm{j}55.1)\,\Omega$ and VSWR=2.67. (b) $Z = (32.7 - \mathrm{j}48.1)\,\Omega$ and VSWR=2.25. (c) $Z = (96.3 + \mathrm{j}65.2)\,\Omega$ and VSWR=1.77.*

P18.25. Match a 25-Ω load to a 50-Ω line using (1) a single quarter-wave section of line, or (2) two quarter-wave line sections. — *Hint: for a quarter-wave section, in Eq. (18.24) $\beta\zeta = \pi/4$, so that the input impedance to the section terminated in Z_L is $Z(\lambda/4) = Z_0^2/Z_L$. With one matching section you need a line of a specific characteristic impedance (which one?). For two matching sections, you have less stringent requirement with respect to the characteristic line impedances (why?).*

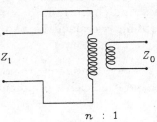

Fig. P18.22. An ideal transformer.

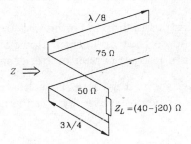

Fig. P18.23. Impedance of a line with a shunt stub.

S **P18.26.** Match a purely capacitive load, $C = 10\,\mathrm{pF}$, to a 50-Ω line at $1\,\mathrm{GHz}$. How many different ways can you think of doing this? — *(a) An open- or a short-circuited stub connected across the capacitor terminals. (b) A line segment of appropriate characteristic impedance between the capacitor and the line. (c) A resistor and a coil connected in series with the capacitor.*

Solution. Eq. (18.24) tells us that a purely reactive load cannot be transformed into a resistive load by any transmission-line section. Therefore, a resistive element must be used for matching, i.e., some power must be lost.

There are many ways one could do the matching. The simplest way is to add to the capacitor a series coil and a series resistor. The coil must have an inductance of $L = 1/(\omega^2 C) = 2.53\,\mathrm{nH}$, and the resistor must have a resistance of $R = 50\,\Omega$.

S **P18.27.** Calculate and plot magnitude and phase of $\rho(f)$ between 1 and $3\,\mathrm{GHz}$ for a 50-Ω open transmission line that is $\lambda/4$ long at $2\,\mathrm{GHz}$. — *Hint: note that, since the line is open-circuited, the reflection coefficient at the load is $\rho_L = 1$. The line being lossless, as one moves by l away from the load the reflection coefficient becomes $\rho = \rho_L\,\mathrm{e}^{-\mathrm{j}2\beta l}$.*

Solution. Since the line is open-circuited, the reflection coefficient at the load is $\rho_L = 1$. The line being lossless, as one moves for l away from the load the reflection coefficient becomes

$$\rho = \rho_L\,\mathrm{e}^{-\mathrm{j}2\beta l},$$

(l is the line length), so that

$$|\rho| = |\rho_L| = 1, \qquad \text{and} \qquad \angle \rho = \angle \rho_L - 2\beta l = -2\beta l = -2 \frac{2\pi}{c/f} \frac{c/f_0}{4} = -\frac{\pi}{f_0} f,$$

where $f_0 = 2\,\text{GHz}$ $[\beta(f_0)\,l = \pi/2]$. It is a simple matter to plot these functions.

P18.28. If you had a cable like the one in problem P18.2 spanning the Atlantic and you sent a continuous signal of 1 MW power from the United States to England, how much power approximately would you get in England? (Look up the approximate distance across the Atlantic in an atlas if you need it.) — *Hint: the distance across the Atlantic (from Boston to Plymouth) is approximately 7,500 km (check in an atlas!). Adopt a cable which introduces an attenuation of 0.5 dB per meter, and calculate the power at the cable end in England.*

S **P18.29.** A printed-circuit board trace in a digital circuit is excited by a voltage $v(t)$, as in Fig. P18.29. Derive an equation for the coupled (cross-talk) signal on an adjacent line, $v_c(t)$, assuming the adjacent line is connected to a load at one end and a scope (infinite impedance) at the other end so that no current flows through it. (*Hint: the coupling is capacitive and you can approximate it by a capacitor between the two traces and use circuit theory.*)

Solution. Referring to Fig. P18.29, let the signal line (represented as a single conductor, the other being the ground) be designated by 1, and the coupled line by 2. Using the coefficients of potential, we can write]

$$v_1 = a_{11}\,q_1' + a_{12}\,q_2', \qquad v_2 = a_{21}\,q_1' + a_{22}\,q_2'.$$

Since $q_2' = 0$, we have $v_2/v_1 = a_{21}/a_{11}$, i.e.,

$$v_c(t) = \frac{a_{21}}{a_{11}}\,v(t).$$

The coefficients of potential are, of course, related to the coefficients of capacitance, C_{ij}.

It is left to the reader as an exercise to prove that for a system with a cross section as shown in Fig. P18.29

$$a_{11} = \frac{1}{2\pi\epsilon} \ln \frac{2h}{R} \qquad \text{and} \qquad a_{21} = \frac{1}{2\pi\epsilon} \ln \frac{\sqrt{d^2 + (2h)^2}}{d}.$$

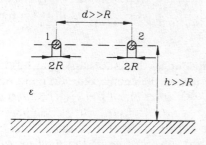

Fig. P18.29. Example of two coupled lines.

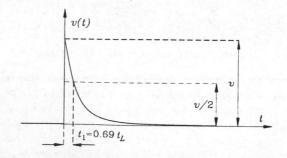

Fig. P18.30. Reflected wave from complex load.

P18.30. Derive the expression $t_1 = 0.69 \, t_L$ when measuring the reflected wave from a complex load, as in Fig. P18.30. This expression shows a practical way to measure the time constant of the reflected wave for the case of complex loads. — *Hint: note that, according to Fig. P18.30,* $v \, e^{-t_1/t_L} = v/2$.

S **P18.31.** Trace the procedure for solving problem P18.8 by means of the Smith chart. — *Hint: start with the normalized load impedance of the first line in the chart, to obtain the actual impedance along the first line. Repeat the procedure for the second line. Finally, determine the corresponding reflection coefficient and VSWR.*

Solution. We start with the normalized load impedance of the first line, $z_{1L} = Z_L/Z_{01}$, in the chart. This point has to be rotated in a clockwise direction (i.e., towards the generator) by ζ_1/λ units of the wavelength scale, where $0 \leq \zeta_1 \leq l_1$, and $\lambda = 2\pi/(\omega\sqrt{\epsilon_0\mu_0})$. We thus obtain the normalized impedance along the first line, $z_1(\zeta_1)$, $0 \leq \zeta_1 \leq l_1$. The actual impedance along the line is hence $Z_1(\zeta_1) = Z_{01}z_1(\zeta_1)$ $(0 \leq \zeta_1 \leq l_1)$.

We then repeat the procedure, starting with the normalized impedance $Z_1(l_1)/Z_{02}$, which represents the normalized load impedance for the second line. By a rotation towards the generator by $(\zeta_2 - l_2)/\lambda$ units of the wavelength scale, we get $z_2(\zeta_2)$, $0 \leq \zeta_2 \leq l_2$. Finally, the impedance along the second line is $Z_2(\zeta_2) = Z_{02}z_2(\zeta_2)$ $(0 \leq \zeta_2 \leq l_2)$.

S **P18.32.** The reciprocals of complex numbers can be determined easily from the Smith chart. Starting with the basic equations defining the chart, deduce how this can be done. — *Hint: note that the reciprocal of a complex number z is obtained as*

$$\frac{1}{z} = \frac{1-\rho}{1+\rho} = \frac{1+\rho \, e^{j\pi}}{1-\rho \, e^{j\pi}} = \frac{1+\rho'}{1-\rho'}.$$

Solution. According to Eq. (18.32), and noting that $-\rho = \rho e^{j\pi}$, the reciprocal of a complex number z is obtained as

$$\frac{1}{z} = \frac{1-\rho}{1+\rho} = \frac{1+\rho \, e^{j\pi}}{1-\rho \, e^{j\pi}} = \frac{1+\rho'}{1-\rho'}.$$

So, for determining $1/z$ we first have to find z on the chart. This locates the complex number ρ in the complex ρ plane. The reciprocal of z, $1/z$, is defined in the chart by $\rho' = \rho e^{j\pi}$. The magnitudes of ρ and ρ' are the same, but their phases differ by π. That is, we simply draw a straight line through z and the origin, and find in the diagram the symmetric point to z, which determines $1/z$.

This conclusion has an important implication. It shows that we can use the Smith chart also as an *admittance chart* $(Y = 1/Z)$, which is particularly valuable when we have two or more lines connected in parallel.

P18.33. A fixed, known complex impedance Z_L is to be connected to a lossless line having a characteristic impedance Z_0. Show that it is possible to eliminate the reflected wave along the line if an appropriate length of the same line, assumed to be short-circuited, is connected at an appropriate place on the line near Z_L (see Fig. P18.33). — *Hint: note that the impedance, and hence also the admittance, of the short-circuited stub can only be imaginary. Next note that, as we move along the line away from the load, the input admittance of the portion of the line terminated by the load varies. At a certain point its real part is equal to $Y_0 = 1/Z_0$, so it* *will be matched to the left of that point if we compensate the imaginary part of admittance by a stub.*

P18.34. What circuit element corresponds to the point on the Smith chart that is defined by the intersection of the circle $r = 1$ and the arc $jx = j1.2$ at 1 GHz, if the normalizing impedance is $50\,\Omega$? — *(a) A 50-Ω resistor. (b) A 60-Ω resistor in series with a 7.82-nH inductor. (c) A 50-Ω resistor in series with a $L = 9.55$-nH inductor.*

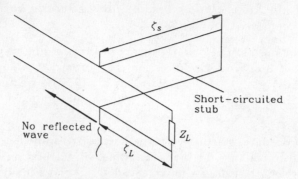

Fig. P18.33. A configuration for matching a load to a transmission line.

P18.35. What circuit element corresponds to the point on the Smith chart that is defined by the intersection of the circle $r = 1$ and the arc $jx = -j0.4$ at 500 MHz, if the normalizing impedance is $50\,\Omega$? — *(a) A 50-Ω resistor. (b) A 50-Ω resistor in series with a $C = 15.92$-pF capacitor. (c) A 100-Ω resistor in series with a $C = 5.63$-pF capacitor.*

P18.36. At the load of a terminated transmission line of characteristic impedance $Z_0 = 100\,\Omega$, the reflection coefficient is $\rho = 0.56 + j0.215$. What is the load impedance? — *(a) $Z_L = (260 - j180)\,\Omega$. (b) $Z_L = (185 + j212)\,\Omega$. (c) $Z_L = (260 + j180)\,\Omega$.*

P18.37. A 50-Ω line is terminated in a load impedance of $Z = 80 - j40\,\Omega$. Find the reflection coefficient of the load and the VSWR. — *(a) $\rho = 0.55\,e^{j23.3°}$, VSWR=3.23. (b) $\rho = 0.37\,e^{-j36.5°}$, VSWR=2.17. (c) $\rho = 0.22\,e^{-j33.2°}$, VSWR=1.32.*

P18.38. A 50-Ω slotted line measurement was done by first placing a short at the place of the unknown load. This results in a large VSWR on the line with sharply defined voltage minima. On an arbitrarily positioned scale along the air-filled coaxial line, the voltage minima are observed at $z_s = 0.1, 1.1$, and 2.1 cm. The short is then replaced by the unknown load, the VSWR is measured to be 2, and the voltage minima (not as sharp as with the short termination) are found at $z = 0.61, 1.61$, and 2.61 cm. Use the Smith chart to find the complex impedance of the load. Explain all your steps. — *(a) $Z_L = 40 - j30\,\Omega$. (b) $Z_L = 30 - j40\,\Omega$. (c) $Z_L = 55 - j48\,\Omega$.*

P18.39. Use a shorted parallel stub to match a 200-Ω load to a 50-Ω transmission line. Include a Smith chart plot with step-by-step explanations. — *The match can be obtained by connecting a short-circuited stub of length a at a distance b from the load along the 50-Ω line, where: (a) $a = 0.176\lambda$, $b = 0.094\lambda$; (b) $a = 0.094\lambda$, $b = 0.176\lambda$; (c) $a = 0.135\lambda$, $b = 0.146\lambda$.*

P18.40. A load consists of a 100-Ω resistor in series with a 10-nH inductor at 1 GHz. Use an open single stub to match it to a 50-Ω line. — *The match can be obtained by connecting* *a short-circuited stub of length a a distance b from the load along the 50-Ω line, where: (a) a = 0.076λ, b = 0.134λ; (b) a = 0.153λ, b = 0.102λ; (c) a = 0.114λ, b = 0.208λ.*

Part 5: Maxwell's Equations and Their Applications

19. Maxwell's Equations

• *Maxwell's equations* are the most general equations of the macroscopic electromagnetic field. Any engineering problem that includes electromagnetic fields is solved starting from these equations, although in some instances it may not be quite obvious. Maxwell's equations can be written in two forms, *integral* and *differential*. Their integral form is

$$\oint_C \boldsymbol{E} \cdot \mathrm{d}\boldsymbol{l} = -\int_S \frac{\partial \boldsymbol{B}}{\partial t} \cdot \mathrm{d}\boldsymbol{S}, \tag{19.1}$$

$$\oint_C \boldsymbol{H} \cdot \mathrm{d}\boldsymbol{l} = \int_S \left(\boldsymbol{J} + \frac{\partial \boldsymbol{D}}{\partial t} \right) \cdot \mathrm{d}\boldsymbol{S}, \tag{19.2}$$

$$\oint_S \boldsymbol{D} \cdot \mathrm{d}\boldsymbol{S} = \int_v \rho \, \mathrm{d}v, \tag{19.3}$$

$$\oint_S \boldsymbol{B} \cdot \mathrm{d}\boldsymbol{S} = 0. \tag{19.4}$$

• The current continuity equation in integral form is important because it must be satisfied by all real currents and charges,

$$\oint_S \boldsymbol{J} \cdot \mathrm{d}\boldsymbol{S} = -\int_v \frac{\partial \rho}{\partial t} \, \mathrm{d}v. \tag{19.5}$$

• The general boundary conditions are specialized forms of the above equations, connecting the electric and magnetic field vectors on the two sides of a boundary surface between two media. They read (the unit vector $\boldsymbol{n}$ is normal to the boundary surface, and is directed into medium 1):

$$E_{1 \text{ tang}} = E_{2 \text{ tang}}; \quad \text{on the surface of a perfect conductor } E_{\text{tang}} = 0. \tag{19.6}$$

$$D_{1 \text{ normal}} - D_{2 \text{ normal}} = \sigma; \quad \text{on the surface of a perfect conductor } D_{\text{norm}} = \sigma. \tag{19.7}$$

$H_{1\text{ tang}} - H_{2\text{ tang}} = J_\text{s} \times n$; on the surface of a perfect conductor $H_\text{tang} = J_\text{s} \times n$. (19.8)

$B_{1\text{ normal}} = B_{2\text{ normal}}$; on the surface of a perfect conductor $B_\text{normal} = 0$. (19.9)

• Differential forms of Eqs. (19.1)-(19.5) are obtained from the integral forms by means of the Stokes's and divergence theorems. They have the form

$$\nabla \times E = -\frac{\partial B}{\partial t}, \tag{19.10}$$

$$\nabla \times H = J + \frac{\partial D}{\partial t}, \tag{19.11}$$

$$\nabla \cdot D = \rho, \tag{19.12}$$

$$\nabla \cdot B = 0. \tag{19.13}$$

$$\nabla \cdot J = -\frac{\partial \rho}{\partial t}. \tag{19.14}$$

• To these equations (as well as to their integral counterparts) it is necessary to add the relationships between vectors (1) D, E, and P; (2) B, H, and M; and (3) J and E. These relations are known as the *constitutive relations*:

$$D = \epsilon_0 E + P, \quad P = P(E), \tag{19.15}$$

$$B = \mu_0(H + M), \quad M = M(B), \tag{19.16}$$

$$J = J(E). \tag{19.17}$$

In linear media,

$$D = \epsilon E, \quad B = \mu H, \quad J = \sigma E. \tag{19.18}$$

• Quantities varying sinusoidally in time are called *time-harmonic*. For time-harmonic currents and charges, *and if the medium is linear*, Eqs. (19.10)-(19.14) can be written with complex (phasor) quantities:

$$\nabla \times \underline{E} = -j\omega \underline{B}, \tag{19.19}$$

$$\nabla \times \underline{H} = \underline{J} + j\omega \underline{D}, \tag{19.20}$$

$$\nabla \cdot \underline{D} = \rho, \tag{19.21}$$

$$\nabla \cdot \underline{B} = 0. \tag{19.22}$$

$$\nabla \cdot \underline{J} = -j\omega \rho. \tag{19.23}$$

The constitutive relations in that case become

$$\underline{D} = \epsilon \, \underline{E}, \qquad \underline{B} = \mu \, \underline{H}, \qquad \underline{J} = \sigma \, \underline{E}. \tag{19.24}$$

Eq.(19.20), combined with $J = \sigma E$ and $D = \epsilon E$, provides the definition of a good conductor ($\sigma \gg \omega\epsilon$) and of a good insulator ($\sigma \ll \omega\epsilon$) in the case of time-harmonic fields and linear media.

- *Poynting's theorem* is the mathematical expression of the law of conservation of energy as applied to electromagnetic fields. It is obtained from Maxwell's equations and has the form

$$\int_v E_i \cdot J \, dv = \int_v \frac{J^2}{\sigma} dv + \frac{\partial}{\partial t} \int_v \left(\frac{1}{2}\epsilon E^2 + \frac{1}{2}\mu H^2 \right) dv + \oint_S (E \times H) \cdot dS, \tag{19.25}$$

where E_i is the "impressed electric field" (representing sources), and v is the domain limited by a closed surface S. According to the law of conservation of energy,

$$\oint_S (E \times H) \cdot dS = \text{power transferred through } S \text{ to region outside } S. \tag{19.26}$$

The cross (vector) product $(E \times H)$ represents the power transferred by the electromagnetic field per unit area, and is known as the *Poynting vector*, $\vec{\mathcal{P}}$:

$$\vec{\mathcal{P}} = E \times H \qquad (\text{W/m}^2). \tag{19.27}$$

- Poynting's theorem in complex form (linear media, sinusoidal time variation) reads

$$\int_v \underline{E}_i \cdot \underline{J}^* dv = \int_v \frac{J^2}{\sigma} dv + 2j\omega \int_v \left(\frac{1}{2}\mu H^2 - \frac{1}{2}\epsilon E^2 \right) dv + \oint_S (\underline{E} \times \underline{H}^*) \cdot dS, \tag{19.28}$$

and the vector

$$\vec{\mathcal{P}} = \underline{E} \times \underline{H}^* \qquad (\text{W/m}^2) \tag{19.29}$$

is the *complex Poynting vector*.

• *Electromagnetic potentials* are auxiliary functions from which the field vectors are obtained with relative ease (most often by differentiation). The most common pair of potentials are the *magnetic vector potential*, **A**, defined by

$$B = \nabla \times A = \text{curl}\,A, \tag{19.30}$$

and the *electric scalar potential*, V, defined by the equation

$$E = -\nabla V - \frac{\partial A}{\partial t} = -\text{grad}V - \frac{\partial A}{\partial t}. \tag{19.31}$$

• The *Lorentz potentials* are a special case of these more general potentials, and are defined by the *Lorentz condition*,

$$\text{div}\,A = \nabla \cdot A = -\epsilon\mu\frac{\partial V}{\partial t}. \tag{19.32}$$

• The Lorentz potentials can be expressed in terms of charges and currents only in *linear, homogeneous media*. In a medium of parameters ϵ, μ, and $\sigma = 0$, the expressions for the potentials in terms of charges and currents are

$$V(\boldsymbol{r},t) = \frac{1}{4\pi\epsilon} \int_{v'} \frac{\rho(\boldsymbol{r'},t - R/c)}{R}\text{d}v', \quad c = \frac{1}{\sqrt{\epsilon\mu}}, \tag{19.33}$$

$$A(\boldsymbol{r},t) = \frac{\mu}{4\pi} \int_{v'} \frac{\boldsymbol{J}(\boldsymbol{r'},t - R/c)}{R}\text{d}v', \quad c = \frac{1}{\sqrt{\epsilon\mu}}. \tag{19.34}$$

In these equations, $\boldsymbol{r}$ is the position vector of the point where the potentials are being determined, $\boldsymbol{r'}$ is the position of the volume element $\text{d}v$, and $\boldsymbol{R} = \boldsymbol{r} - \boldsymbol{r'}$. These expressions tell us that the potential at point $\boldsymbol{r}$ at a time t is not due to the elemental source at time t, but to that at an earlier time, $(t - R/c)$. In other words, the potentials propagate with a finite velocity, $c = 1/\sqrt{\epsilon\mu}$, i.e., they are *retarded* in their reaching the field point. In a vacuum, $c = c_0$, the velocity of light.

• For time-harmonic sources of the electromagnetic field, the retarded potentials can be written as complex (phasor) quantities,

$$\underline{V}(\boldsymbol{r}) = \frac{1}{4\pi\epsilon} \int_{v'} \frac{\rho(\boldsymbol{r'})\text{e}^{-\text{j}\omega R/c}}{R}\text{d}v', \quad c = \frac{1}{\sqrt{\epsilon\mu}}, \tag{19.35}$$

$$\underline{A}(r) = \frac{\mu}{4\pi} \int_{v'} \frac{\underline{J}(r')e^{-j\omega R/c}}{R} dv', \quad c = \frac{1}{\sqrt{\epsilon\mu}}. \tag{19.36}$$

The retardation can be neglected provided that the largest dimension of the field domain we consider, d_{max}, is determined by the inequality $\omega d_{max}/c \ll 1$, or

$$d_{max} \ll \frac{c}{\omega} = \frac{1}{\omega\sqrt{\epsilon\mu}}. \tag{19.39}$$

If this is satisfied for an electromagnetic field, the field is said to be a *quasi-static* (nearly-static) field.

QUESTIONS

Q19.1. Why (and when) it is allowed to move the time derivative in the equation

$$\oint_S J \cdot dS = -\frac{d}{dt} \oint_S D \cdot dS$$

to act on D only? — *(a) There are no restrictions whatsoever. (b) D must be time-varying. (c) S must be constant in time.*

S **Q19.2.** Does Eq. (19.1) tell us that a time-varying magnetic field is the source of a time-varying electric field? Explain. — *(a) Yes, because the integral on the left-hand side depends on the time variation of B. (b) No, because the sources of the electric field are only electric charges. (c) No, because if B is time-varying, Eq. (19.1) does not require that E be also time-varying.*

Answer. Yes. If the magnetic field is time constant, the electric field vector has only the component which satisfies the relation $\oint E \cdot dl = 0$. If the magnetic field varies in time, the electric field vector has also a time-varying component which does not satisfy this equation (i.e., the induced electric field).

Q19.3. Why would an electric field inside a perfect conductor produce a current of infinite density? Would such a current be physically possible? Explain. — *Hint: recall the constitutive relation $J = \sigma E$, and the expression for the density of energy in the magnetic field.*

S **Q19.4.** Why surface currents are possible on surfaces of perfect conductors, when a nonzero tangential electric field there is not possible? Is this a current of finite volume density? — *(a) The E-field cannot exists inside a perfect conductor, but a tangential E-field can exist on its very surface. (b) In a perfect conductor, no field is needed to create a current. (c) Surface currents exist because there can be no currents inside.*

Answer. Surface currents and charges over perfect conductors are precisely a kind of shield preventing the field to enter a perfect conductor. Their distribution is such that their field exactly cancels the external field inside the conductor.

Q19.5. Write the full set of Maxwell's equations in differential form for the special case of a static electric field, assuming that the dielectric is linear, but inhomogeneous. — *Hint: all time derivatives are zero, there is no magnetic field.*

Q19.6. Write the full set of Maxwell's equations in differential form for the special case of a static electric field produced by the charges on a set of conducting bodies situated in a vacuum. — *Hint: use the answer to the preceding question and note that in this case $\rho = 0$.*

Q19.7. Write the full set of Maxwell's equations in differential form for the special case of a steady current flow in a homogeneous conductor of conductivity σ, with no impressed electric field. — *Convince yourself that the correct answer is:* $\nabla \times E = 0$, $\nabla \cdot E = 0$, $\nabla \cdot J = 0$, $J = \sigma E$.

S **Q19.8.** Write the full set of Maxwell's equations in differential form for the special case of a steady current flow in an inhomogeneous poor dielectric, with impressed electric field E_i present. — *Convince yourself that the correct answer is:* $\nabla \times E = 0$, $\nabla \cdot D = \rho$, $\nabla \cdot J = 0$, $D = \epsilon E$, $J = \sigma(E + E_i)$.

Answer. $\nabla \times E = 0$, $\nabla \cdot D = \rho$, $\nabla \cdot J = 0$, $D = \epsilon E$, $J = \sigma(E + E_i)$.

Q19.9. Write the full set of Maxwell's equations in differential form for the special case of a time-constant magnetic field in a linear medium of permeability μ, produced by a steady current flow. — *Find three errors in the following answer:* $\nabla \times H = J + \partial E/\partial t$, $\nabla \cdot B = 0$, $B = \mu H + M$.

Q19.10. Write the full set of Maxwell's equations in differential form for the special case of a time-constant magnetic field, produced by a permanent magnet of magnetization M (a function of position). — *Find three errors in the following answer:* $\nabla \times H = J$, $\nabla \cdot B = -\partial E/\partial t$, $B = \mu(H + M)$.

Q19.11. Write the full set of Maxwell's equations in differential form for the special case of a time-constant magnetic field produced by both steady currents and magnetized matter, if the medium is not linear. — *Convince yourself that the following answer is correct:* $\nabla \times H = J$, $\nabla \cdot B = 0$, $B = \mu_0(H + M)$.

Q19.12. Write the full set of Maxwell's equations in differential form for the special case of a quasi-static electromagnetic field, produced by quasi-static currents in nonferromagnetic conductors. — *Find three errors in the following answer:* $\nabla \times (\epsilon E) = -\partial B/\partial t$, $\nabla \times H = J + \partial D/\partial t$, $\nabla \cdot D = \rho/\epsilon_0$, $\nabla \cdot H = 0$, $\nabla \cdot J = -\partial \rho/\partial t$, $D = \epsilon E$, $B = \mu H$.

S **Q19.13.** Write Maxwell's equations in differential form for an arbitrary electromagnetic field in a vacuum, no free charges being present. — *Find three errors in the following answer:* $\nabla \times E = -\mu \partial H/\partial t$, $\nabla \times H = \epsilon \partial E/\partial t$, $\nabla \cdot E = \rho$, $\nabla \cdot H = 0$.

Answer. $\nabla \times E = -\mu_0 \partial H/\partial t$, $\nabla \times H = \epsilon_0 \partial E/\partial t$, $\nabla \cdot E = 0$, $\nabla \cdot H = 0$.

Q19.14. Write Maxwell's equations for an arbitrary electromagnetic field in a homogeneous perfect dielectric of permittivity ϵ and permeability μ. — *Find three errors in the following answer:* $\nabla \times E = -\mu_0 \partial H/\partial t$, $\nabla \times H = \epsilon_0 \partial E/\partial t$, $\nabla \cdot E = \rho$, $\nabla \cdot H = 0$.

Q19.15. Write the full set of differential Maxwell's equations *in scalar form* in the rectangular coordinate system. Note that *eight* simultaneous, partial differential equations result. Write these equations neatly and save them for future reference. — *Hint: use the expressions for curl and divergence in the rectangular coordinate system.*

Q19.16. Repeat question Q19.15 for the cylindrical coordinate system. — *Hint: use the expressions for curl and divergence in the cylindrical coordinate system.*

Q19.17. Repeat question Q19.15 for the spherical coordinate system. — *Hint: use the expressions for curl and divergence in the spherical coordinate system.*

Q19.18. Write differential Maxwell's equations in scalar form for the particular case of an electromagnetic field in a vacuum ($J = 0$, $\rho = 0$), if the field vectors are only functions of the cartesian coordinate z and of time t. — *Hint: use the scalar equations you obtained in the answer to Q19.15. Note that the partial derivatives with respect to y and z are zero, that the medium is a vacuum, and that $J = 0$ and $\rho = 0$.*

Q19.19. Write differential Maxwell's equations in scalar form for a good conductor, for the particular case of an axially symmetrical system with dependence of the field vectors only on the cylindrical coordinate r and time t. Assume that $J = J_z u_z$ and $\rho = 0$. — *Hint: use the scalar equations you obtained in the answer to Q19.16. Note that the partial derivatives with respect to cylindrical coordinates ϕ and z are zero, that the current density vector is assumed to have only the z-component, and that $\rho = 0$. Note also that, since the conductor is assumed to be good, the displacement current density, $\partial D/\partial t$, should be omitted.*

Q19.20. Repeat question Q19.19 for $B = B_z u_z$. — *Hint: the procedure is the same, with one difference only. What is this difference?*

Q19.21. Write differential Maxwell's equations in complex form for an arbitrary electromagnetic field in a very good conductor, of conductivity σ and permeability μ. — *Find three errors in the following answer:* $\nabla \times E = -j\omega^2 \mu H$, $\nabla \times H = J + j\epsilon E$, $\nabla \cdot E = 0$, $\nabla \cdot H = 0$, $J = \sigma D$.

S   **Q19.22.** Write differential Maxwell's equations in complex form for a quasi-static electromagnetic field. — *Find three errors in the following answer:* $\nabla \times E = -\partial B/\partial t$, $\nabla \times H = J + j\omega D$, $\nabla \cdot D = \rho/\epsilon$, $\nabla \cdot B = 0$, $\nabla \cdot J = -j\omega\rho$, $D = \epsilon E$, $B = \mu H$, $J = \sigma E$.

Answer. $\nabla \times E = -j\omega B$, $\nabla \times H = J$, $\nabla \cdot D = \rho$, $\nabla \cdot B = 0$, $\nabla \cdot J = -j\omega\rho$, $D = \epsilon E$, $B = \mu H$, $J = \sigma E$.

Q19.23. Write differential Maxwell's equations in complex form for an arbitrary electromagnetic field in a perfect dielectric of permittivity ϵ and permeability μ, no free charges being present. — *Convince yourself that the following answer is correct:* $\nabla \times E = -j\omega\mu H$, $\nabla \times H = j\omega\epsilon E$, $\nabla \cdot E = 0$, $\nabla \cdot H = 0$.

S   **Q19.24.** Write the most general integral Maxwell's equations in complex form. — *Find three errors in the following answer:*

$$\oint_C E \cdot dl = -j\omega \int_S H \cdot dS, \qquad \oint_C H \cdot dl = \int_S (J + j\omega E) \cdot dS,$$

$$\oint_S D \cdot dS = \frac{1}{\epsilon} \int_v \rho dv, \qquad \oint_S B \cdot dS = 0.$$

Answer.

$$\oint_C E \cdot dl = -j\omega \int_S B \cdot dS, \qquad \oint_C H \cdot dl = \int_S (J + j\omega D) \cdot dS,$$

$$\oint_S \boldsymbol{D} \cdot \mathrm{d}\boldsymbol{S} = \int_v \rho \mathrm{d}v, \qquad \oint_S \boldsymbol{B} \cdot \mathrm{d}\boldsymbol{S} = 0.$$

Q19.25. The current intensity through a resistor of resistance R is I. What is the flux of the Poynting vector through any closed surface enclosing the resistor? — *(a)* $-RI^2$. *(b)* RI^2. *(c) It cannot be determined without knowing the exact resistor geometry.*

Q19.26. A capacitor, of capacitance C, is charged with a charge Q. What is the flux of the Poynting vector through any surface enclosing the capacitor, if the charge Q (1) is constant in time, or (2) varies in time as $Q = Q_\mathrm{m} \cos \omega t$? — *(a) Zero,* $-\omega Q_\mathrm{m}^2 \sin 2\omega t/(2C)$. *(b) Zero,* $-\omega Q_\mathrm{m}^2 \cos 2\omega t/(2C)$. *(c)* $Q^2/2C$, $-\omega Q_\mathrm{m}^2 \sin \omega t/(2C)$.

Q19.27. A coil, of inductance L, carries a current $i(t)$. What is the flux of the Poynting vector through any surface enclosing the coil? — *(a) It cannot be determined without knowing the exact geometry of the coil. (b)* $Li(\mathrm{d}i/\mathrm{d}t)$. *(c)* $-Li(\mathrm{d}i/\mathrm{d}t)$.

Q19.28. A dc generator of emf $\mathcal{E}$ is open-circuited. What is the flux of the Poynting vector through any surface enclosing the generator? — *(a) Undefined. (b) Zero. (c)* $-\mathcal{E}^2$.

Q19.29. Repeat question Q19.28 assuming that a current $i(t)$ flows through the generator, and its internal resistance is R. — *(a)* $\mathcal{E}i(t) + Ri^2(t)$. *(b) Zero. (c)* $\mathcal{E}i(t) - Ri^2(t)$.

Q19.30. What is the time-average value of the Poynting vector, if complex rms values are known for the electric and magnetic field strength, $\boldsymbol{E}$ and $\boldsymbol{H}$. — *(a)* $Re(\boldsymbol{E} \times \boldsymbol{H}^*)$. *(b)* $(\boldsymbol{E} \times \boldsymbol{H}^*)$. *(c)* $Re(\boldsymbol{E}^* \times \boldsymbol{H})$.

S **Q19.31.** The largest dimension of a coil at a very high frequency is on the order of $(\omega \sqrt{\epsilon \mu})^{-1}$. Is it possible at such high frequencies to define the inductance of the coil in the same way as in a quasi-static case? Explain. — *(a) Yes, the inductance does not depend on frequency. (b) Yes, provided that interturn capacitance can be neglected. (c) No, because the current along the coil is not the same at all points.*

Answer. It is not possible, since the current along the coil at an instant of time is not the same at all points.

Q19.32. The length of a long 60-Hz power transmission line is equal to $0.5(\omega \sqrt{\epsilon_0 \mu_0})^{-1}$. Is this a quasi-static system? What is the length of the line? — *(a) It is a quasi-static system, the line length is 398 km. (b) It is not a quasi-static system, the line length is 137 km. (c) It is not a quasi-static system, the line length is 398 km.*

Q19.33. A parallel-plate capacitor has plates of linear dimensions comparable with $(\omega \sqrt{\epsilon \mu})^{-1}$, where ω is the operating angular frequency. Is it possible to determine the capacitance of such a capacitor in the same way as in the static and quasi-static case? Explain. — *(a) It is not, since the charge over the capacitor plates is not distributed uniformly. (b) It is, because the capacitance does not depend on frequency. (c) It is not, because there are currents over the capacitor plates.*

S **Q19.34.** An electric circuit operates at a high frequency f. The largest linear dimension of the circuit is $2(\omega\sqrt{\epsilon\mu})^{-1}$. Are Kirchhoff's laws applicable in this case for analyzing the circuit? Explain. — *(a) The Kirchhoff Current Law is valid, the Kirchhoff Voltage Law is not. (b) None of the two Laws is valid. (c) The Kirchhoff Current Law is not valid, the Kirchhoff Voltage Law is valid.*

Answer. The Kirchhoff Current Law remains valid for the circuit nodes, provided that they are small enough. However, the current intensities along branches of such a circuit are not constant along a branch. Therefore, in contrast to circuit theory, the equations according to the Kirchhoff Current Law are independent *for all N nodes of the circuit* (not for $N - 1$ nodes).

The Kirchhoff Voltage Law is not valid, since there is substantial coupling between the branches due to the induced electric field, and the induced voltages depend on the circuit shape.

Q19.35. Write the Lorentz condition in complex form. — *(a)* $\nabla \cdot \mathbf{A} = \operatorname{div}\mathbf{A} = \mathrm{j}\omega\epsilon\mu V$. *(b)* $\nabla \cdot \mathbf{A} = \operatorname{div}\mathbf{A} = -\mathrm{j}\omega\epsilon\mu V$. *(c)* $\nabla \cdot \mathbf{A} = \operatorname{div}\mathbf{A} = -\mathrm{j}\omega(\epsilon/\mu)V$.

Q19.36. A current pulse of duration $\Delta t = 10^{-9}$ s was excited in a small wire loop. After how many Δt's is the magnetic and induced electric field of this pulse going to be detected at a point $r = 10\,\mathrm{m}$ from the loop? — *(a)* $T/\Delta t = 66.6$. *(b)* $T/\Delta t = 44.4$. *(c)* $T/\Delta t = 33.3$.

PROBLEMS

S **P19.1.** A current $i(t) = I_\mathrm{m}\cos\omega t$ flows through the leads of a parallel-plate capacitor of plate area S and distance between them d. If the permittivity of the dielectric of the capacitor is ϵ, prove that the displacement current through the capacitor dielectric is exactly $i(t)$. Ignore fringing effect. — *Hint: find $q(t)$ on the plates, hence $D(t)$ and $\partial D(t)/\partial t$.*

Solution. The charge on the capacitor plates is given by

$$Q = \int i\,\mathrm{d}t = \frac{I_\mathrm{m}}{\omega}\sin\omega t.$$

The electric field intensity in the capacitor is hence

$$E = \frac{Q}{\epsilon S} = \frac{I_\mathrm{m}}{\epsilon\omega S}\sin\omega t,$$

and the displacement current through the capacitor dielectric is

$$\frac{\partial D}{\partial t}S = \epsilon\frac{\partial E}{\partial t}S = I_\mathrm{m}\cos\omega t = i.$$

P19.2. A spherically symmetrical charge distribution disperses under the influence of mutually repulsive forces. Suppose that the charge density $\rho(r,t)$, as a function of the distance r from the center of symmetry and of time, is known. Prove that the total current density at any point is zero. — *Hint: to determine $J(r,t)$, find first $i(r,t)$ as an integral, and divide by $4\pi r^2$. To determine $\partial D(r,t)/\partial t$, use the generalized Gauss' law to determine $D(r,t)$ as an integral.*

P19.3. Determine the magnetic field as a function of time for the dispersing charge distribution in problem P19.2. — *Hint: note that the total current density at all points is zero.*

S *P19.4. Small-scale models are used often in engineering practice, including electrical engineering. Starting from differential Maxwell's equations for a linear medium, derive the necessary conditions for the electromagnetic field in a small-scale model to be similar to the field in a real, n times larger model. (These conditions are usually referred to as the conditions of the *electrodynamic similitude*.) (*Hints:* (1) Write the first two differential Maxwell's equations for the full-scale system, and for the model. (2) Note that the coordinates in the latter are n times smaller, and find the conditions under which, in spite of that, the two sets of equations will be the same.) — *Suppose we made a small-scale model geometrically similar to a real system. An example of a full-scale system and a small-scale model is shown in Fig. P19.4. The parameters ϵ, μ, and σ of the full-scale system are known functions of the coordinates. Frequency $f = \omega/(2\pi)$ of the generators in the full-scale system is also known. It is necessary to determine the parameters ϵ', μ', and σ' as functions of the coordinates in the small-scale model, and the corresponding frequency $f' = \omega'/(2\pi)$ of the generators so that the electromagnetic fields in the two cases are the same. These parameters are given by: (a) $\omega'\mu' = n^2\omega\mu$, $\sigma' = \sigma$, $\omega'\epsilon' = n^2\omega\epsilon$. (b) $\omega'\mu' = n\omega\mu$, $\sigma' = n\sigma$, $\omega'\epsilon' = n\omega\epsilon$. (c) $\omega'\mu' = \omega\mu/n$, $\sigma' = \sigma/n$, $\omega'\epsilon' = \omega\epsilon/n$. What these conditions reduce to if $\epsilon' = \epsilon$ and $\mu' = \mu$?*

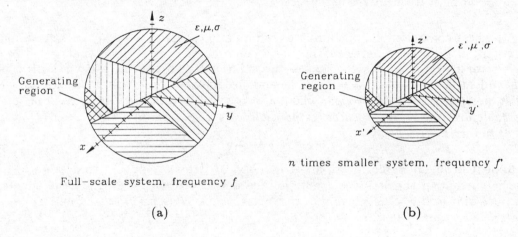

Full—scale system, frequency f n times smaller system, frequency f'

(a)   (b)

Fig. P19.4. Two electrodynamically similar systems. (a) The full-scale system, (b) the small-scale model.

Solution. Suppose we made a small-scale model geometrically similar to a real system. An example of a full-scale system and a small-scale model is shown in Fig. P19.4. The parameters ϵ, μ, and σ of the full-scale system are known functions of the coordinates. The frequency $f = \omega/(2\pi)$ of the generators in the full-scale system is also known. It is necessary to determine the parameters ϵ', μ', and σ' as the functions of coordinates in the small-scale model, and the corresponding frequency $f' = \omega'/(2\pi)$ of generators in the model, so that the electromagnetic fields in the two cases are the same.

The first two differential Maxwell's equations in the complex form for the full-scale system have the form

$$\nabla \times \boldsymbol{E} = \operatorname{curl}\boldsymbol{E} = -\mathrm{j}\omega\mu\boldsymbol{H}, \qquad \nabla \times \boldsymbol{H} = \operatorname{curl}\boldsymbol{H} = \sigma\boldsymbol{E} + \mathrm{j}\omega\epsilon\boldsymbol{E}.$$

In the small-scale model all the lengths are n times smaller. In order that the curl at the respective points of the model be the same as in the real system, the length coordinates, with respect to which differentiation is performed, must be n times as small. For example, $\partial E'_x / \partial y' = \partial E'_x / \partial (y/n) = n \partial E'_x / \partial y$, which means that the first two Maxwell's equations for the model must have the form

$$n \operatorname{curl} \boldsymbol{E'} = -j\omega' \mu' \boldsymbol{H'}, \qquad n \operatorname{curl} \boldsymbol{H'} = \sigma' \boldsymbol{E'} + j\omega' \epsilon' \boldsymbol{E'}.$$

The differentiation implied by the curl in these equations is performed with respect to coordinates of the same "size" as in the equations for the original system. The boundary conditions in the two cases are identical. Hence the solutions will also be equal, provided that respective pairs of Maxwell's equations above are equal. This is the case if

$$\omega' \mu' = n\omega\mu, \qquad \sigma' = n\sigma, \qquad \omega' \epsilon' = n\omega\epsilon.$$

In practice, models are usually made of the same dielectric, i.e., $\epsilon' = \epsilon$ and $\mu' = \mu$. In that case, both the first and the third conditions are satisfied if $\omega' = n\omega$. This means that the *frequency of the generators in an n times smaller model must be n times as large.* The second condition ($\sigma' = n\sigma$) often cannot be satisfied (for example, if the conductors in the full-scale system are made of silver). Fortunately, the conductivity usually is *not* a decisive parameter, and the impossibility of fulfilling the condition $\sigma' = n\sigma$ most frequently is not critical.

P19.5. A lossless coaxial cable, of conductor radii a and b, carries a steady current of intensity I. The potential difference between the cable conductors is V. Prove that the flux of the Poynting vector through a cross section of the cable is VI, using the known expressions for vectors $\boldsymbol{E}$ and $\boldsymbol{H}$ in the cable. Sketch the dependence of the magnitude of the Poynting vector on the distance r from the cable axis, where $a < r < b$. — *Hint: note that the lines of vector $\boldsymbol{E}$ in the cable are radial with respect to the cable axis, while the lines of vector $\boldsymbol{H}$ are circles centered at the axis.*

P19.6. Repeat problem P19.5 for an air stripline with strips of width a that are a distance d apart, if the current in the strips is I and voltage between them V. Neglect the edge effects. — *Hint: note that vector $\boldsymbol{E}$ is normal to the strips, while $\boldsymbol{H}$ is parallel to them. Both vectors lie in the cross section of the line (Fig. P19.6).*

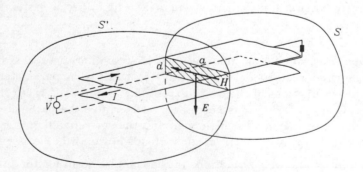

Fig. P19.6. Calculation of the flux of the Poynting vector through the cross section of a strip line.

S **P19.7.** The stripline from the preceding problem is connected to a sinusoidal generator of emf $\mathcal{E}$ and angular frequency ω. The other end of the line is connected to a capacitor of capacitance

C. Apply Poynting's theorem in complex form to a closed surface enclosing (1) the generator, or (2) the capacitor. — *Hint: see Fig. P19.6. The results are those predicted by circuit theory.*

Solution. In complex notation

$$E = \frac{\mathcal{E}}{d} \quad \text{and} \quad H = \frac{I}{a} = \frac{j\omega Q}{a} = \frac{j\omega C \mathcal{E}}{a},$$

where Q stands for the charge on the capacitor plates. Note that the vector product $\boldsymbol{E} \times \boldsymbol{H}$ is directed from the generator away (see Fig. P19.6).

(1) When the closed surface encloses *only the generator* (surface S' in Fig. P19.6), on the left-hand side of the power-balance equation, which represents Poynting's theorem in complex form, is the complex power of the generator, $\mathcal{E}I^*$. On the right-hand side of the equation is the flux of the complex Poynting vector through this closed surface. Since the field exists only between the strips,

$$\oint_{S'} (\boldsymbol{E} \times \boldsymbol{H}^*) \cdot \mathrm{d}\boldsymbol{S}' = EH^* ad = \mathcal{E}I^*.$$

(2) The flux of the complex Poynting vector through the surface which encloses *only the load* (surface S in Fig. P19.6, except that the resistor should be replaced by a capacitor) is given by

$$\oint_{S} (\boldsymbol{E} \times \boldsymbol{H}^*) \cdot \mathrm{d}\boldsymbol{S} = -EH^* ad = -\frac{\mathcal{E}}{d} \frac{-j\omega C \mathcal{E}}{a} ad = 2j\omega \frac{C\mathcal{E}^2}{2}$$

(note that $\mathrm{d}\boldsymbol{S} = -\mathrm{d}\boldsymbol{S}'$). This is precisely $2j\omega$ times the average energy stored in the electric field of the capacitor. According to Poynting's theorem in complex form, this should be so.

P19.8. Repeat problem P19.7 assuming that the load is an inductor of inductance L, instead of a capacitor. — *Hint: see Fig. P19.6. The results are those predicted by circuit theory.*

P19.9. Repeat problem P19.7, assuming that the line is a lossless coaxial line of conductor radii a and b. — *The result is the same as in problem P19.7.*

P19.10. Starting from Maxwell's equations in differential form for linear, homogeneous dielectric media, derive the partial differential equations the solutions of which are given in Eqs. (19.33) and (19.34). — *Hint: use the Lorentz condition in Eq. (19.32).*

P19.11. Derive the retarded potentials in Eqs. (19.35) and (19.36) from Eqs. (19.33) and (19.34). — *Hint: assume that the sources, ρ and $\boldsymbol{J}$, and the potentials, V and $\boldsymbol{A}$, vary in time as $e^{j\omega t}$.*

P19.12. Suppose a system is regarded as approximately quasi-static if its largest dimension d satisfies the inequality $d\omega\sqrt{\epsilon\mu} \leq 0.1$. Determine the largest value of d thus defined for the electrodynamic systems in a vacuum if the frequency of the generators is (1) 60 Hz, (2) 10 MHz, or (3) 10 GHz. — *(a) $d_1 = 79.6\,km$, $d_2 = 0.477\,m$, $d_3 = 0.477\,mm$. (b) $d_1 = 9.26\,km$, $d_2 = 0.277\,m$, $d_3 = 0.277\,mm$. (c) $d_1 = 3.2\,km$, $d_2 = 0.177\,m$, $d_3 = 0.177\,mm$.*

P19.13. Compare the rms values of vectors $\boldsymbol{J}$ and $\partial \boldsymbol{D}/\partial t$ in copper, seawater and wet ground, for frequencies f of (1) 60 Hz, (2) 10 kHz, (3) 100 MHz, or (4) 10 GHz. For copper, assume

$\epsilon = \epsilon_0$, $\sigma = 56 \cdot 10^6$ S/m. For seawater, adopt $\epsilon = 10\epsilon_0$, $\sigma = 4$ S/m, and for the ground $\epsilon = 10\epsilon_0$ and $\sigma = 10^{-2}$ S/m. — *Hint: note that, in fact, we need to determine the ratio $k = \sigma/(2\pi f \epsilon)$.*

20. The Skin Effect

• A time-varying current has a tendency to concentrate near the surfaces of conductors. If the frequency is very high, the current is restricted to a very thin layer near the conductor surfaces, and this effect is therefore known as the *skin effect*. The cause of the skin effect is electromagnetic induction. The skin effect is of considerable practical importance. For example, at very high frequencies a very thin layer of conductor carries most of the current, so we can coat any conductor with silver (which is the best, but expensive, conductor), and have practically the entire current flow through this thin silver coating.

• The skin effect is much more pronounced for a ferromagnetic conductor than for a nonferromagnetic conductor of the same conductivity. For example, for iron at 60 Hz the thickness of this layer is on the order of only 0.5 mm. Therefore, ferromagnetic cores for alternating current electrical motors, generators, transformers, etc., are made of thin, mutually insulated sheets. At very high frequencies, ferrites (ferrimagnetic materials) are used, because they have very low conductivity.

• It is simplest to analyze the skin effect in a homogeneous conducting half-space with current parallel to the interface. Let the current be z-directed, and the y axis be normal to the interface (Fig. P20.2, p.215). If the angular frequency of the current is ω, and the medium has a conductivity σ and permeability μ, the complex current density in the conducting half space is found to be

$$J_z(y) = J_z(0)\mathrm{e}^{-ky}\mathrm{e}^{-\mathrm{j}ky}, \qquad k = \sqrt{\frac{\omega\mu\sigma}{2}} \quad (1/\mathrm{m}). \qquad (20.1)$$

The intensity of the current density vector decreases exponentially with increasing y. At a distance

$$\delta = \frac{1}{k} = \sqrt{\frac{2}{\omega\mu\sigma}} \qquad (\mathrm{m}), \qquad (20.2)$$

the amplitude of the current density vector decreases to $1/e$ of its value $J_z(0)$ at the boundary surface. This distance is known as the *skin depth*. Although derived for a special case of currents in a half-space, the above analysis is valid for a current distribution in any conductor whose radius of curvature is much larger than the skin depth. Table 20.1 summarizes the values of the skin depth for copper ($\sigma = 57 \cdot 10^6$ S/m, $\mu = \mu_0$), iron ($\sigma = 10^7$ S/m, $\mu_\mathrm{r} = 1000$), sea water ($\sigma = 4$ S/m, $\mu = \mu_0$) and wet soil ($\sigma = 0.01$ S/m, $\mu = \mu_0$), at 60 Hz (power-line frequency), 10^3 Hz, 10^6 Hz and 10^9 Hz.

TABLE 20.1. **Skin depth δ for some common materials**

Material	$f = 60$ Hz	$f = 10^3$ Hz	$f = 10^6$ Hz	$f = 10^9$ Hz
Copper	8.61 mm	2.1 mm	0.067 mm	2.11 μm

Iron	0.65 mm	0.16 mm	5.03 μm	0.016 μm
Sea water	32.5 m	7.96 m	0.25 m	7.96 mm
Wet soil	650 m	159 m	5.03 m	0.16 m

• Using Poynting's theorem, the power of Joule's losses and the reactive power in a conductor per unit area of the boundary surface are found to be

$$P_J = \int_S R_s |H_0|^2 dS = (P_{\text{reactive}})_{\text{internal}} \qquad \text{(W)}, \qquad (20.3)$$

where H_0 is the complex rms value of the tangential component of the vector H *on the conductor surface*, and

$$R_s = \sqrt{\frac{\omega\mu}{2\sigma}} \qquad (\Omega) \qquad (20.4)$$

is known as the *surface resistance* of the conductor.

• Using Eq. (20.3), it is possible to evaluate the resistance and internal reactance of many structures at high frequencies. For example, for a round wire the resistance per unit length at high frequencies is found to be

$$R' = \frac{R_s}{2\pi a} \qquad (\Omega/\text{m}), \qquad (20.5)$$

and it is equal to the *internal* (inductive) reactance of the wire per unit length.

• The term *proximity effect* refers to the influence of an alternating current in one conductor on the current distribution in another, close-by conductor. It is also due to electromagnetic induction.

QUESTIONS

S **Q20.1.** Three long parallel wires a distance d apart are in one plane. At their ends they are connected together. These common ends are then connected by a large loop to a generator of sinusoidal emf. Are the currents in the three wires the same? Explain. [*Hint:* Have in mind Eq. (14.3), where $J dv$ is substituted by $i dl$.] — *(a) The currents are the same. (b) The current in the middle conductor is greater than in the side conductors. (c) The current in the middle conductor is less than in the side conductors.*

Answer. The induced electric field along the middle wire is different from that along the two side wires. Therefore the current in the middle wire differs from that in the two side wires, and it is smaller in intensity.

Q20.2. N long parallel thin wires are arranged uniformly around a circular cylinder. At their ends the wires are connected by a large loop to a generator of sinusoidal emf. Are the currents in the N wires the same? Explain. — *(a) The currents are the same. (b) Those in the wires close to the generator are the greatest. (c) Those in the wires further from the generator are the greatest.*

Q20.3. Another wire is added in question Q20.2 along the axis of the cylinder. Is the current in the added wire the same as in the rest? Is it smaller or greater? Explain having in mind Eq. (14.3). — *The answers are the same as in question Q20.1.*

Q20.4. A thin metallic strip of width d carries a sinusoidal current of a high frequency. What do you expect the distribution of current in the strip to be like? — *(a) It will be uniform. (b) The current density will be greater along the middle of the strip. (c) The current density will be greater along the edges of the strip.*

S **Q20.5.** The two conductors of a coaxial line are connected in parallel to a generator of sinusoidal emf. Is the current intensity in the two conductors the same? If it is not, does the difference depend on frequency? Explain. — *(a) The larger part is in the outer conductor; depends on frequency. (b) The larger part is in the inner conductor; depends on frequency. (c) The currents are the same, because the resistances of the two conductors are approximately the same.*

Answer. It is not the same — the larger part of current is in the outer line conductor. With increasing frequency, this effect becomes progressively more pronounced, because the skin effect becomes progressively more pronounced. Compare also the answer to Q20.3.

Q20.6. A metal coin is situated in a time-harmonic uniform magnetic field, with faces normal to the field lines. What are the lines of eddy currents in the coin like? What are the lines of the induced electric field of these currents? — *(a) There are no eddy currents. (b) The lines of eddy currents and of the induced electric field are circles. (c) The lines of eddy currents and of the induced electric field are not easy to determine.*

Q20.7. So-called induction furnaces are used for melting iron by producing large eddy currents in iron pieces. Assume that the iron in the furnace is first in the form of small ferromagnetic objects (nails, screws, etc.). What do you expect to happen if they are exposed to a very strong time-harmonic magnetic field? What happens once they melt? — *Hint: note that in the beginning the objects are ferromagnetic, but practically separated, and when melted the entire volume of the liquid metal is paramagnetic, but in one large piece.*

S **Q20.8.** Two parallel coplanar thin strips carry equal time-harmonic currents. What do you think the current distribution in the strips is like if the currents in the strips are (1) in the same direction, and (2) in opposite directions? — *(a) For currents in the same direction the skin effect is pronounced, for currents in the opposite directions it is not. (b) For currents in the same direction the skin effect is not pronounced, for currents in the opposite directions it is pronounced. (b) The skin effect is pronounced in both cases.*

Answer. (1) The skin effect is very pronounced. The current density is the greatest near the strip edges, and the smallest near their centers. (2) The current is distributed very nearly uniformly over the cross section of both strips, since the induced electric field of one strip due to the current in it is practically canceled by the induced field of the opposite current in the nearby strip.

S **Q20.9.** A thick copper conductor of square cross section carries a large time-harmonic current. Where do you expect the most intense Joule's heating of the conductor? Explain. — *(a) Along the conductor axis. (b) Along the middle of conductor sides. (c) Along the conductor edges.*

Answer. Along the conductor edges, where the current density, due to the skin effect, is the greatest.

Q20.10. A ferromagnetic core of a solenoid is made of thin sheets. If the current in the solenoid is time-harmonic, where to you expect the strongest heating of the core due to eddy currents? — *(a) At the middle point of the core. (b) At the solenoid end. (c) Since it is made of thin sheets, it is uniform.*

Q20.11. Describe the procedure of determining the resistance and internal inductance per unit length of a stripline at high frequencies. Neglect edge effect. — *Hint: use Ampère's law to find H for a given current I in the strips, use Eq. (20.3), and equate the result with $R'I^2$.*

Q20.12. When compared with current density on the surface, what is the magnitude of current density in a thick conducting sheet one skin depth below the surface, and what it is at two skin depths below the surface? — *(a) 0.268, 0.105. (b) 0.568, 0.235. (c) 0.368, 0.135.*

PROBLEMS

P20.1. Check all skin depth values given in Table 20.1. — *Hint: use Eq. (20.2).*

S **P20.2.** Starting from Eq. (20.1), prove that the total current in the half-space in Fig. P20.2 is the same as if a current of constant density $J_z(0)/(1+j)$ exists in a slab $0 \leq y \leq \delta$. — *Hint: evaluate the current through a strip x wide, from $y = 0$ to $y = \infty$.*

Solution. Referring to Fig. P20.2,

$$\frac{dI}{dx} = \int_0^\infty J_z(y)\,dy = \int_0^\infty J_z(0)\,e^{-ky}\,e^{-jky}\,dy = \frac{J_z(0)}{(1+j)\,k} = \frac{J_z(0)}{1+j}\,\delta.$$

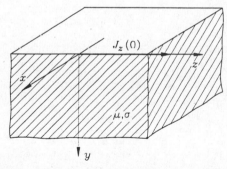

Fig. P20.2. A thick slab with time-harmonic currents.

P20.3. Determine the total Joule's losses per unit area of the half-space in Fig. P20.2 by integrating the density of Joule's losses. Compare the result with Eq. (20.3). — *Hint: integrate $|J_z(y)|^2/\sigma S\,dy$ from $y = 0$ to $y = \infty$.*

P20.4. Using Poynting's theorem in complex form, prove that for any conductor with two close terminals, at very high frequencies the conductor resistance and internal reactance are equal. Find the (integral) expression for these quantities. — *Hint: use Poynting's theorem and Eq. (20.3), to obtain*

$$R = X_{\text{internal}} = \frac{1}{|I_0|^2} \oint_S R_s |H_{\text{tangential to } S}|^2 dS.$$

The integration is over the conductor surface, S.

S **P20.5.** A stripline of strip width $a = 2\,\mathrm{cm}$, distance between them $d = 2\,\mathrm{mm}$, and the thickness of the strips $b = 1\,\mathrm{mm}$ carries a time-harmonic current of rms value $I = 0.5\,\mathrm{A}$ and frequency $f = 1\,\mathrm{GHz}$. The strips are made of copper. Neglecting fringing effect, determine the line resistance and total inductance per unit length. — (a) $R' = 0.44\,\Omega/m$, $L' = L'_{\text{internal}} + L'_{\text{external}} = 25.8\,nH/m$. (b) $R' = 0.84\,\Omega/m$, $L' = 125.8\,nH/m$. (c) $R' = 0.55\,\Omega/m$, $L' = 83.4\,nH/m$.

Solution. Neglecting the fringing effects, the magnetic field intensity on the surface of each strip is $H(0) = I/a$, so the Joule's losses per unit length of the line are

$$P_J' = 2R_{\mathrm{s}}\left(\frac{I}{a}\right)^2 a = \frac{2R_{\mathrm{s}}}{a}\,I^2.$$

Hence, at high frequencies, the resistance and internal reactance per unit length of the line are given by

$$R' = \omega L'_{\text{internal}} = \frac{2R_{\mathrm{s}}}{a} = 0.84\,\Omega/m,$$

L'_{internal} being the high-frequency internal inductance per unit length. Since $L'_{\text{external}} = \mu_0 d/a$ (see problem P15.15), the total inductance per unit length of the strip line is

$$L' = L'_{\text{internal}} + L'_{\text{external}} = \frac{2R_{\mathrm{s}}}{\omega a} + \frac{\mu_0 d}{a} = 125.8\,nH/m.$$

S **P20.6.** Starting from the equations

$$\frac{\mathrm{d}J_z}{\mathrm{d}y} = -\mathrm{j}\omega\sigma B_x, \qquad -\frac{\mathrm{d}B_x}{\mathrm{d}y} = \mu J_z$$

for current distribution parallel to the interface of a flat sheet (Fig. P20.2), determine the distribution of current in a flat conducting sheet of thickness d. The sheet conductivity is σ, permeability μ, and angular frequency of the current is ω. Set the origin of the y coordinate at the sheet center, and assume that the rms value of the current density at the center is $J_z(0)$. Plot the resulting current distribution. — *The correct result is:* $J_z(y) = J_z(0)(\cos ky \cosh ky + \mathrm{j}\sin ky \sinh ky)$.

Solution. The complex rms value of the current density in the sheet is given by Eq. (20.5). Since J_z is an *even* function of y, i.e. $J_z(-y) = J_z(y)$, we find that $J_1 = J_2$, from which

$$J_z(y) = J_1\left(e^{Ky} + e^{-Ky}\right), \qquad K = (1+\mathrm{j})k.$$

Now, $J_z(0) = 2J_1$, so that we can rewrite the last equation as

$$J_z(y) = \frac{1}{2}J_z(0)\left(e^{Ky} + e^{-Ky}\right) = J_z(0)\cosh Ky = J_z(0)\left(\cos ky \cosh ky + \mathrm{j}\sin ky \sinh ky\right).$$

***P20.7.** The current density in the slab shown in Fig. P20.2 is given in Eq. (20.1). (1) Find $E_z(y)$ in the slab. (2) Starting from Maxwell's equation in complex form and rectangular coordinate system, find $H_x(y)$. (3) Use the expressions for $E_z(y)$ and $H_x(y)$ and Poynting's theorem to prove Eq. (20.3). — *Hint: in step (2), note that, due to the direction of current, H has only the H_x component, which is a function of y alone.*

P20.8. Starting from Eq. (20.1), derive the expression for the instantaneous value of the current density, $J_z(y,t)$. — *Hint: recall that the instantaneous value is obtained as the real part of the current density multiplied by $\sqrt{2}e^{j\omega t}$.*

P20.9. Calculate the resistance per unit length of a round copper wire of radius $a = 1\,\text{mm}$, from the frequency for which the skin depth is one tenth of the wire radius, to the frequency $f = 10\,\text{GHz}$. Plot this resistance as a function of frequency. — *Let the starting frequency be f_1, and the end frequency $f = f_2$. (a) $R_1' = 0.0184\,\Omega/m$, $R_2' = 3.226\,\Omega/m$. (b) $R_1' = 0.0284\,\Omega/m$, $R_2' = 4.226\,\Omega/m$. (c) $R_1' = 0.0384\,\Omega/m$, $R_2' = 5.226\,\Omega/m$.*

S **P20.10.** Assume that in a ferromagnetic round wire of radius a, conductivity σ, and permeability μ, there is an axial magnetic field of angular frequency ω and of rms flux density B practically constant over the wire cross section. Find the expressions for eddy currents in the wire and eddy current losses in the wire per unit length. — *Hint: eddy currents in the wire are circulating currents, i.e., the lines of the current density vector are circles centered along the wire axis and normal to the axis. The complex rms value of the density of these currents is $J = -j\omega Br/2$ ($r \leq a$).*

Solution. See problem P15.25. The eddy currents in the wire are circulating currents, i.e., the lines of the current density vector are circles centered on the wire axis. The complex rms value of the density of these currents is $J = -j\omega Br/2$ ($r \leq a$). The time-average power of Joule's losses due to eddy currents per unit length of the wire is

$$(P'_{\text{eddy currents}})_{\text{ave}} = \frac{1}{8}\pi\sigma\omega^2 a^4 B^2.$$

P20.11. A bunch of N insulated round wires of radius a, conductivity σ, and permeability μ is exposed to an axial time-harmonic magnetic field of angular frequency ω. The frequency is sufficiently low that the field can be considered uniform over the cross section of the wires. If the rms value of the magnetic flux density is B_0, determine the time-average eddy current power losses in the bunch, per unit volume *of the wires*. Use the result of the preceding problem. Specifically, calculate the losses per unit volume assuming $B_0 = 0.1\,\text{T}$, $a = 0.5\,\text{mm}$, $\sigma = 10^7\,\text{S/m}$, $\mu = 1000\mu_0$, and $f = 60\,\text{Hz}$. — *(a)* $dP_{\text{eddy currents}}/dv_{\text{wires}} = 444\,W/m^3$. *(b)* $dP_{\text{eddy currents}}/dv_{\text{wires}} = 644\,W/m^3$. *(c)* $dP_{\text{eddy currents}}/dv_{\text{wires}} = 844\,W/m^3$.

***P20.12.** Consider a straight wire of radius a, conductivity σ, and permeability μ. Let the wire axis be the z axis of a cylindrical coordinate system. Assume there is a current in the wire of rms value I and angular frequency ω. Starting from Maxwell's equations in cylindrical coordinates, derive the differential equation for the only existing, J_z component of the current density vector in the wire. Note that, by symmetry, the only component of H is H_ϕ. Do *not* attempt to solve the equation you obtain. (If your equation is correct, it is known as a Bessel differential equation, and its solutions are known as Bessel functions.) — *The correct result is*

$$\frac{\mathrm{d}^2 J_z}{\mathrm{d}u^2} + \frac{1}{u}\frac{\mathrm{d}J_z}{\mathrm{d}u} + J_z = 0, \quad u = r\sqrt{-\mathrm{j}\omega\mu\sigma}.$$

21. Uniform Plane Electromagnetic Waves

• Maxwell's theory predicts the existence of specific electromagnetic fields, known as *electromagnetic waves*. These fields, once created by time-varying currents and charges, continue to move with a finite velocity independently of the sources that produced them.

• The *wave equations* are second-order partial differential equations derived from Maxwell's equations for fields in *homogeneous linear media*. They are of the same form for vectors E and H,

$$\nabla^2 E - \epsilon\mu\frac{\partial^2 E}{\partial t^2} - \mu\sigma\frac{\partial E}{\partial t} = 0, \tag{21.1}$$

$$\nabla^2 H - \epsilon\mu\frac{\partial^2 H}{\partial t^2} - \mu\sigma\frac{\partial H}{\partial t} = 0. \tag{21.2}$$

• If the field is time-harmonic, in complex notation this becomes

$$\nabla^2 \underline{E} + (\omega^2\epsilon\mu - \mathrm{j}\omega\mu\sigma)\underline{E} = 0, \tag{21.3}$$

$$\nabla^2 \underline{H} + (\omega^2\epsilon\mu - \mathrm{j}\omega\mu\sigma)\underline{H} = 0. \tag{21.4}$$

These equations are known as the *Helmholtz equations*.

• The *uniform plane electromagnetic wave* is the simplest of all waves. It is assumed that it exists in a *homogeneous, lossless medium* of parameters ϵ, μ and $\sigma = 0$, and that both vectors E and H depend only on one coordinate (e.g., z), and on time. The solution of the wave equation (i.e., of Maxwell's equations) in that case requires that the z-components of both fields be zero. Assuming, for simplicity, that only the $E_x(x,t)$ exists, the final solution is

$$E_x(z,t) = E_1 f_1\left(t - \frac{z}{c}\right) + E_2 f_2\left(t + \frac{z}{c}\right), \qquad c = \frac{1}{\sqrt{\epsilon\mu}} \quad \text{(m/s)}, \tag{21.5}$$

where E_1 and E_2 are constants, and f_1 and f_2 are *any* functions of the arguments $(t - z/c)$ and $(t + z/c)$, respectively. We know from Chapter 18 that $f_1(t - z/c)$ is an incident (forward) travelling (electric field) wave, and $f_2(t + z/c)$ is a reflected (backward) travelling wave (with respect to the z axis).

- The associated magnetic field is given by

$$H_y(z,t) = \sqrt{\frac{\epsilon}{\mu}} E_x(z,t). \tag{21.6}$$

Consequently, for a *single* incident plane wave, the energy density of the magnetic field, $\mu H^2/2$, at all points and at all instants, equals the energy density of the electric field, $\epsilon E^2/2$. Since the vectors $\boldsymbol{E}$ and $\boldsymbol{H}$ are transversal to the direction of propagation, this kind of uniform plane wave is known as a *transverse electromagnetic wave*, or TEM wave.

- The square root in Eq. (21.6) has a dimension of impedance, and is known as the *intrinsic impedance* (η or Z) of the medium:

$$\eta = \sqrt{\frac{\mu}{\epsilon}} \qquad (\Omega). \tag{21.7}$$

The most important (and frequent) waves in practice are those propagating in a vacuum (or air). In that case,

$$\eta_0 = \sqrt{\frac{\mu_0}{\epsilon_0}} \simeq 120\pi \simeq 377\,\Omega, \qquad c_0 = \frac{1}{\sqrt{\epsilon_0 \mu_0}} \simeq 3 \cdot 10^8\,\text{m/s}. \tag{21.8}$$

- Note that the cross product of vectors $\boldsymbol{E}$ and $\boldsymbol{H}$, i.e., the Poynting vector, *is in the direction of propagation of the wave*:

$$\boldsymbol{E} \times \boldsymbol{H} = E_x(z,t) H_y(z,t) \boldsymbol{u}_z = \mathcal{P}(z,t) \boldsymbol{u}_z \qquad (\text{W/m}^2). \tag{21.9}$$

This means that the wave *transports electromagnetic energy*. If the E-field has an y-component (instead of an x-component), the H-field will have a $-x$-component, since the cross product $\boldsymbol{E} \times \boldsymbol{H}$ must be in the $+z$-direction (the direction of propagation of the wave). The relations between the two is

$$H_x(z,t) = -\sqrt{\frac{\epsilon}{\mu}} E_y(z,t). \tag{21.10}$$

- The complex form of the components of a time-harmonic plane wave propagating in the $+z$-direction is

$$\underline{E}_x(z) = \underline{E}\mathrm{e}^{-\mathrm{j}\beta z}, \quad \underline{H}_y(z) = \frac{1}{\eta}\underline{E}\mathrm{e}^{-\mathrm{j}\beta z}, \quad \underline{E} = E\mathrm{e}^{\mathrm{j}\theta}, \tag{21.11}$$

where

$$\beta = \frac{\omega}{c} = \frac{2\pi}{\lambda} \qquad \text{(radian/m)} \qquad\qquad (21.12)$$

is the *phase coefficient*, and λ is the *wavelength* of the wave.

•   If the tip of the $\boldsymbol{E}$-vector at a point traces in the course of time a line segment, the polarization of the wave is *linear*. If it traces an ellipse, the polarization is *elliptical*, and if it traces a circle, it is *circular*.

•   For a wave propagating in a medium that has a permittivity and/or permeability which is a function of frequency, the velocity of wave propagation is a function of frequency. The same is true if the medium is lossy. Such media are known as *dispersive media*. If a non-sinusoidal signal propagates in such a medium, it is distorted. This effect is known as *dispersion*. For a wave in a dispersive medium two velocities are defined. The *phase velocity* is given by

$$v_{\text{ph}}(\omega) = \frac{\omega}{\beta(\omega)} \qquad \text{(m/s)}, \qquad\qquad (21.13)$$

and represents the velocity of the phase of the wave. The *group velocity* is the velocity of the signal in a dispersive medium, and can be defined *only if dispersion is small*. The energy travels at the group velocity, given by

$$v_{\text{g}}(\omega) = \frac{1}{\text{d}\beta(\omega)/\text{d}\omega} \qquad \text{(m/s)}. \qquad\qquad (21.14)$$

QUESTIONS

S **Q21.1.** What would Eqs. (21.1) and (21.2) be like for an inhomogeneous perfect dielectric? Would it be possible to obtain the wave equation in that case? — *Hint: try to derive the wave equations for such a medium.*

Answer. Only the second of Eqs. (22.1) would be changed, and it would read $\nabla \cdot (\epsilon \boldsymbol{E}) = \epsilon \nabla \cdot \boldsymbol{E} + \boldsymbol{E} \cdot \nabla \epsilon = 0$. This would make it impossible to obtain the wave equations, and in fact to solve the equations at all. (Note that ϵ is a function of coordinates.)

Q21.2. Derive the Helmholtz equations, (21.3) and (21.4), from the wave equations, (21.1) and (21.2). — *Hint: recall that a partial derivative with respect to time needs to be replaced by* $j\omega$.

Q21.3. Write the expressions for at least three functions representing forward and backward travelling waves. — *Which of the following four functions* does not *represent a forward or backward travelling wave?* (a) $f_1(z,t) = \sin \omega(t \mp z/c)$. (b) $f_2(z,t) = \cos \omega(t \mp z/c)$, (c) $f_3(z,t) = A(t \mp z/c)$ *if* $0 < t \mp z/c < a/c$, *else* $f_3(z,t) = 0$. (d) $f_4(z,t) = (t \mp z/c)(t \pm z/c)$.

Q21.4. Is a plane electromagnetic wave with a component of the electric or magnetic field in the direction of propagation possible? Explain. — *(a) It is possible, but the other components are then different. (b) It is possible, and nothing else is changed. (c) It is not possible, since then Maxwell's divergence equations cannot be satisfied.*

Q21.5. A perfect-dielectric medium is not homogeneous, but ϵ is a smooth function of position, $\epsilon = \epsilon(x,y,z)$. Is a uniform plane wave possible in such a medium? Explain. — *Hint: see the answer to question Q21.1.*

S **Q21.6.** Write the equation $\partial^2 E_x / \partial z^2 - \epsilon\mu \partial^2 E_x / \partial t^2 = 0$ in complex form and find its solutions. — *Hint: recall how we obtain an equation in complex form from that in time domain.*

Answer. The equation in complex form reads $\partial^2 E_x / \partial z^2 + \omega^2 \epsilon\mu E_x = 0$. Its solution is $E_x(z) = E_0 e^{\mp j\beta z}$, where $\beta = \omega\sqrt{\epsilon\mu}$.

Q21.7. What is the ratio of the wavelengths of a sinusoidal plane wave of frequency f if it propagates in perfect dielectrics of permittivities ϵ_1 and ϵ_2, and permeability μ_0? — *(a)* $\lambda_1/\lambda_2 = \sqrt{\epsilon_1/\epsilon_2}$. *(b)* $\lambda_1/\lambda_2 = \sqrt{\epsilon_2/\epsilon_1}$. *(c)* $\lambda_1/\lambda_2 = \epsilon_2/\epsilon_1$.

Q21.8. What is the wavelength in a vacuum corresponding to the following frequencies of a plane wave: (1) 60 Hz, (2) 10 kHz, (3) 1 MHz, (4) 100 MHz, (5) 1 GHz, (6) 10 GHz, (7) 100 GHz? — *Which three of the following wavelengths do not correspond to any of the given frequencies? (1) 2000 km, (2) 5000 km, (3) 30 km, (4) 300 m, (5) 600 m, (6) 3 m, (7) 30 cm, (8) 3 cm, (9) 3 mm, (10) 300 μm.*

Q21.9. Does the expression $E = E_1\cos(\omega t - \beta z)u_z$ represent a possible electric field of a plane wave? Explain. — *(a) It does, because it has the proper dependence on t and z. (b) It does, although this is not a standard plane wave. (c) It does not, because it implies that the component of E in the direction of propagation varies in space and time.*

S **Q21.10.** Does the expression $E = E_1 e^{j\beta x}u_z$ represent a possible phasor expression for the electric field of a plane wave? Explain. — *(a) No, because it is z-directed. (b) No, because it propagates in the $-x$-direction. (c) Yes (why?).*

Answer. It does — it represents a wave propagating in the $-x$-direction, and having a z-component of the electric field vector, which is admissible.

S **Q21.11.** A circular loop of radius a is situated in the field of a plane electromagnetic wave of wavelength $\lambda = a$. Is it possible in principle to evaluate the emf induced in the loop? If you think it is, can it be used for the evaluation of current intensity in the loop by means of circuit theory? Explain. — *Hint: note that the Maxwell's first equation in integral form is valid for any size of the loop. So, the emf can be calculated. Can the current be evaluated by means of circuit theory?*

Answer. We can evaluate the emf (Maxwell's first equation in integral form is valid for any size of the loop). The current intensity in the loop cannot be evaluated by means of circuit theory, because the loop is "electrically large", and the current intensity is not constant around the loop.

Q21.12. Does the concept of linear wave polarization make sense if the wave is not time-harmonic? What about the concept of circular and elliptical polarization? Explain. — *(a) All three concepts are completely general. (b) The concept of linear polarization does not make sense. (c) The concepts of circular and elliptical polarization do not make sense.*

Q21.13. A time-harmonic, linearly polarized plane wave propagates along the z axis. Located along the z axis is a row of small free charges. How do the charges move in time? Sketch their approximate position over a few time intervals. — *Hint: note that the charges experience a force due to the electric field of the wave.*

Q21.14. Repeat question Q21.13 assuming that the polarization of the wave is circular. — *Hint: same as in the preceding question.*

Q21.15. What is the complex representation of the electric field of an elliptically polarized plane wave defined by the equations $E_x(z,t) = E_1 \cos(\omega t - \beta z)$, $E_y(z,t) = E_2 \sin(\omega t - \beta z)$? — *Convince yourself that the correct answer is $E_x(z) = E_1 e^{-j\beta z}$, $E_y(z) = jE_2 e^{-j\beta z}$.*

Q21.16. Assuming that the wave propagates into the paper, is the polarization of the wave represented in the two sketches in Fig. Q21.16 right-handed or left-handed? Explain. — *Hint: note that for the direction of propagation into the paper, a right-hand polarized wave should rotate as a right-hand screw progressing in that direction.*

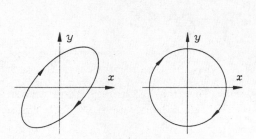

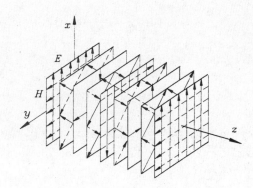

Fig. Q21.16. Elliptic and circular polarization.

Fig. Q21.17 Elliptically polarized wave.

S   **Q21.17.** Is the polarization of the wave represented in Fig. Q21.17 right-handed or left-handed? Explain. — *Hint: the same as in the preceding question.*

Answer. A wave has a right-handed polarization if, looking down the direction of propagation, the tip of the vector **E** rotates in the clockwise direction, else it is left-handed. The wave in Fig. Q21.17 propagates in the z-direction, and the **E**-vector rotates with respect to that direction in the clockwise direction. Therefore the polarization of the wave is right-handed.

PROBLEMS

P21.1. Assuming an E_y-component of the electric field, *derive* the corresponding component of the magnetic field in Eq. (21.10), following the same procedure as when E_x was assumed. — *Hint: you need to reconsider the expression for the curl in a rectangular coordinate system.*

S   **P21.2.** Prove that for a plane wave with both E_x and E_y components the vector **E** is normal to vector **H**. Evaluate the Poynting vector in that case. — *Hint: prove that the dot product of the two vectors is zero. To evaluate the Poynting vector, start from its definition and express the two field vectors as the sum of their two components.*

Solution. The dot product of vectors **E** and **H** is

$$\mathbf{E} \cdot \mathbf{H} = [E_x(z,t)\,\mathbf{u}_x + E_y(z,t)\,\mathbf{u}_y] \cdot [H_x(z,t)\,\mathbf{u}_x + H_y(z,t)\,\mathbf{u}_y]$$

$$= \sqrt{\frac{\epsilon}{\mu}}\,[E_x(z,t)\,\mathbf{u}_x + E_y(z,t)\,\mathbf{u}_y] \cdot [-E_y(z,t)\,\mathbf{u}_x + E_x(z,t)\,\mathbf{u}_y]$$

$$= \sqrt{\frac{\epsilon}{\mu}}\,[-E_x(z,t)E_y(z,t) + E_y(z,t)E_x(z,t)] = 0,$$

which means that these vectors are mutually orthogonal.

The Poynting vector is given by

$$\vec{\mathcal{P}} = \mathbf{E} \times \mathbf{H} = [E_x(z,t)\,\mathbf{u}_x + E_y(z,t)\,\mathbf{u}_y] \times [H_x(z,t)\,\mathbf{u}_x + H_y(z,t)\,\mathbf{u}_y]$$

$$= \sqrt{\frac{\epsilon}{\mu}}\,[E_x(z,t)\,\mathbf{u}_x + E_y(z,t)\,\mathbf{u}_y] \times [-E_y(z,t)\,\mathbf{u}_x + E_x(z,t)\,\mathbf{u}_y]$$

$$= \sqrt{\frac{\epsilon}{\mu}}\,\left[E_x^2(z,t) + E_y^2(z,t)\right]\mathbf{u}_z = \sqrt{\frac{\epsilon}{\mu}}\,E^2(z,t)\,\mathbf{u}_z.$$

P21.3. Repeat the entire derivation of the plane waves for time-harmonic plane waves and starting from complex forms of all the equations. — *Hint: start from the Helmholtz equation, Eq. (21.3).*

S **P21.4.** A time-harmonic plane wave with an rms value of the electric field vector $E = 10\,\mathrm{mV/m}$ propagates in a vacuum, and is normally incident on a screen that totally absorbs the energy of the wave. Find the absorbed energy per square meter of the screen in one hour. — *(a)* $W_{\mathrm{abs}} \simeq 1.95\,mJ.$ *(b)* $W_{\mathrm{abs}} \simeq 2.95\,mJ.$ *(c)* $W_{\mathrm{abs}} \simeq 0.95\,mJ.$

Solution. We know that the average Poynting vector represents the *power flow per unit area* normal to the direction of propagation of the wave. The absorbed energy on the area of $S = 1\,\mathrm{m}^2$ of the screen in $t = 1\,\mathrm{h} = 3600\,\mathrm{s}$ is

$$W_{\mathrm{abs}} = \mathcal{P}_{\mathrm{ave}} S t = E H S t = \sqrt{\frac{\epsilon_0}{\mu_0}}\,E^2 S t \simeq 0.95\,\mathrm{mJ}.$$

P21.5. By measurements it was found that the time-average power of the sun's radiation on the surface of the earth is about $1.35\,\mathrm{kW/m}^2$, for normal incidence of the plane waves from the sun. This radiation is composed of a very wide band of frequencies, and the components of different frequencies are generally polarized elliptically. Assuming, for simplicity, that the entire radiation is a linearly polarized wave of a single frequency, determine the rms value of its electric and magnetic field. — *(a)* $E = 713\,V/m$, $H = 1.9\,A/m$. *(b)* $E = 356\,V/m$, $H = 0.95\,A/m$. *(c)* $E = 178\,V/m$, $H = 0.48\,A/m$.

P21.6. The radius of the earth is about $6350\,\mathrm{km}$. Assuming the entire energy of the sun's radiation reaching the earth is absorbed by the earth, calculate the time-average power of the absorbed energy, and the energy absorbed by the earth in one day. Compare this with the total man-produced energy, assuming that the time-average power of this energy during the day is about 12,500 GW. — (a) $W_{\text{abs}} \simeq 1.5 \cdot 10^{12}$ J. (b) $W_{\text{abs}} \simeq 1.5 \cdot 10^{16}$ J. (c) $W_{\text{abs}} \simeq 1.5 \cdot 10^{22}$ J.

P21.7. Due to various effects, human exposure to electromagnetic radiation is considered to be harmful above a certain time-average value of the Poynting vector. This estimated value depends on frequency, and differs greatly among different countries in the world. Assuming that above 10 GHz this value is on the order of $10 \, \text{mW/cm}^2$, compute the corresponding rms value of the electric and magnetic field of the plane wave with this time-average value of the Poynting vector. Compare this value of the electric field with the rms value of TV and broadcasting stations, which is on the order of mV/m. — (a) $E = 194 \, V/m$, $H = 0.515 \, A/m$. (b) $E = 392 \, V/m$, $H = 1.04 \, A/m$. (c) $E = 784 \, V/m$, $H = 2.08 \, A/m$.

P21.8. A circular wire loop of radius a is situated in a vacuum in the electromagnetic field of a plane wave, of wavelength λ ($\lambda \gg a$), and the rms value of the electric field strength E. How should the loop be positioned in order that the emf induced in it be maximal? Determine the rms value of the emf in that case. — (a) $\mathcal{E}_{\text{max}} = 2\pi a^2 E/\lambda$. (b) $\mathcal{E}_{\text{max}} = 2\pi^2 a^2 E/\lambda$. (c) $\mathcal{E}_{\text{max}} = \pi^2 a^2 E/\lambda$.

S **P21.9.** A rectangular wire loop with sides a and b is situated in a vacuum in the electromagnetic field of a time-harmonic plane wave. The amplitude of the electric field strength of the wave is E, and its wavelength is λ ($\lambda \gg a, b$). The loop is oriented so that the maximal emf is induced in it. In case (1) the sides a are parallel to the electric field of the wave, and in case (2) the sides b are parallel to the electric field of the wave. Evaluate in both cases the emf (a) starting from Faraday's law of electromagnetic induction in its usual form ($e = -\mathrm{d}\Phi/\mathrm{d}t$), and (b) as an integral of the electric field strength of the wave around the contour. — *Referring to Fig. P21.9:* (a) $e = \mu_0 \omega a b H_{\text{m}} \sin \omega t$. (b) $e = 2\pi \mu_0 \omega a b H_{\text{m}} \sin \omega t$. (c) $e = \pi \mu_0 \omega a b H_{\text{m}} \sin \omega t$.

Solution. (a) The loop with sides a parallel to the electric field intensity vector of the wave is shown in Fig. P21.9. Suppose that in the region where the loop is situated $H = H_{\text{m}} \cos \omega t$. Using Faraday's law we then obtain

$$e = -\mu_0 \frac{\mathrm{d}H}{\mathrm{d}t} \, ab = \mu_0 \omega a b H_{\text{m}} \sin \omega t.$$

(b) According to Fig. P21.9, it must also be

$$e = [-E_1 + E_2] \, a.$$

In applying Faraday's law we assumed that H is approximately constant at all points on the loop. A similar approximation for E would result in $e = 0$, so we must take into account the (very small) difference between E_1 and E_2. Let $E_1 = E_{\text{m}} \cos \omega t$. Then

$$E_2 = E_{\text{m}} \cos(\omega t - \beta b) = E_{\text{m}} \cos\left(\omega t - \frac{2\pi b}{\lambda}\right).$$

Since $b \ll \lambda$, we have that $\cos(2\pi b/\lambda) \simeq 1$ and $\sin(2\pi b/\lambda) \simeq 2\pi b/\lambda$. So

$$e = a E_{\text{m}} \left[-\cos \omega t + \cos\left(\omega t - \frac{2\pi b}{\lambda}\right)\right]$$

$$= a E_\mathrm{m} \left(- \cos \omega t + \cos \omega t \cos \frac{2 \pi b}{\lambda} + \sin \omega t \sin \frac{2 \pi b}{\lambda} \right) \simeq a E_\mathrm{m} \frac{2 \pi b}{\lambda} \sin \omega t.$$

This is the same as in (a), because $H_\mathrm{m} = \sqrt{\epsilon_0/\mu_0}\, E_\mathrm{m}$ and $\omega = 2\pi/(\lambda\sqrt{\epsilon_0\mu_0})$.

The result for e remains the same when the sides b of the loop are parallel to the electric field of the wave.

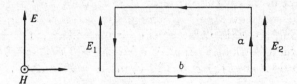

Fig. P21.9. A rectangular loop in the field of a plane wave.

P21.10. A sinusoidal plane wave, of frequency f and time-average value of the Poynting vector $\mathcal{P}$, propagates through distilled water ($\mu = \mu_0$, $\epsilon = \epsilon_\mathrm{r}\epsilon_0$). Find the rms value of the emf induced in a small circular loop of radius a, oriented so that the emf is maximal. What condition must be met in order that the loop can be considered as a quasi-static system? — *(a) $\mathcal{E} = \pi^2 f \mu_0 \sqrt{\mathcal{P}/\eta}\, a^2$. (b) $\mathcal{E} = 2\pi f \mu_0 \sqrt{\mathcal{P}/\eta}\, a^2$. (c) $\mathcal{E} = 2\pi^2 f \mu_0 \sqrt{\mathcal{P}/\eta}\, a^2$.*

S **P21.11.** Prove that an elliptically polarized wave can be represented as a sum of two circularly polarized waves. — *Hint: start from the expressions for an elliptically polarized wave in question Q21.15, and by mathematical manipulations make them represent two circularly polarized waves.*

Solution. Starting from the expressions for the components of the E-field of an elliptically polarized plane wave in Eqs. (21.30) and (21.31), we can write

$$\boldsymbol{E} = E_1 \cos(\omega t - \beta z)\, \boldsymbol{u}_x + E_2 \sin(\omega t - \beta z)\, \boldsymbol{u}_y.$$

The constants E_1 and E_2 can always be represented as

$$E_1 = E' + E'', \qquad E_2 = E' - E'',$$

where $E' = (E_1 + E_2)/2$ and $E'' = (E_1 - E_2)/2$. We can now represent vector $\boldsymbol{E}$ as a sum

$$\boldsymbol{E} = \boldsymbol{E}_\mathrm{I} + \boldsymbol{E}_\mathrm{II},$$

where

$$\boldsymbol{E}_\mathrm{I} = E' \cos(\omega t - \beta z)\, \boldsymbol{u}_x + E' \sin(\omega t - \beta z)\, \boldsymbol{u}_y$$

and

$$\boldsymbol{E}_\mathrm{II} = E'' \cos(\omega t - \beta z)\, \boldsymbol{u}_x - E'' \sin(\omega t - \beta z)\, \boldsymbol{u}_y$$

are circularly polarized vectors. Note that of the two vectors, E_I and E_{II}, the polarization of the first is left-handed, and of the second right-handed (with respect to the $+z$-direction of propagation of the wave). Convince yourself that this is indeed so!

It is left to the reader to write down the corresponding expressions for the H-field.

S **P21.12.** Determine the time-average Poynting vector of a circularly polarized plane wave (1) starting from the expressions for the plane wave in time domain, and (2) starting from the phasor expressions for the plane wave. — *Note that $\vec{\mathcal{P}}$ is constant in time.* (a) $\vec{\mathcal{P}}_{\text{ave}} = (E_{\text{m}}^2/2\eta)\,u_z$. (b) $\vec{\mathcal{P}}_{\text{ave}} = (E_{\text{m}}^2/\eta)\,u_z$. (c) $\vec{\mathcal{P}}_{\text{ave}} = 2(E_{\text{m}}^2/\eta)\,u_z$.

Solution. (1) In time domain,

$$E = E_{\text{m}}\cos(\omega t - \beta z)\,u_x + E_{\text{m}}\sin(\omega t - \beta z)\,u_y,$$

$$H = \frac{E_{\text{m}}}{\eta}\cos(\omega t - \beta z)\,u_y - \frac{E_{\text{m}}}{\eta}\sin(\omega t - \beta z)\,u_x,$$

so that

$$\vec{\mathcal{P}} = E \times H = \frac{E_{\text{m}}^2}{\eta}\left[\cos^2(\omega t - \beta z) + \sin^2(\omega t - \beta z)\right]u_z = \frac{E_{\text{m}}^2}{\eta}\,u_z.$$

Note that $\vec{\mathcal{P}}$ is *constant* in time. The time-average Poynting vector is hence $\vec{\mathcal{P}}_{\text{ave}} = (E_{\text{m}}^2/\eta)\,u_z$.

(2) In complex (phasor) domain,

$$\underline{E} = \frac{E_{\text{m}}}{\sqrt{2}}e^{-j\beta z}\,u_x - j\frac{E_{\text{m}}}{\sqrt{2}}e^{-j\beta z}\,u_y, \qquad \underline{H} = \frac{E_{\text{m}}}{\eta\sqrt{2}}e^{-j\beta z}\,u_y + j\frac{E_{\text{m}}}{\eta\sqrt{2}}e^{-j\beta z}\,u_x,$$

from which

$$\underline{\vec{\mathcal{P}}} = \underline{E} \times \underline{H}^* = \left(\frac{E_{\text{m}}}{\sqrt{2}}e^{-j\beta z}\,u_x - j\frac{E_{\text{m}}}{\sqrt{2}}e^{-j\beta z}\,u_y\right) \times \left(\frac{E_{\text{m}}}{\eta\sqrt{2}}e^{j\beta z}\,u_y - j\frac{E_{\text{m}}}{\eta\sqrt{2}}e^{j\beta z}\,u_x\right)$$

$$= \frac{E_{\text{m}}^2}{2\eta}\,u_z + \frac{E_{\text{m}}^2}{2\eta}\,u_z = \frac{E_{\text{m}}^2}{\eta}\,u_z.$$

Note that the complex Poynting vector has *no imaginary component*. Finally, $\vec{\mathcal{P}}_{\text{ave}} = \text{Re}\{\underline{\vec{\mathcal{P}}}\} = (E_{\text{m}}^2/\eta)\,u_z$.

P21.13. The electric field of a plane wave in complex (phasor) form is given by $E(z) = (E_x u_x + E_y u_y)e^{-j\beta z}$. The components E_x and E_y are arbitrary complex numbers. Assuming that the wave propagates in the $+z$ direction, discuss the polarization of the wave, stating whether it is right-handed or left-handed for circular and elliptic polarization, if (1) $E_x = 1$, $E_y = 0$; (2) $E_x = 0$, $E_y = 5$; (3) $E_x = j$, $E_y = -j$; (4) $E_x = j$, $E_y = 2$; (5) $E_x = (1 + j)$,

$E_y = 0$; (6) $E_x = 1$, $E_y = \mathrm{j}$; or (7) $E_x = (1 + 2\mathrm{j})$, $E_y = (1 - \mathrm{j})$. — *Convince yourself that the polarization of the wave is: (1) linear; (2) linear; (3) linear; (4) right-hand elliptical; (5) linear; (6) left-hand circular; (7) right-hand elliptical.*

P21.14. Two plane waves of equal frequencies and phases propagate in the same, z direction. Both are circularly polarized, but in opposite directions (one is right-handed, the other left-handed). The amplitudes of the electric field strength of the two waves are E_1 and E_2. Find the polarization of the resultant wave in terms of E_1 and E_2, starting from the expressions of the waves in time domain. — *Refer to Fig. P21.14, and prove the following: if $E_1 \neq E_2$, the resultant wave is polarized elliptically (if $E_1 > E_2$ the polarization is right-hand elliptical, if $E_1 < E_2$ it is left-hand elliptical). If $E_1 = E_2$, the wave becomes linearly polarized. Circular polarization is obtained only if $E_1 = 0$ or $E_2 = 0$.*

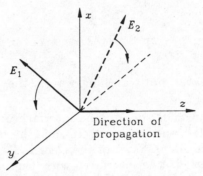

Fig. P21.14. Two circularly polarized plane waves propagating in the same direction.

P21.15. Two linearly polarized sinusoidal waves of the same frequency propagate in the z-direction. The electric field vectors of the two waves, $\boldsymbol{E}_1$ and $\boldsymbol{E}_2$, are along the x- and y axis, respectively. Plot the trace that the tip of the resulting vector, $\boldsymbol{E} = \boldsymbol{E}_1 + \boldsymbol{E}_2$, traces at $z = 0$ in time, as a function of the ratio of amplitudes E_1 and E_2, and their relative phase, ϕ. — *Hint: you need to find the x- and y-components of the total wave at $z = 0$, and to plot the vector sum of these two components as a function of time.*

22. Reflection and Refraction of Plane Waves

• If a plane wave encounters an obstacle, it induces conduction or polarization currents in it. The field that is then produced by these currents is known as the *scattered field*, the process itself as *scattering of electromagnetic waves*, and such obstacles are called *scatterers*.

• Generally, the determination of scattered fields is a difficult problem. However, plane waves incident on a *plane* boundary between two *homogeneous* media result in *plane* scattered waves, the *reflected wave* into the medium from which the original wave is incident, and the *refracted wave* or *transmitted wave* into the other medium.

• The simplest case is when a uniform plane wave is incident normally onto the planar interface between a perfect dielectric, of parameters ϵ and μ, and a perfect conductor. If the interface is at $z = 0$, and if the incident wave is of angular frequency ω, has E_x and H_y

components, and propagates in the $+z$-direction, the complex components of the resultant wave in front of the interface are

$$E_{\text{total}}(z) = -2\mathrm{j}E\sin\beta z\ \boldsymbol{u}_x, \tag{22.1}$$

$$H_{\text{total}}(z) = 2H\cos\beta z\ \boldsymbol{u}_y. \tag{22.2}$$

The instantaneous values of the two vectors are

$$E_{\text{total}}(z,t) = 2E\sqrt{2}\sin\beta z\sin\omega t\ \boldsymbol{u}_x, \tag{22.3}$$

$$H_{\text{total}}(z,t) = 2H\sqrt{2}\cos\beta z\cos\omega t\ \boldsymbol{u}_y. \tag{22.4}$$

The total waves are *standing waves*, not travelling waves (they do not contain a factor $\mathrm{e}^{\pm\mathrm{j}\beta z}$).

• If a plane wave is incident normally on a plane interface between two lossless media, there are two scattered waves, reflected and transmitted. Let the parameters of the two media be ϵ_1, μ_1, and ϵ_2, μ_2, and let the incident wave, of the electric field $E_{1\mathrm{i}}$ and of angular frequency ω, propagate in medium 1 towards the interface, with the vector $\boldsymbol{E}$ parallel to the x axis. Adopting the reference directions of the incident and two scattered electric field vectors to be in the $+x$-direction, the reflected and transmitted electric fields are given by

$$E_{1\mathrm{r}} = \frac{\eta_2 - \eta_1}{\eta_1 + \eta_2}E_{1\mathrm{i}}, \qquad E_2 = \frac{2\eta_2}{\eta_1 + \eta_2}E_{1\mathrm{i}}. \tag{22.5}$$

The ratio $E_{1\mathrm{r}}/E_{1\mathrm{i}}$ is known as the *reflection coefficient*, and the ratio $E_2/E_{1\mathrm{i}}$ the *transmission coefficient* (note the similarity to the transmission line ρ and τ):

$$\rho = \frac{\eta_2 - \eta_1}{\eta_1 + \eta_2}, \qquad \tau = \frac{2\eta_2}{\eta_1 + \eta_2} \qquad \text{(dimensionless)}. \tag{22.6}$$

• Note that ρ and τ are defined with respect to the same reference directions of all three components of the *electric field* of the three waves. In medium 2 there is only the progressive transmitted wave, while in medium 1 there are the incident and the reflected waves. The total electric field in medium 1 is

$$\boldsymbol{E}_1(z) = \boldsymbol{E}_{\mathrm{i}}(z) + \boldsymbol{E}_{\mathrm{r}}(z) = E_{1\mathrm{i}}\mathrm{e}^{-\mathrm{j}\beta_1 z}\left(1 + \rho\mathrm{e}^{+\mathrm{j}2\beta_1 z}\right)\ \boldsymbol{u}_x. \tag{22.7}$$

If $\rho > 0$ (i.e., if $\eta_2 > \eta_1$), the expression in the parenthesis is the largest, equal to $(1 + \rho)$, in planes defined by the equation (note that medium 1 occupies the half-space $z < 0$)

$$2\beta_1 z_{\max} = -2n\pi, \text{ or } z_{\max} = -\frac{n\lambda_1}{2}, \qquad n = 0, 1, \ldots. \tag{22.8}$$

This expression is minimal, equal to $(1 - \rho)$, in planes

$$z_{\min} = -(2n + 1)\frac{\lambda_1}{4}, \qquad n = 0, 1, \ldots . \tag{22.9}$$

If $\rho < 0$ (i.e., if $\eta_2 < \eta_1$), $z_{\max}$ and $z_{\min}$ simply exchange places. The ratio

$$\mathrm{SWR} = \frac{|E_1(z)|_{\max}}{|E_1(z)|_{\min}} = \frac{1 + |\rho|}{1 - |\rho|} \qquad \text{(dimensionless)} \tag{22.10}$$

is known as the *standing wave ratio* (SWR).

• If a plane wave is obliquely incident on a perfectly conducting plane, the plane containing the vector n and the directions of propagation of the incident and reflected waves is known as the *plane of incidence*. Any incident plane wave can be represented as a superposition of two plane waves, one with the vector E normal to the plane of incidence (*normal* or *horizontal polarization*), and the other with the vector E parallel to it (*parallel* or *vertical polarization*). In the case of normal polarization, with the vector E_{inc} in the direction of the x axis, the total electric field has only an x-component, given by

$$E_{\mathrm{total}}(y, z) = 2\mathrm{j}E \sin(\beta z \cos\theta)\, \mathrm{e}^{-\mathrm{j}\beta y \sin\theta}. \tag{22.11}$$

This is a standing wave in the z-direction, and a traveling wave in the y-direction. The wavelength in the z-direction is

$$\lambda_z = \frac{2\pi}{\beta \cos\theta} = \frac{\lambda}{\cos\theta}. \tag{22.12}$$

The vector E is zero in the planes

$$z_{E=0} = \frac{n\lambda}{2\cos\theta}, \qquad n = 0, 1, 2, \ldots . \tag{22.13}$$

In the direction of the y axis, the total field behaves as a traveling (progressing) wave, with a phase velocity and a wavelength along the y axis

$$v_{\mathrm{ph}} = \frac{\omega}{\beta_y} = \frac{c}{\sin\theta}, \qquad c = \frac{1}{\sqrt{\epsilon\mu}}, \tag{22.14}$$

$$\lambda_y = \frac{2\pi}{\beta_y} = \frac{\lambda}{\sin\theta}. \tag{22.15}$$

The total H-field has two components, H_y and H_z, given by

$$H_{\mathrm{total}\, y}(y, z) = -2\frac{E}{\eta} \cos\theta \cos(\beta z \cos\theta)\, \mathrm{e}^{-\mathrm{j}\beta y \sin\theta}, \tag{22.16}$$

$$H_{\text{total } z}(y,z) = 2\text{j}\frac{E}{\eta}\sin\theta\sin(\beta z\cos\theta)\text{e}^{-\text{j}\beta y\sin\theta}. \tag{22.17}$$

- For parallel polarization, the incident and the reflected waves have the y-component and the z-component of the electric field. The total components of the electric field are

$$E_{\text{total } y}(y,z) = 2\text{j}E\cos\theta\sin(\beta z\cos\theta)\text{e}^{-\text{j}\beta y\sin\theta}, \tag{22.18}$$

$$E_{\text{total } z}(y,z) = 2E\sin\theta\cos(\beta z\cos\theta)\text{e}^{-\text{j}\beta y\sin\theta}. \tag{22.19}$$

The total H-field has only the x-component,

$$H_{\text{total } x}(y,z) = 2\frac{E}{\eta}\cos(\beta z\cos\theta)\text{e}^{-\text{j}\beta y\sin\theta}. \tag{22.20}$$

- Let a plane wave be obliquely incident from medium 1 (of parameters ϵ_1 and μ_1) on a planar interface between medium 1 and medium 2 (of parameters ϵ_2 and μ_2). Assume that the direction of propagation of the incident wave makes an angle θ_1 with the normal to the interface. Then the direction of propagation of the reflected wave makes the same angle with respect to the normal. The wave in medium 2 is refracted, its angle being given by *Snell's law*,

$$\frac{\sin\theta_1}{\sin\theta_2} = \frac{c_1}{c_2} = \sqrt{\frac{\epsilon_2\mu_2}{\epsilon_1\mu_1}}. \tag{22.21}$$

The ratio $c_1/c_2 = n_{12}$ is the *index of refraction*. If $\mu_1 = \mu_2 = \mu_0$,

$$\frac{c_1}{c_2} = \sqrt{\frac{\epsilon_2}{\epsilon_1}} \qquad (\mu_1 = \mu_2). \tag{22.22}$$

For $\epsilon_1 < \epsilon_2$, we have $\theta_1 > \theta_2$. In the other case $\theta_1 < \theta_2$, so that if

$$\sin\theta_1 = \sqrt{\frac{\epsilon_2}{\epsilon_1}}, \tag{22.23}$$

there is no refracted wave, a phenomenon known as *total reflection*, with θ_1 being *the angle of total reflection.*

- The reflection and transmission coefficients for normal and parallel polarization of the incident wave are known as the *Fresnel coefficients*, and are given respectively by

$$\rho_{\text{n}} = \left(\frac{E_{1\text{r}}}{E_{1\text{i}}}\right)_{\text{n}} = \frac{\eta_2\cos\theta_1 - \eta_1\cos\theta_2}{\eta_2\cos\theta_1 + \eta_1\cos\theta_2}, \qquad \tau_{\text{n}} = \left(\frac{E_2}{E_{1\text{i}}}\right)_{\text{n}} = \frac{2\eta_2\cos\theta_1}{\eta_2\cos\theta_1 + \eta_1\cos\theta_2}, \tag{22.24}$$

$$\rho_{\text{p}} = \left(\frac{E_{1\text{r}}}{E_{1\text{i}}}\right)_{\text{p}} = \frac{\eta_1 \cos\theta_1 - \eta_2 \cos\theta_2}{\eta_1 \cos\theta_1 + \eta_2 \cos\theta_2}, \qquad \tau_{\text{p}} = \left(\frac{E_2}{E_{1\text{i}}}\right)_{\text{p}} = \frac{2\eta_2 \cos\theta_1}{\eta_1 \cos\theta_1 + \eta_2 \cos\theta_2}. \quad (22.25)$$

In these expressions, $\cos\theta_2$ must be calculated from Snell's law.

• If the two media are perfect dielectrics of equal permeabilities, the Fresnel's coefficients become

$$\rho_{\text{n}} = \frac{\cos\theta_1 - \sqrt{\epsilon_2/\epsilon_1}\,\cos\theta_2}{\cos\theta_1 + \sqrt{\epsilon_2/\epsilon_1}\,\cos\theta_2}, \qquad \tau_{\text{n}} = \frac{2\cos\theta_1}{\cos\theta_1 + \sqrt{\epsilon_2/\epsilon_1}\,\cos\theta_2}, \quad (22.26)$$

$$\rho_{\text{p}} = \frac{\sqrt{\epsilon_2/\epsilon_1}\,\cos\theta_1 - \cos\theta_2}{\sqrt{\epsilon_2/\epsilon_1}\,\cos\theta_1 + \cos\theta_2}, \qquad \tau_{\text{p}} = \frac{2\cos\theta_1}{\sqrt{\epsilon_2/\epsilon_1}\,\cos\theta_1 + \cos\theta_2}. \quad (22.27)$$

The reflection coefficient in the *parallel* polarization case is *zero* if

$$\tan\theta_1 = \sqrt{\epsilon_2/\epsilon_1}. \quad (22.28)$$

This angle θ_1 is known as the *Brewster*, or *polarization* angle.

QUESTIONS

S **Q22.1.** For what orientation and position of a small wire loop, Fig. Q22.1, is the emf induced in it maximal? — (a) *On the conducting plane, parallel to the plane.* (b) *On the conducting plane, normal to it, normal to vector E.* (c) *On the conducting plane, normal to it, normal to vector H.*

Answer. One of possible places is close to the conductor, where the magnetic field is maximal. The loop should be oriented so that the magnetic field is normal to its plane.

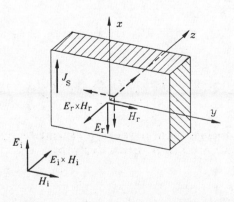

Fig. Q22.1. Normal incidence of plane wave.

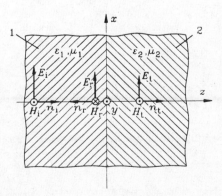

Fig. Q22.5. Interface between two dielectrics.

S **Q22.2.** Prove that the time-average value of the Poynting vector at any point in Fig. Q22.1 is zero. — *Hint: note that the total electric and magnetic field differ in phase by $\pi/2$.*

Answer. The electric and magnetic field differ in phase by $\pi/2$, and therefore the time-average of their product is zero. Another way of reaching this conclusion is to evaluate the time-average of the Poynting vector as the real part of the complex Poynting vector. It is zero because one of the vectors is multiplied by j, and the other is not.

Q22.3. If waves are represented in phasor form, how can you distinguish a standing wave from a traveling wave? — *(a) A standing wave does not contain a function of βz. (b) A standing wave does not have a factor of the form $e^{\mp j\beta z}$. (c) A standing wave contains a factor of the form $f(\beta z)$.*

Q22.4. If waves are represented in the time domain, how can you distinguish a standing wave from a traveling wave? — *(a) A standing wave has a time dependence of the form $(t - z/c)$, while a traveling wave does not. (b) A standing wave does not have a time dependence of the form $(t - z/c)$, while a traveling wave does. (c) A standing wave has a time dependence of the form $(t + z/c)$, while a traveling wave does not.*

S **Q22.5.** Does the emf induced in a small loop of area S placed in Fig. Q22.5 at a coordinate $z > 0$ depend on z? Does it depend on z if $z < 0$? Explain. — *(a) For $z > 0$ it does, for $z < 0$ it does not. (b) Depends on z in both cases. (c) For $z > 0$ it does not, for $z < 0$ it does.*

Answer. The points $z > 0$ are in medium 2, where we have only one forward wave. Therefore the emf induced in the loop does not depend on z if $z > 0$. If $z < 0$ (in medium 1), we have a superposition of two waves traveling in opposite directions, adding up to a forward plus a standing wave, and the emf induced in the loop depends on z for $z < 0$.

Q22.6. Is there a position of a small loop in Fig. Q22.6 in which the emf induced in it is practically zero irrespective of the orientation of the loop? — *(a) At the surface of the plane. (b) At $z = \lambda/(4\cos\theta)$. (c) Such a position of the loop does not exist.*

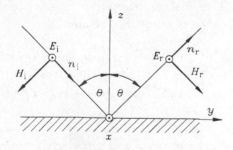

Fig. Q22.6 Oblique incidence of plane wave.

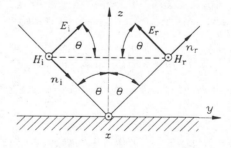

Fig. Q22.7. Oblique incidence of plane wave.

Q22.7. Repeat question Q22.6 for Fig. Q22.7. — *The answers are the same as for the preceding question.*

Q22.8. Is total reflection possible if a wave is incident from air onto a dielectric surface? Explain. — *(a) It is not. (b) It is. (c) It is, but only if the incident angle is greater than the angle of total reflection.*

Q22.9. Why is there no counterpart of the Brewster angle for a wave with vector E normal to the plane of incidence? — *(a) Because in that case the transmission coefficient is nonzero for all incident angles. (b) Because the reflection coefficient is nonzero for all incident angles. (c) Because the transmission coefficient is zero for all incident angles.*

S **Q22.10.** A linearly polarized plane wave is incident from air onto a dielectric half-space, with the vector E at an angle α $(0 < \alpha < \pi/2)$ with respect to the plane of incidence. Is the polarization of the transmitted and reflected wave linear? If not, what is the polarization of the two waves? Does it depend, for a given α, on the properties of the dielectric medium? — *Hint: use superposition. (a) It is linear. (b) It is circular. (c) It is elliptical.*

Answer. We can represent such a wave as a sum of two waves, one polarized normal, and the other parallel, with respect to the plane of incidence. The reflection and transmission coefficients for the two component waves are different. If the two media are perfect dielectrics, all the coefficients are *real*, and the two components are in phase. Consequently, they combine in a wave whose polarization is linear. If one of the media is lossy, or if both are lossy, this is not the case, and both the reflected and transmitted waves are elliptically polarized.

PROBLEMS

S **P22.1.** A linearly polarized plane wave of rms electric field strength E and angular frequency ω is normally incident from a vacuum on a large, perfectly conducting flat sheet. Determine the induced surface charges and currents on the sheet. — *With reference to Fig. Q22.1, convince yourself that $\sigma(t) = 0$, and $J_s(t) = 2\,(E/\eta_0)\,\sqrt{2}\cos\omega t\,\mathbf{u}_x$,.*

Solution. With reference to Fig. Q22.1, the instantaneous values of resultant electric and magnetic field intensity vectors in a vacuum, $E_{\text{total}}(z,t)$ and $H_{\text{total}}(z,t)$, which represent a standing wave, are given in Eqs.(22.5) and (22.6). The induced surface charges and currents on the sheet surface are obtained by using the corresponding boundary conditions,

$$\sigma(t) = \epsilon_0 \mathbf{n}\cdot E_{\text{total}}(0,t) = 0, \qquad J_s(t) = \mathbf{n}\times H_{\text{total}}(0,t) = 2\,\frac{E}{\eta_0}\,\sqrt{2}\cos\omega t\,\mathbf{u}_x,$$

where $\mathbf{n} = -\mathbf{u}_z$ is the unit vector normal to the sheet, and directed towards the vacuum.

P22.2. Note that the induced surface currents in problem P22.1 are situated in the magnetic field of the incident wave. Determine the time-average force per unit area (the pressure) on the sheet. (This is known as *radiation pressure*). — *(a) $p_{\text{ave}} = \epsilon_0 E^2$. (b) $p_{\text{ave}} = \epsilon_0 E^2/2$. (c) $p_{\text{ave}} = 2\epsilon_0 E^2$.*

P22.3. Repeat problems P22.1 and P22.2 assuming the wave is polarized circularly. — *(a) $p_{\text{ave}} = 4\mu_0 H^2$. (b) $p_{\text{ave}} = 2\mu_0 H^2$. (c) $p_{\text{ave}} = \mu_0 H^2$.*

S **P22.4.** If the conductivity σ of the sheet in problem P22.1 is large, but finite, its permeability is μ, and the frequency of the wave is ω, find the time-average power losses in the sheet per unit area. Specifically, find these losses if $f = 1\,\text{MHz}$, $E = 1\,\text{V/m}$, $\sigma = 56\cdot 10^6\,\text{S/m}$ (copper), and $\mu = \mu_0$. — *(a) $dP_J/dS = 15.0\,nW/m^2$. (b) $dP_J/dS = 7.5\,nW/m^2$. (c) $dP_J/dS = 22.5\,nW/m^2$.*

Solution. According to Eq. (20.9), the only quantity necessary for determining the time-average power of Joule's losses in the conducting sheet is the rms value H_0 of the total tangential magnetic field at the conductor surface. Since the conductor is assumed to be very good, H_0 is approximately the same as *if the conductor were perfect*. Hence, the time-average power losses per unit area are given by

$$\frac{\mathrm{d}P_\mathrm{J}}{\mathrm{d}S} = R_\mathrm{s}|H_0|^2 = R_\mathrm{s}\left(\frac{2E}{\eta_0}\right)^2 = 4\,\frac{\epsilon_0}{\mu_0}\sqrt{\frac{\pi\mu f}{\sigma}}\,E^2 = 7.5\,\mathrm{nW/m^2},$$

where R_s stands for the surface resistance of the sheet.

P22.5. A plane wave, of wavelength λ, is normally incident from a vacuum on a large, perfectly conducting sheet. A circular loop of radius a ($a \ll \lambda$) should be at a location at which the induced emf is maximal, as near as possible to the sheet. If the electric field of the incident wave is E, calculate this maximal emf. — (a) $\mathcal{E}_\mathrm{max} = 4\pi^2 a^2 E/\lambda$. (b) $\mathcal{E}_\mathrm{max} = 2\pi^2 a^2 E/\lambda$. (c) $\mathcal{E}_\mathrm{max} = \pi^2 a^2 E/\lambda$.

P22.6. Assume that a time-harmonic surface current of density $J_{\mathrm{s}x} = J_{\mathrm{s}0}\cos\omega t$ exists over an infinitely large plane sheet. Write the integral expression for the electric field strength vector at a distance z from the sheet. Do not evaluate the integral, but reconsider problem P22.1 to see if you know what the result must be. — *Referring to Fig. Q22.1, convince yourself that the electric field strength at a point $(0,0,z)$, where $z < 0$, due to the currents over the sheet, is*

$$\boldsymbol{E}(z,t) = -\mu_0 \int_{-\infty}^{\infty}\int_{-\infty}^{\infty} \frac{\partial J_{\mathrm{s}y}(t - R/c_0)/\partial t}{R}\,\mathrm{d}x\,\mathrm{d}y\,\boldsymbol{u}_x, \qquad R = \sqrt{x^2 + y^2 + z^2}.$$

What must the value of this integral be?

P22.7. A linearly polarized plane wave, of frequency $f = 1\,\mathrm{MHz}$, is normally incident from a vacuum on the plane surface of distilled water ($\mu = \mu_0$, $\epsilon = 81\epsilon_0$, $\sigma \simeq 0$). The rms value of the electric field strength of the incident wave is $E = 100\,\mathrm{mV/m}$. A loop of area $S = 100\,\mathrm{cm^2}$ wound with $N = 5$ turns is situated in water so that the emf induced in it is maximal. Determine the rms value of the emf. — (a) $\mathcal{E} = 3.88\,\mathrm{V}$. (b) $\mathcal{E} = 3.88\,\mathrm{V}$. (c) $\mathcal{E} = 1.88\,\mathrm{V}$.

S **P22.8.** A plane wave propagating in dielectric 1, of permittivity ϵ_1 and permeability μ_1, impinges normally on a dielectric slab 2, of permittivity ϵ_2, permeability μ_2, and thickness d. To the right of the slab there is a semi-infinite medium of permittivity ϵ_3 and permeability μ_3. Determine the reflection coefficient at the interface between media 1 and 2. Plot the reflection coefficient as a function of the slab thickness, d, for given permittivities. Consider cases when (1) $\epsilon_1 > \epsilon_2 > \epsilon_3$, (2) $\epsilon_3 > \epsilon_2 > \epsilon_1$, (3) $\epsilon_2 > \epsilon_1 > \epsilon_3$, and (4) $\epsilon_2 > \epsilon_3 > \epsilon_1$. — *Hint: refer to Fig. P22.8, and write the boundary conditions on the two boundaries. The starting equations are relatively simple, but the evaluation of the coefficients is not straightforward.*

Solution. With reference to Fig. P22.8, the boundary conditions in Eqs. (19.10) and (19.12) tell us that

$$E_\mathrm{i} + E_\mathrm{r} = E_{2\mathrm{i}} + E_{2\mathrm{r}}, \qquad \frac{E_\mathrm{i}}{\eta_1} - \frac{E_\mathrm{r}}{\eta_1} = \frac{E_{2\mathrm{i}}}{\eta_2} - \frac{E_{2\mathrm{r}}}{\eta_2},$$

$$E_{2\mathrm{i}}\,e^{-j\beta_2 d} + E_{2\mathrm{r}}\,e^{j\beta_2 d} = E_3, \qquad \frac{E_{2\mathrm{i}}}{\eta_2}\,e^{-j\beta_2 d} - \frac{E_{2\mathrm{r}}}{\eta_2}\,e^{j\beta_2 d} = \frac{E_3}{\eta_3}.$$

After a somewhat lengthy, but elementary, calculation we find that the reflection coefficient at the interface between medium 1 and the *equivalent medium* consisting of media 2 and 3 is

$$\rho_{1,e} = \frac{\eta_e - \eta_1}{\eta_1 + \eta_e},$$

where η_e is the *equivalent intrinsic impedance* of media 2 and 3 (with respect to medium 1), given by

$$\eta_e = \eta_2 \frac{\eta_3 \cos \beta_2 d + j \eta_2 \sin \beta_2 d}{\eta_2 \cos \beta_2 d + j \eta_3 \sin \beta_2 d}.$$

P22.9. Assume that in the preceding problem the thickness of the slab is (1) half a wavelength, and (2) a quarter of a wavelength in the slab. Determine the relationship between the intrinsic impedances of the three media for which in the two cases there will be no reflected wave into medium 1. (The first of these conditions is used in antenna covers, called radomes. The second is used in optics, for so-called anti-reflection, or AR, coatings. The thickness and relative permittivity of a thin transparent layer over lenses can be designed in this way so that the reflection of light from the lens is minimized.) — *(a) In case 1, $\eta_1 = \eta_3$, in case 2 $\eta_2 = \sqrt{\eta_1 \eta_3}$. (b) $\eta_1 = \eta_2$, $\eta_2 = \eta_3$. (c) $\eta_1 = \sqrt{\eta_2 \eta_3}$, $\eta_2 = \eta_3$.*

P22.10. Find the reflection and transmission coefficients for the interface between air and fresh water ($\epsilon = 81\epsilon_0$, $\sigma \simeq 0$), in the case of perpendicular incidence. — *(a) $\rho = -0.6$, $\tau = 0.4$. (b) $\rho = -0.68$, $\tau = 0.32$. (c) $\rho = -0.8$, $\tau = 0.2$.*

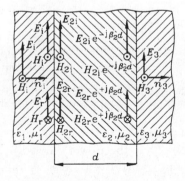

Fig. P22.8. A dielectric slab in plane wave.

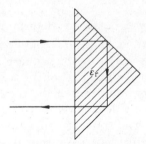

Fig. P22.14. Reflection of light by a prism.

P22.11. A plane wave is normally incident on the interface between air and a dielectric having a permeability $\mu = \mu_0$, and an unknown permittivity ϵ. The measured standing-wave ratio in air is 1.8. Determine ϵ. — *(a) $\epsilon = 2.58\,\epsilon_0$. (b) $\epsilon = 3.24\,\epsilon_0$. (c) $\epsilon = 2.20\,\epsilon_0$.*

P22.12. What is the position of a small loop of area S in Fig. Q22.7 in order that the emf induced in it be maximal? If the electric field of the wave is E and its frequency f, calculate this maximal emf. — *Hint: see Eqs. (22.16) and (22.17). The plane of the loop can be normal to either the y axis, or to the z axis. Assume it is normal to the z axis. (a) $z_n = n\lambda/(2\cos\theta)$, $\mathcal{E}_{max} = 4\pi f\sqrt{\epsilon\mu}\,ES$. (b) $z_n = n\lambda/(\cos\theta)$, $\mathcal{E}_{max} = 2\pi f\sqrt{\epsilon\mu}\,ES$. (c) $z_n = n\lambda/(4\cos\theta)$, $\mathcal{E}_{max} = 4\pi f\sqrt{\epsilon\mu}\,ES$. Solve the problem also if the loop is normal to the y axis!*

P22.13. Repeat problem P22.12 for a small loop placed in the wave in Fig. Q22.6. — *Hint: start from Eqs. (22.16) and (22.17).*

P22.14. Determine the minimal relative permittivity of a dielectric medium for which the critical angle of total reflection from the dielectric into air is less than 45 degrees. Is it possible to make from such a dielectric a right-angled isosceles triangular prism that returns the light wave as in Fig. P22.14? Is there reflection of the light wave when it enters the prism? — *(a)* $\epsilon_r = 2.2$. *(b)* $\epsilon_r = 2.1$. *(c)* $\epsilon_r = 2$. *Is there a relatively small reflection when the wave enters the prism and leaves it?*

S **P22.15.** A plane wave with parallel polarization is incident at an angle of $\pi/4$ from air on a perfect dielectric with $\epsilon_r = 4$ and $\mu = \mu_0$. Find the Fresnel coefficients. What fraction of the incident power is reflected, and what is transmitted into the dielectric? More generally, plot the Fresnel coefficients and the reflected and transmitted power as a function of ϵ_r, assuming its value is between 1 and 80. — *Hint: use Eq. (22.27).*

Solution. The Fresnel coefficients are given in Eq. (22.46). For $\epsilon_1 = \epsilon_0$ the coefficients become

$$\rho_p = \frac{\sqrt{\epsilon_r}\cos\theta_1 - \cos\theta_2}{\sqrt{\epsilon_r}\cos\theta_1 + \cos\theta_2}, \qquad \tau_p = \frac{2\cos\theta_1}{\sqrt{\epsilon_r}\cos\theta_1 + \cos\theta_2} \qquad (\mu_1 = \mu_2).$$

The angle θ_1 is given, and $\cos\theta_2$ is obtained from $\sin\theta_2$ determined from Snell's law:

$$\sin\theta_2 = \frac{1}{\sqrt{\epsilon_r}}\sin\theta_1, \qquad \cos\theta_2 = \sqrt{1 - \sin^2\theta_2}.$$

It is a relatively simple matter to plot the Fresnel coefficients as functions of ϵ_r.

The power balance is determined by means of the Poynting vector. The Poynting vector of the incident wave is $\mathcal{P}_i = E_i^2/\eta_1$, that of the reflected wave is $\rho_p^2 \mathcal{P}_i$, and that of the transmitted wave is $\eta_1 \tau_p^2 \mathcal{P}_i/\eta_2$. Consider one square meter of the wave front of the incident wave. The reflected wave corresponding to this part of the wave front has the same area (of one square meter). The area of the corresponding wavefront of the transmitted wave, however, is $\cos\theta_2/\cos\theta_1$ square meters. (A square meter of the incident wave front illuminates $1/\cos\theta_1$ square meters of the interface, and it is projected to the wave front of $\cos\theta_2$ this area for the transmitted wave.) Thus the power carried by a square meter of the incident wave front is split between the reflected and transmitted waves as follows:

$$P_{\text{reflected}} = \rho_p^2 P_{\text{incident}}, \qquad P_{\text{transmitted}} = \frac{\eta_1}{\eta_2}\frac{\cos\theta_2}{\cos\theta_1}\tau_p^2 P_{\text{incident}}.$$

For $\epsilon_r = 4$ the Fresnel coefficients are $\rho_p = 0.204$ and $\tau_p = 0.602$. Since $\cos\theta_2 = 0.935$, we have

$$P_{\text{reflected}} = 0.0416 P_{\text{incident}}, \qquad P_{\text{transmitted}} = 0.9585 P_{\text{incident}}.$$

It is seen that the two add up to the power of the incident wave (except for a small roundoff error), as they should.

P22.16. Repeat problem P22.15 for a normally polarized wave. — *Hint: use Eq. (22.26). The solution parallels that of the preceding problem.*

P22.17. A plane wave with normal polarization is incident at an angle of 60° from air onto deep fresh water with $\epsilon_r = 81$ ($\sigma = 0$). The rms value of the incident electric field is $1\,\mathrm{V/m}$. Find the rms value of the reflected and transmitted electric field. — (a) $\rho_n = 0.8943$, $\tau_n = 0.0529$. (b) $\rho_n = -0.8943$, $\tau_n = 0.0529$. (c) $\rho_n = -0.7943$, $\tau_n = 0.1529$.

P22.18. Repeat problem P22.17 for parallel polarization. — (a) $\rho_p = 0.6377$, $\tau_p = 0.1820$. (b) $\rho_p = 0.5422$, $\tau_p = 0.2731$. (c) $\rho_p = -0.7023$, $\tau_p = 0.1065$.

P22.19. Is there an incident angle in problems P22.17 and P22.18 for which the reflected wave is eliminated? If so, calculate this angle for the two polarizations. — (a) *Such an angle exists for both polarizations, and* $\theta_{1p} = 76.4°$, $\theta_{1n} = 67.2°$. (b) *The angle exists only for parallel polarization, and* $\theta_{1p} = 83.7°$. (c) *The angle exists only for normal polarization, and* $\theta_{1n} = 78.3°$.

23. Waveguides and Resonators

● Maxwell's equations predict that electromagnetic waves can be guided *through hollow metallic tubes*, known as *waveguides*. It turns out that the components of the field transverse to the direction of the waveguide can be expressed in terms of the components of vectors E and H along the waveguide. Referring to Fig. P23.5a (p. 246), the x- and y-components of the two vectors are given by

$$E_x = -\frac{1}{K^2}\left(\gamma\frac{\partial E_z}{\partial x} + j\omega\mu\frac{\partial H_z}{\partial y}\right), \tag{23.1}$$

$$E_y = -\frac{1}{K^2}\left(\gamma\frac{\partial E_z}{\partial y} - j\omega\mu\frac{\partial H_z}{\partial x}\right), \tag{23.2}$$

$$H_x = -\frac{1}{K^2}\left(-j\omega\epsilon\frac{\partial E_z}{\partial y} + \gamma\frac{\partial H_z}{\partial x}\right), \tag{23.3}$$

$$H_y = -\frac{1}{K^2}\left(j\omega\epsilon\frac{\partial E_z}{\partial x} + \gamma\frac{\partial H_z}{\partial y}\right), \tag{23.4}$$

where

$$K^2 = \gamma^2 + \beta^2, \qquad \beta^2 = \omega^2\epsilon\mu. \tag{23.5}$$

The propagation coefficient γ (and hence also the parameter K) are determined from boundary conditions for a specific waveguide.

● The solutions of these equations are possible only for certain distinct values of the parameter K, which in turn depend on the boundary conditions. These values of K are known as its *eigenvalues*, or *characteristic values*.

• The most common waveguides are those of rectangular cross section. While transmission lines can support TEM (transverse electromagnetic) waves, hollow metallic waveguides support only the *transverse electric*, or TE, waves (with a nonzero component of vector $\boldsymbol{H}$ in the direction of propagation), and *transverse magnetic*, or TM, waves. TE and TM waves can propagate only if their frequency is above a certain critical frequency, and their velocity depends on frequency. Among numerous applications, closed sections of waveguides are used as electromagnetic *cavity resonators*.

• If a rectangular waveguide extends along the z axis (Fig. P23.5a, p.246), a TE wave propagating in the $+z$-direction has a propagation coefficient of the form

$$\gamma = \mathrm{j}\beta, \qquad \beta = \omega\sqrt{\epsilon\mu}\sqrt{1 - \frac{f_c^2}{f^2}}, \tag{23.6}$$

where

$$f_c = \frac{c}{2}\sqrt{\left(\frac{m}{a}\right)^2 + \left(\frac{n}{b}\right)^2}, \qquad m, n = 0, 1, 2, \ldots, \qquad c = \frac{1}{\sqrt{\epsilon\mu}}, \tag{23.7}$$

is the *cutoff frequency* of the wavetype corresponding to a particular pair (m, n). This wavetype is known as the TE$_{mn}$ mode. Note that the wave propagates without attenuation (assuming no losses) only if γ is imaginary, i.e., if $f > f_c$. Such modes are known as *propagating modes*. If γ is real (i.e., if $f > f_c$), the field amplitudes are attenuated along the z axis exponentially. Such modes are called *evanescent modes*.

• All components of TE waves are derived from the z-component of vector $\boldsymbol{H}$. The final expressions for the TE$_{mn}$ wave at $z = 0$ are:

$$H_z(x, y) = H_0 \cos\left(\frac{m\pi}{a}x\right)\cos\left(\frac{n\pi}{b}y\right) \quad (\text{at } z = 0), \tag{23.8}$$

$$E_x(x, y) = \frac{\mathrm{j}\omega\mu}{K^2}\frac{n\pi}{b}H_0 \cos\left(\frac{m\pi}{a}x\right)\sin\left(\frac{n\pi}{b}y\right) \quad (\text{at } z = 0), \tag{23.9}$$

$$E_y(x, y) = -\frac{\mathrm{j}\omega\mu}{K^2}\frac{m\pi}{a}H_0 \sin\left(\frac{m\pi}{a}x\right)\cos\left(\frac{n\pi}{b}y\right) \quad (\text{at } z = 0), \tag{23.10}$$

$$H_x(x, y) = \frac{\gamma}{K^2}\frac{m\pi}{a}H_0 \sin\left(\frac{m\pi}{a}x\right)\cos\left(\frac{n\pi}{b}y\right) \quad (\text{at } z = 0), \tag{23.11}$$

$$H_y(x, y) = \frac{\gamma}{K^2}\frac{n\pi}{b}H_0 \cos\left(\frac{m\pi}{a}x\right)\sin\left(\frac{n\pi}{b}y\right) \quad (\text{at } z = 0). \tag{23.12}$$

The constant H_0 depends on the level of excitation, and the values of the wave components for any z are obtained by multiplying the above expressions by $\mathrm{e}^{-\gamma z} = \mathrm{e}^{-\mathrm{j}\beta z}$.

- The *phase* and *group velocity* of the TE$_{mn}$ mode are given by

$$v_{\text{ph}} = \frac{\omega}{\beta} = \frac{c}{\sqrt{1 - f_c^2/f^2}}, \qquad (23.13)$$

$$v_{\text{g}} = c\sqrt{1 - f_c^2/f^2}. \qquad (23.14)$$

Since the phase (and group) velocity depend on frequency, rectangular waveguides are *dispersive structures*. Only waves of frequencies $f > f_c$ can propagate along a waveguide.

- Travelling TEM, TE and TM waves have a common property that the ratio of transverse components of the electric and magnetic field is constant in the cross section of the system. This ratio is known as the *wave impedance*, and for the direct wave of the three wave types is given by

$$Z_{\text{TEM,TE,TM}} = \frac{E_x}{H_y} = -\frac{E_y}{H_x}. \qquad (23.15)$$

In specific,

$$Z_{\text{TEM}} = \sqrt{\frac{\mu}{\epsilon}}, \qquad Z_{\text{TE}} = \frac{j\omega\mu}{\gamma}, \qquad Z_{\text{TM}} = \frac{\gamma}{j\omega\epsilon}. \qquad (23.16)$$

- The TE$_{10}$ mode is known as the *dominant mode*, because it has the lowest cutoff frequency and is the only mode that can propagate in the frequency range from its cutoff frequency to the next higher cutoff frequency. It is the most commonly used mode in rectangular waveguides, and its field components are

$$H_z(x,y) = H_0 \cos\left(\frac{\pi}{a}x\right) \quad \text{(TE$_{10}$ mode)}, \qquad (23.17)$$

$$E_x(x,y) = E_z(x,y) = H_y(x,y) = 0 \quad \text{(TE$_{10}$ mode)}, \qquad (23.18)$$

$$E_y(x,y) = -j\omega\mu\frac{a}{\pi}H_0\sin\left(\frac{\pi}{a}x\right) \quad \text{(TE$_{10}$ mode)}, \qquad (23.19)$$

$$H_x(x,y) = j\beta\frac{a}{\pi}H_0\sin\left(\frac{\pi}{a}x\right) \quad \text{(TE$_{10}$ mode)}. \qquad (23.20)$$

The cutoff frequency of the TE$_{10}$ mode is

$$f_{c\ \text{TE10}} = \frac{c}{2a}, \qquad (23.21)$$

and the phase velocity, the wave impedance, etc., for the TE$_{10}$ mode are obtained from the general expressions using this cutoff frequency.

- The wavelength along the z axis of a waveguide is $\lambda_z = 2\pi/\beta$, or

$$\lambda_z = \frac{\lambda_0}{\sqrt{1 - f^2/f_c^2}}, \tag{23.22}$$

where λ_0 is the wavelength of a plane wave of the same frequency in the medium of parameters ϵ and μ.

- The power transmitted by a forward wave through any waveguide is obtained as the flux of the Poynting vector over the waveguide cross section. For the TE_{10} mode this yields

$$P_{TE10} = \frac{ab}{2}\sqrt{\frac{\mu}{\epsilon}}\frac{f^2}{f_c^2}\sqrt{1 - \frac{f_c^2}{f^2}}\,|H_0|^2. \tag{23.23}$$

- All TM waves in a rectangular waveguide are obtained as derivatives with respect to the x- and y-coordinates from their E_z-component. The expressions for the field components are similar to those for TE modes, and the critical frequencies and the propagation coefficients are the same as for TE modes. The lowest TM mode is the TM_{11} mode.

- Rectangular waveguides can support either TE modes, or TM modes (or, of course, a combination of these). However, some guiding structures can support *only* wave types that are a combination of TEM, TE and/or TM modes. These wave types are called *hybrid* modes. An example of such a waveguide is the *microstrip line*. It consists of a dielectric substrate with metallization on one side, and a metallic strip on the other. It can be shown that in such a waveguide both E_z and H_z are always present, so a microstrip line supports only hybrid modes. However, the components of the electric and magnetic field vectors along the direction of propagation are small compared to the other components. It is therefore said that a microstrip line supports a "quasi-TEM" mode (almost a TEM mode). Thus it is possible to define a characteristic impedance and a propagation constant, and then use TEM mode, or transmission-line equations. These line parameters are expressed in terms of an *effective dielectric constant*, which depends on the relative permittivity and thickness of the substrate.

- Classical resonant circuits with lumped elements cannot be used above about 100 MHz, due to Joule's losses and radiation. Instead, between about 500 MHz and 3 GHz, resonators are usually in the form of shorted segments of shielded transmission lines, and above about 3 GHz to few tens of GHz in the form of metallic boxes of various shapes, known as *cavity resonators*. At still higher frequencies we use so-called Fabry-Perot resonators, consisting of two parallel, highly polished metal plates. The basic parameters of an electromagnetic resonator are its *resonant frequency*, f_r, the type of wave inside it, and its *quality factor, Q*.

- The Q-factor is defined in terms of the electromagnetic energy contained in the resonator, W_{em}, and the total energy lost in one cycle, $W_{lost/cycle}$, in the resonator containing this energy, as

$$Q = 2\pi\frac{W_{em}}{W_{lost/cycle}} = \omega_r\frac{W_{em}}{P_{losses}} \qquad \text{(dimensionless)}. \tag{23.24}$$

This general definition is valid also for resonant circuits.

• The simplest resonant cavity is in the form of a parallelepiped (rectangular box). It can be obtained by introducing appropriate short circuits (transverse metallic walls) into a rectangular waveguide, and the electromagnetic field in it as a sum of appropriate incident and reflected waves in the waveguide. In the case of incident TE_{mn} wave, if the short circuits are p half wavelenghts along the z axis apart, we say that there is a TE_{mnp} mode in the cavity. For the TE_{101} mode in the cavity, the field components are found to be

$$E_{y \ \text{res}}(x, y, z) = -2\omega\mu\frac{a}{\pi}H_0 \sin\left(\frac{\pi}{a}x\right)\sin\left(\frac{\pi}{d}z\right), \tag{23.25}$$

$$H_{x \ \text{res}}(x, y, z) = 2\text{j}\frac{a}{d}H_0 \sin\left(\frac{\pi}{a}x\right)\cos\left(\frac{\pi}{d}z\right), \tag{23.26}$$

$$H_{z \ \text{res}}(x, y, z) = -2\text{j}H_0 \cos\left(\frac{\pi}{a}x\right)\sin\left(\frac{\pi}{d}z\right). \tag{23.27}$$

QUESTIONS

S **Q23.1.** Write the instantaneous value of the E_{total}-field given by the expression $E_{\text{total}}(x, y, z) = E(x, y)e^{-\gamma z}$, where $\gamma = \alpha + \text{j}\beta$. — (a) $E(x,y)e^{\alpha z}\cos(\omega t + \beta z)$. (b) $E(x,y)e^{-\alpha z}\cos(\omega t - \beta z)$. (c) $E(x,y)e^{-\alpha z}\cos(\omega t + \beta z)$.

Answer. $E_{\text{total}}(x, y, z, t) = E(x, y)e^{-\alpha z}\cos(\omega t - \beta z)$.

Q23.2. Complete the derivation of Eqs. (23.15) and (23.16) for TEM waves. — *Hint: use Eqs. (23.1)-(23.4), the expression for the propagation coefficient of TEM waves.*

Q23.3. Define in your own words the TEM, TE, and TM waves. What does "mode" mean? *Hint: refer to the definitions in the introduction to this chapter.*

Q23.4. Can the complex propagation coefficient γ in Eq. (23.6) be real? Can it have a real part? — *(a) It can never be real. (b) It is real if $f > f_c$. (c) It is real if $f < f_c$.*

Q23.5. What are eigenvalues (characteristic values) of a parameter in a boundary-value problem? What do they depend on? — *(a) Values of a parameter for which boundary conditions can be satisfied. (b) Values of a parameter typical for a problem. (c) Values of a parameter defined by the user.*

S **Q23.6.** The wave impedance of a TEM wave is always real. Are the wave impedances of TE and TM wave also always real? Explain. — *(a) Only if the propagation coefficient is real. (b) Only if the propagation coefficient is imaginary. (c) It is always imaginary.*

Answer. They are real only if the propagation coefficient is purely imaginary, i.e., if in the guide there is only a wave propagating without attenuation.

Q23.7. Under which conditions is the relation $Z_{\text{TE}}Z_{\text{TM}} = Z_{\text{TEM}}^2 = \mu/\epsilon$ valid? — *(a) It is valid in all cases. (b) It is valid if the propagation coefficients of the TE and TM waves are the same, and if both are either forward or backward waves. (c) The only condition is that the propagation coefficients of the TE and TM waves are the same.*

Q23.8. What is the physical meaning of the coefficients m and n in the field components inside a rectangular waveguide in Eqs. (23.8) to (23.12)? — *(a) There is no physical meaning. (b) They tell us how many half-wavelengths there are in the x- and y-directions. (c) They tell us how many wavelengths there are in the x- and y-directions.*

S **Q23.9.** What is the phase and group velocity in a rectangular waveguide in these three cases? (1) $f < f_c$, (2) $f = f_c$, and (3) $f > f_c$. — *Convince yourself that: (1) Both are imaginary, meaning no propagation. (2) Phase velocity is infinite, group velocity zero. (3)* $v_{\mathrm{ph}} = c/\sqrt{1 - f_c^2/f^2}$, $v_{\mathrm{g}} = c\sqrt{1 - f_c^2/f^2}$.

Answer. (1) Both are imaginary, meaning no propagation. (2) Phase velocity is infinite, group velocity zero. (3) $v_{\mathrm{ph}} = c/\sqrt{1 - f_c^2/f^2}$, $v_{\mathrm{g}} = c\sqrt{1 - f_c^2/f^2}$.

Q23.10. What is the attenuation constant in a rectangular waveguide in these three cases? (1) $f < f_c$, (2) $f = f_c$, and (3) $f > f_c$. — *Convince yourself that: (1)* $\alpha = \omega\sqrt{\epsilon\mu}\sqrt{f_c^2/f^2 - 1}$. *(2) Zero. (3) Zero.*

Q23.11. What are the parameters that determine the cutoff frequency in a waveguide? — *(a) The numbers m and n. (b) The numbers m and n, and the properties of the dielectric in the waveguide. (c) The numbers m and n, the properties of the dielectric, the lengths of the two waveguide sides.*

S **Q23.12.** A signal consisting of frequencies in the vicinity of a frequency f_1, and a signal consisting of frequencies in the vicinity of a frequency f_2, propagate unattenuated along a rectangular waveguide in the TE_{10} mode. If $f_1 < f_2$, which is faster? — *(a) That of higher frequency. (b) That of lower frequency. (c) The velocity of the two is the same.*

Answer. Since the group velocity is the one determining the velocity of the signals, the signal at the higher frequency is faster.

Q23.13. What will eventually happen with the signals from the preceding question if the waveguide is long? — *(a) They will only be attenuated. (b) They will become broader and distorted. (c) They will become compressed and distorted.*

Q23.14. A signal consisting of frequencies in the vicinity of a frequency f_1 propagates along a rectangular waveguide as a TE_{10} mode. What happens if the bandwidth of the signal is relatively large? — *(a) Its shape will change considerably. (b) Its magnitude will change, but not its shape. (c) It does not matter how large the bandwidth is.*

Q23.15. What are *propagating modes* and *evanescent modes* in a waveguide? — *(a) Propagating modes are those that propagate, evanescent modes are transients after the field has been turned on. (b) Propagating modes are attenuated only due to possible losses, evanescent modes are those that are attenuated due to the waveguide geometry. (c) Propagating modes are those that propagate attenuated, evanescent modes are transients after the field has been switched off.*

S **Q23.16.** You would like to have openings for airing a shielded room (a Faraday's cage) without enabling electromagnetic energy to enter or leave the cavity. You are aware that a field of a certain microwave frequency is particularly pronounced around the room, but you do not know its polarization. Can you make the openings in the form of waveguide sections? What profile of the waveguide would you use? — *Convince yourself in the following: it is possible to make such* *openings using waveguide sections with evanescent mode. A circular waveguide is preferable (why?).*

Answer. It is possible to make such openings, with the transverse dimension of the waveguide and its length such that the signal could only be an evanescent mode, and that the attenuation of the adopted waveguide length is sufficient. A circular waveguide is best to use, since we do not know the polarization of the undesired field.

Q23.17. You are using a square waveguide that is bent and twisted along its way. The waveguide is excited with the TE_{10} mode (the E field parallel to the y axis). Can you be certain about the polarization of the wave at the receiving point? Explain. — *(a) Yes, it will remain the same. (b) No, the TE_{01} mode will also be excited. (c) No, an infinite number of higher-order modes will also be excited and propagate along the guide.*

Q23.18. What is the physical meaning of the *dominant mode* in a waveguide? — *(a) The simplest mode. (b) The one which dominates the others, i.e., is of the highest amplitude. (c) The only one which can propagate unattenuated along a waveguide in a certain frequency range.*

Q23.19. A rectangular waveguide along which waves of many frequencies and modes propagate is terminated in a large metal box. Can you extract from the box a signal of a specific frequency and a desired mode by connecting a section of the same waveguide at another point of the box? — *(a) Not with the same waveguide. (b) Yes. (c) Not with any waveguide.*

Q23.20. How would you construct a high-pass filter (i.e., a filter transmitting only frequencies above a certain frequency), using sections of rectangular waveguides? — *Find three errors in the following answer: the waveguide sections should operate in their first dominant mode, and the size of the waveguides should be such that the TE_{00} cutoff frequency corresponds to half the specified low frequency limit.*

Q23.21. Propose a method for exciting the TE_{11} mode in a rectangular waveguide. — *Convince yourself that the following is correct: a possibility is to use two coaxial feeders connected at the centers of two adjacent waveguide sides, about a quarter of a wavelength along the guide from the guide short circuit. The outer cable conductors should be connected to the waveguide, and the inner protrude slightly through a small hole drilled in one and the other waveguide wall. The two signals need to be in phase.*

S **Q23.22.** You would like for a rectangular waveguide with a TE_{10} wave *radiate* (leak) from a series of narrow slots you made in its walls. For this, you need slots that would force the internal waveguide currents to appear on its outer surface. How do the slots need to be oriented to accomplish this? — *(a) Along the center line of the guide broad side. (b) Transverse to the guide, on its narrow side. (c) On its narrow side, along the z axis or inclined at an angle with respect to the transversal plane.*

Answer. To answer this question, it is necessary to inspect the picture of the surface currents on the waveguide walls. It is seen that on the narrower side the slots must be along the z axis, or at least inclined; a slot normal to the z axis on that side would radiate very little. On the wider side, the slots must not be along the side axis.

Q23.23. Sketch the electric and magnetic field lines for two microstrip lines, one with a substrate twice the thickness of the other, but with the same permittivity. In which case is the quasi-TEM approximation more accurate? Explain. —*Hint: note that the fringing field is more pronounced for the thicker substrate.*

Q23.24. Sketch the electric and magnetic field lines for two microstrip lines on substrates of equal thickness, but where one has a permittivity two times higher than the other. In which case is the quasi-TEM approximation more accurate. Explain. —*Hint: the fringing field is less pronounced for the substrate of higher permittivity.*

Q23.25. Explain why it is hard to achieve a large Q factor in a classical resonant circuit (parallel C and an L, with losses). Why do losses go up as the value of inductance and the frequency increase? — *Hint: recall that the greater the inductance, the longer is the wire, and that the skin effect increases with frequency.*

S **Q23.26.** You would like to have a coaxial-line resonator with as large a Q-factor as possible for a given outer resonator size. What would you do? — *(a) Reduce the radius of the inner conductor. (b) Increase the radius of the inner conductor. (c) Reduce the radius of the outer conductor.*

Answer. Much of the loss comes from the relatively small radius of the inner conductor, so that the surface current there is relatively large. If we increase the inner conductor radius as much as allowed by other restrictions (mechanical, maximal voltage, etc.), the Q-factor would be maximal possible.

Q23.27. Propose two methods for the excitation and energy extraction from a coaxial resonator. — *Convince yourself that the following answer is correct: a coaxial probe in the middle of the resonator outer conductor, or a loop probe at one end of the resonator.*

Q23.28. Find the energy contained in coaxial resonators of lengths λ, $\frac{3}{2}\lambda$, and 2λ. What is the Q factor of these resonators? — *Hint: you need to integrate the energy of a standing wave in resonators of these lengths.*

Q23.29. Sketch the current and voltage along an open-ended microstrip line resonator that is half of a guided wavelength long. What is the impedance at the center of the resonator, and what at the two ends? — *Hint: note that at the resonator ends the current is zero, and the voltage maximal.*

Q23.30. What loss mechanisms can you think of in an open-ended microstrip line resonator? — *Hint: think of losses in conductors and in the substrate, and think of other possible losses (there are two more).*

S **Q23.31.** A rectangular waveguide with a TE_{10} mode is terminated in a large rectangular cavity (e.g., of a microwave oven). Describe qualitatively what happens. — *Hint: think in terms of the reflected and transmitted energies.*

Answer. The wave traveling along the waveguide will partly be transmitted to the cavity, and partly reflected from this discontinuity in the waveguide. If inside the cavity there are losses (e.g., objects that absorb microwave energy and are thereby heated up), energy will be continuously transmitted to the cavity.

Q23.32. Propose two methods for the excitation and energy extraction from a cavity resonator with a TE_{101} wave type in it. — *Convince yourself that the following answer is correct: a coaxial probe at the top or bottom cavity wall, or a loop probe at the middle of a side wall, normal to the top and bottom cavity walls.*

PROBLEMS

S **P23.1.** Prove that for any TEM wave the electric and magnetic field vectors are normal to each other at all points. — *Hint: check the dot product of the transverse components of the two fields.*

Solution. Starting from Eq. (23.7) we obtain

$$\mathbf{E} \cdot \mathbf{H} = (E_x\,\mathbf{u}_x + E_y\,\mathbf{u}_y) \cdot (H_x\,\mathbf{u}_x + H_y\,\mathbf{u}_y)$$

$$= \pm Z_{\mathrm{TEM}}\,(H_y\,\mathbf{u}_x - H_x\,\mathbf{u}_y) \cdot (H_x\,\mathbf{u}_x + H_y\,\mathbf{u}_y) = \pm Z_{\mathrm{TEM}}\,(H_y H_x - H_x H_y) = 0,$$

which means that the vectors $\mathbf{E}$ and $\mathbf{H}$ are normal to each other.

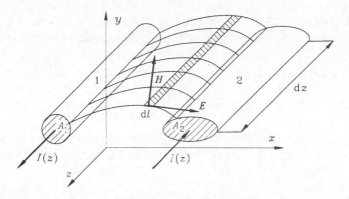

Fig. P23.2. Determination of the magnetic flux per unit length of a two-conductor transmission line of arbitrary cross section.

P23.2. Prove that at any cross section of a two-conductor transmission line with a forward traveling wave, the ratio of the voltage between the conductors and the current in them equals Z_{TEM} in Eq. (23.16). Show that for two-conductor transmission lines, $C'L' = \epsilon\mu$. — *Hint: use Fig. P23.2, and find the external inductance per unit length.*

P23.3. Prove that since at any cross section of a multiconductor transmission line $\sum Q'(z) = 0$, it follows that also $\sum I(z) = 0$, where the sum refers to all the conductors of the line. — *Hint: note that the transverse electric field is the same as in electrostatics, and that the energy of a nonzero charge on an infinite conductor is infinite.*

P23.4. Prove that Eqs. (23.8) to (23.12) imply that the electric and magnetic field vectors of a TE wave are normal to each other at all points. — *Hint: see problem P23.1.*

S **P23.5.** Write the instantaneous values of all the components of the TE_{10} wave in a rectangular waveguide. From these equations, sketch the distribution of the E-field and H-field in the waveguide at $t = 0$. — *Hint: refer to Fig. P23.5a, find the time-domain TE_{10} field components, and check if the sketch of the field in Fig. P23.5b is what you expect.*

Solution. With reference to Fig. P23.5a, the field components of the TE_{10} mode in time domain are

$$E_y(x, y, z, t) = \sqrt{2}\,\omega\mu\,\frac{a}{\pi}\,|H_0|\,\sin\left(\frac{\pi}{a}\,x\right)\sin(\omega t - \beta z + \alpha),$$

$$H_x(x, y, z, t) = -\sqrt{2}\,\beta\,\frac{a}{\pi}\,|H_0|\,\sin\left(\frac{\pi}{a}\,x\right)\sin(\omega t - \beta z + \alpha),$$

$$H_z(x, y, z, t) = \sqrt{2}\,|H_0|\,\cos\left(\frac{\pi}{a}\,x\right)\cos(\omega t - \beta z + \alpha),$$

$$E_x(x, y, z, t) = E_z(x, y, z, t) = H_y(x, y, z, t) = 0,$$

where $\alpha = \angle\{H_0\}$.

The field distributions at $t = 0$ are shown in Fig. P23.5b.

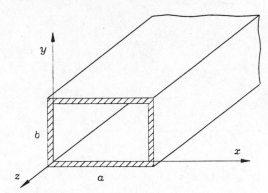

Fig. P23.5a. A rectangular waveguide.

P23.6. Determine the cutoff frequencies of an air-filled waveguide with $a = 2.5$ cm and $b = 1.25$ cm, for the following wave types: (1) TE_{01}, (2) TE_{10}, (3) TE_{11}, (4) TE_{21}, (5) TE_{12}, and (6) TE_{22}. — *Find three answers which do not have their pair: (1) 9.5 GHz, (2) 12 GHz, (3) 6 GHz, (4) 5 GHz (5) 13.42 GHz, (6) 16.97 GHz, (7) 24.74 GHz, (8) 26.83 GHz, (9) 36.5 GHz.*

P23.7. Plot the mode impedances between 8 and 12 GHz for an air-filled rectangular waveguide with $a = 2.5$ cm and $b = 1.25$ cm, for the following wave types: (1) TE_{01}, (2) TE_{10}, (3) TE_{11}, (4) TE_{21}, (5) TE_{12}, and (6) TE_{22}. — *Hint: use Eqs. (23.6), (23.7), and (23.16).*

P23.8. Plot the wavelength λ_z along a rectangular waveguide with $a = 2$ cm, $b = 1$ cm, and air as the dielectric, if the wave is of the TE_{10} type, for frequencies between 8 GHz and 12 GHz. Is the wavelength shorter or longer than in an air-filled coaxial line? — *At (a) 8 GHz, and at (b) 10 GHz: (a) $\lambda_{z1} = 11.78$ cm, $\lambda_{z2} = 6.53$ cm. (b) $\lambda_{z1} = 12.78$ cm, $\lambda_{z2} = 7.53$ cm. (c) $\lambda_{z1} = 10.78$ cm, $\lambda_{z2} = 4.53$ cm.*

P23.9. Plot the phase and group velocity in problem P23.8. — *At (a) 8 GHz, and at (b) 10 GHz: (a) $v_{ph1} = 8.32 \cdot 10^8$ m/s, $v_{g1} = 1.24 \cdot 10^8$ m/s, $v_{ph2} = 4.13 \cdot 10^8$ m/s, $v_{g2} = 2.28 \cdot 10^8$ m/s.*

(b) $v_{\mathrm{ph}1} = 8.62 \cdot 10^8 \ m/s$, $v_{\mathrm{g}1} = 1.04 \cdot 10^8 \ m/s$, $v_{\mathrm{ph}2} = 4.53 \cdot 10^8 \ m/s$, $v_{\mathrm{g}2} = 1.98 \cdot 10^8 \ m/s$. *(c)* $v_{\mathrm{ph}1} = 6.62 \cdot 10^8 \ m/s$, $v_{\mathrm{g}1} = 1.44 \cdot 10^8 \ m/s$, $v_{\mathrm{ph}2} = 4.03 \cdot 10^8 \ m/s$, $v_{\mathrm{g}2} = 2.45 \cdot 10^8 \ m/s$.

S **P23.10.** In a rectangular waveguide from problem P23.8, two signals are launched at the same instant. The frequency range of the first is in the vicinity of $f_1 = 10 \ \mathrm{GHz}$, and of the second in the vicinity of $f_2 = 12 \ \mathrm{GHz}$. Both signals propagate as TE_{10} waves. Find the time intervals the two signals need to cover a distance $L = 10 \ \mathrm{m}$, and the difference between the two intervals. Which signal is faster? — *(a)* $\Delta t = t_1 - t_2 = 3.8 \ ns$. *(b)* $\Delta t = 5.8 \ ns$. *(c)* $\Delta t = 7.8 \ ns$.

Solution. The signal propagates with velocity v_{g}, so that the time interval is given by $t = L/v_{\mathrm{g}}$. For the first signal $t_1 = 50.5 \ \mathrm{ns}$, and for the second $t_2 = 42.7 \ \mathrm{ns}$, whence $\Delta t = t_1 - t_2 = 7.8 \ \mathrm{s}$. The second, higher-frequency, signal is faster.

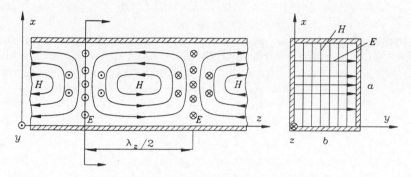

Fig. P23.5b. Distribution of the electric and magnetic fields of a TE_{10} wave in a rectangular waveguide.

P23.11. Consider the microstrip in Fig. P23.12. Derive the equation

$$\epsilon_{\mathrm{r}} \frac{\partial H_{z,\mathrm{air}}}{\partial y} - \frac{\partial H_{z,\mathrm{diel}}}{\partial y} = (\epsilon_{\mathrm{r}} - 1) \frac{\partial H_y}{\partial y}$$

starting from the boundary condition for the tangential electric field. — *Hint: start from Maxwell's equation* $\nabla \times \boldsymbol{H} = j\omega\epsilon\boldsymbol{E}$ *in rectangular coordinates, valid for two close points on the two sides of the air-dielectric boundary in Fig. P23.12. Note that the tangential component to the boundary is* E_x. *When equating the two tangential E-field components, note that the normal components of vector* $\boldsymbol{H}$ *on the air/dielectric boundary are the same.*

P23.12. The *effective dielectric constant* of a microstrip line depends on its dimensions approximately as

$$\epsilon_e = \frac{\epsilon_r + 1}{2} + \frac{\epsilon_r - 1}{2} \frac{1}{\sqrt{1 + 12 \, h/w}},$$

where the parameters are explained in Fig. P23.12. Plot the effective dielectric constant for h/w ratios between 0.1 and 10 (this is the approximate range for practical use), and for substrates that have relative permittivities of 2.2 (teflon-based Duroid), 4.6 (FR4 laminate),

9 (aluminum nitride), 12 (high-resistivity silicon), and 13 (gallium arsenide). — *Note: the effective dielectric constant allows the microstrip line to be considered as if the ground plane and the strip are situated in a homogeneous dielectric of that permittivity, i.e., to analyze it as a transmission line with a TEM wave along it.*

P23.13. The approximate formulas for microstrip line impedance and propagation constant based on the quasi-TEM approximation are given by

$$\beta = \omega\sqrt{\epsilon_0\mu_0}\sqrt{\epsilon_e}$$

$$Z_0 = \begin{cases} \dfrac{60}{\sqrt{\epsilon_e}}\ln\left(\dfrac{8h}{w} + \dfrac{w}{4h}\right) & , \quad \dfrac{w}{h} \leq 1 \\[3mm] \dfrac{120\pi}{\sqrt{\epsilon_e}\left[\frac{w}{h}+1.393+0.667\ln\left(\frac{w}{h}+1.444\right)\right]} & , \quad \dfrac{w}{h} > 1 \end{cases}$$

Plot the characteristic impedance as a function of the ratio w/h (between 0.1 and 10), and for the relative permittivities from problem P23.12. What can you conclude about the impedance as the line gets narrower? — *Note: this problem is intended to give you an idea of the parameters of a microstrip line assumed to support a quasi TEM mode.*

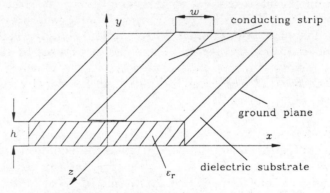

Fig. P23.12. A microstrip line.

P23.14. Plot the current, voltage, and impedance along a half-wavelength coaxial resonator short-circuited at both ends. If you want to feed the resonator with another piece of the same kind of cable, at which place along the resonator would you do it and why? — *Hint: if a resonator is lossless, current and voltage in it are shifted in phase by $\pi/2$, but since they are always lossy, there is a current component (what is its distribution?) which is in phase with the voltage.*

P23.15. Plot the current, voltage, and impedance along a half-wavelength microstrip line resonator open-circuited at both ends. You want to feed the resonator with a 50-Ω microstrip line. Propose (sketch) a way to do it, and explain. — *See hint for problem P23.14.*

S **P23.16.** Determine the maximum possible energy stored in a cubical resonant air-filled cavity with $a = b = d = 10$ cm, at a resonant frequency corresponding to the TE_{101} wave. The electric strength of air is 30 kV/cm. — *Hint: integrate the maximal energy density using Eq. (23.25).* $W_{em\,maximal} = 0.01\,J$. (b) $W_{em\,maximal} = 0.015\,J$. (c) $W_{em\,maximal} = 0.02\,J$.

Solution. The electric field intensity is maximal for $x = a/2$ and $z = d/2$. The maximal *instantaneous* value of E_y is limited by the dielectric strength of air, $E_{\text{maximal}} = 30\,\text{kV/cm}$, i.e.,

$$\sqrt{2}\left(2\omega\mu_0\frac{a}{\pi}\,|H_0|_{\text{maximal}}\right) = E_{\text{maximal}},$$

so that $|H_0|_{\text{maximal}} = \pi E_{\text{maximal}}/(2\sqrt{2}\omega\mu_0 a)$. The energy stored in the cavity is obtained as the integral of the electric energy density over the cavity volume. The result is

$$W_{\text{em maximal}} = \frac{1}{\pi^2}\,\epsilon_0\mu_0^2 a^3 bd\omega^2|H_0|_{\text{maximal}}^2 = \frac{1}{8}\cdot\epsilon_0 abd E_{\text{maximal}}^2 = 0.01\,\text{J}.$$

24. Fundamentals of Electromagnetic Wave Radiation and Antennas

• The process of producing electromagnetic waves is known as *electromagnetic radiation*. Structures that are designed to efficiently radiate electromagnetic waves are referred to as *transmitting antennas*. Transmitting antennas do not radiate equally in all directions, i.e., they have certain *directional radiation properties*. The same structures as transmitting antennas are used for extracting energy from an electromagnetic wave, in which case they are called *receiving antennas*. Transmitting and receiving antennas are vocal cords and ears of radio-communication systems.

• A transmitting antenna takes energy from a source and radiates a part of this energy in the form of a free electromagnetic wave. The source is usually connected to the antenna via a transmission line (the antenna *feeder* or *feed*). Looking from the source, a transmitting antenna is just a receiver of energy. If the source is time-harmonic, it "sees" the transmitting antenna as a complex impedance Z_A, known as the (transmitting) *antenna impedance*.

• A receiving antenna transforms a part of energy carried by an electromagnetic wave into the voltage between two antenna terminals connected to a receiver. Thus, a receiving antenna behaves as a voltage generator. In frequency domain and in complex notation, it has an emf and an internal impedance. The generator internal impedance equals the antenna impedance in the transmitting mode. The emf equals the open-circuit voltage across the antenna terminals, and depends on the antenna shape and size, *and on the direction and polarization of the incident wave*. The *directional properties* of a receiving antenna are known if they are known for the same antenna in transmitting mode.

• Relatively frequently, antennas are close to approximately flat conducting surfaces. If the surface is approximated by a perfectly conducting plane, the antenna can be analyzed by image theory. The images of the antenna at respective points should have opposite charges, and currents in opposite directions with respect to the conducting plane. An antenna in the form of a metal rod excited at such a plane is known as a *monopole antenna*. An antenna with two arms and a generator between them is a *dipole antenna*. The impedance of a monopole antenna is exactly one half the impedance of the dipole antenna (we can view the dipole antenna as the monopole antenna and its image.

• The *Hertzian dipole* is the simplest of all radiating systems. It consists of a straight, thin wire conductor of length l with two small conducting spheres or disks at the ends. If a generator of sinusoidal emf exciting the dipoles produce a current I in it, the electric and magnetic field far from the dipole (in the *radiation* or *far zone*), with respect to a spherical coordinate system with the origin at the dipole center and with the z axis along the dipole, is given by

$$E_\theta(r,\theta) = \frac{\mathrm{j}\beta Il\sin\theta}{4\pi r}\sqrt{\frac{\mu}{\epsilon}}\,\mathrm{e}^{-\mathrm{j}\beta r}, \tag{24.1}$$

$$H_\phi(r,\theta) = \frac{\mathrm{j}\beta Il\sin\theta}{4\pi r}\,\mathrm{e}^{-\mathrm{j}\beta r}. \tag{24.2}$$

In these equations, r is the distance from the dipole ($r \gg \lambda \gg l$), and θ is the angle between r and the dipole. Note that the ratio of the amplitudes of the two vectors is the same as for a plane wave,

$$\frac{E_\theta(r,\theta)}{H_\phi(r,\theta)} = \eta = \sqrt{\frac{\mu}{\epsilon}}. \tag{24.3}$$

• Both far-field components of the field of the Hertzian dipole depend on the distance r from the dipole as $1/r$. No static field has this dependence on r. This type of field is the *radiation (far) field*. Since all antennas can be considered as large assemblies of Hertzian dipoles, the far field of all antennas also depends on r as $1/r$, and the radiated power density as $1/r^2$.

• To characterize the distribution of the radiated field in different directions, a plot is made of the radiated electric field normalized to its maximal value. Such a graph is known as the *antenna radiation pattern*. It is a three-dimensional plot but, more often, it is plotted in two-dimensional cuts of the three-dimensional pattern. Usual planes for these cuts are those containing the E vector (the *E-plane pattern*), or the H vector (the *H-plane pattern*).

• A frequently used antenna is a straight wire dipole of total length equal to about half of a wavelength. For such a *half-wave dipole*, the current distribution is approximately a sine function, with the maximum at the generator. The impedance of a half-wave dipole is, *roughly*, $Z_A = (73 + \mathrm{j}0)\,\Omega$. Its radiation pattern is similar to that of the Hertzian dipole, except that the E-plane pattern consists of two slightly flattened circles, instead of true circles.

• A descriptor of the antenna directional properties used more frequently than the radiation pattern is the *antenna directivity*, defined as

$$D(\theta,\phi) = \frac{4\pi r^2|\vec{\mathcal{P}}(r,\theta,\phi)|}{P_{\mathrm{rad}}} = \sqrt{\frac{\epsilon}{\mu}}\frac{4\pi r^2|\boldsymbol{E}(r,\theta,\phi)|^2}{P_{\mathrm{rad}}}, \qquad P_{\mathrm{rad}} = R_{\mathrm{rad}}|I_0|^2. \tag{24.4}$$

In this equation, $\vec{\mathcal{P}}(r,\theta,\phi)$ is a time-average Poynting vector at a point in the far field, and P_{rad} is the power radiated by the antenna. The directivity is usually given in decibels,

$$[D(\theta,\phi)]_{\mathrm{dB}} = 10\log\{D(\theta,\phi)\} \quad (\mathrm{dB}). \tag{24.5}$$

• A hypothetical antenna radiating equally in all directions is known as an *isotropic*, or *omni-directional antenna*. For an isotropic antenna the directivity is 0 dB. Therefore, the directivity in Eqs. (24.4) and (24.5) is defined with respect to the isotropic antenna as a reference. The plot of the directivity in space (or in desired cuts) is known as the *antenna power pattern*. If the direction (defined by angles θ and ϕ) is not specified, by convention this means that *the maximum value of the directivity is implied*,

$$D = [D(\theta, \phi)]_{\text{max}}. \tag{24.6}$$

For example, the directivity of the Hertzian dipole is given by $[D(\theta)]_{\text{Hertzian dipole}} = 1.5 \sin^2 \theta$, so that its maximal directivity $D_{\text{Hertzian dipole}} = 1.5$. The directivity of the half-wave dipole is given by

$$D(\theta) = \sqrt{\frac{\mu}{\epsilon}} \frac{1}{\pi R_{\text{rad}}} \frac{\cos^2\left(\frac{\pi}{2}\cos\theta\right)}{\sin^2\theta} \qquad (R_{\text{rad}} \simeq 73\,\Omega), \tag{24.7}$$

and its maximal directivity is $D = D(\pi/2) \simeq 1.64$.

• The receiving pattern of a receiving antenna is the same as the radiation pattern of the antenna in transmitting mode. The antenna *effective area* is used frequently for describing *any* receiving antenna. It is defined by the equation

$$P_{\text{rec., matched load, optimal reception}} = A_{\text{eff}}(\theta, \phi)\mathcal{P}, \tag{24.8}$$

where $\mathcal{P}$ is the magnitude of the Poynting vector at the location of the antenna. Note the conditions described in the subscript which must hold (the receiver matched to the antenna, the receiving antenna positioned for optimal reception). In addition, the wave incident on the antenna from the direction considered must be of the same polarization as that of the wave the antenna would radiate in that direction when transmitting. The effective area can be expressed in terms of the antenna directivity as

$$A_{\text{eff}}(\theta, \phi) = \frac{\lambda^2}{4\pi} D(\theta, \phi). \tag{24.9}$$

For example, the effective area of a half-wave dipole with sinusoidal current distribution, in the direction normal to the dipole, is about $0.13\lambda^2$.

• The *Friis transmission formula* describes the power transmission in a line-of-sight (i.e., with no reflections) radio link between a transmitting antenna 1 and a receiving antenna 2. It reads

$$P_{2 \text{ matched load, optimal reception}} = \frac{D_1(\theta_1, \phi_1) A_2(\theta_2, \phi_2)}{4\pi r^2} P_{1 \text{ rad}}. \tag{24.10}$$

"Matched load" means that the formula is valid only if the receiving antenna is matched to the load (the receiver input). "Optimal reception" means that, in addition, the receiving antenna must be positioned so with respect to the incident field that the emf induced in it is the maximal possible.

QUESTIONS

S **Q24.1.** You have a black box with two terminals. You connect a generator to these terminals and find out that the black box behaves as an impedance. Can you check by observing the measured impedance whether the terminals belong to a transmitting antenna inside the box? Explain. — *(a) This is not possible. (b) Yes, because the antenna impedance can be greatly influenced by a nearby object. (c) Yes, because the antenna impedance differs from any circuit-theory impedance.*

Answer. Yes. The antenna impedance is greatly influenced by nearby object. Assume you put a large metal object close to the box. If inside the box there is an antenna, its impedance will change appreciably. If inside the box there is an ordinary electric circuit, this will have practically no effect on the impedance.

Q24.2. You have a black box with two terminals. You connect a load to these terminals and find out that the black box behaves as a generator. Can you check by observing the measured current in the load whether the terminals belong to a receiving antenna inside the box? Explain. — *(a) No, because a receiving antenna is not a normal circuit-theory generator. (b) Yes, if we screen the black box. (c) No, because the properties of the receiving antenna as a generator do not differ from those of a circuit-theory generator.*

Q24.3. Why are the images of antennas in Fig. Q24.3 as indicated? — *Hint: note that boundary conditions on the perfectly conducting plane must be satisfied if it is replaced by images.*

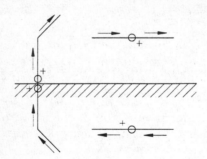

Fig. Q24.3. Examples of images of antennas above a perfectly conducting ground plane.

Q24.4. On many short antennas there are small conducting balls at each end. What are these balls for? — *(a) They serve as a decoration. (b) They improve mechanical properties of the antenna. (c) They make the current at the antenna end nonzero, and thus the current along the entire antenna greater.*

S **Q24.5.** Assume that a short wire dipole (the Hertzian dipole) has no end spheres. Will there be a current in the two short wire segments? If the answer is yes, what do you expect this current distribution to be like? — *(a) The absence of the end spheres changes nothing. (b) The absence of the end spheres makes the current in the dipole zero. (c) The absence of the end spheres force the current to become zero at the wire segment ends.*

Answer. There will be a current, which goes to zero at the wire segment ends.

Q24.6. What is the relationship between the phasor current I in the Hertzian dipole, and the charges Q and $-Q$ on the dipole end spheres? — *(a) No relationship, because the current is along the wire, and the charge is at the spheres. (b) $I = j\omega Q$. (c) $I = -dQ/dt$.*

S **Q24.7.** Take a pencil and assume it is a Hertzian dipole. What is its radiation pattern in space like? — *(a) Like a doughnut with zero inner radius and circular cross section, with the pencil along the doughnut axis. (b) A sphere with a diameter along the pencil. (c) A flattened ellipsoid, with its smaller axis along the pencil.*

Answer. Like a doughnut with zero inner radius and circular cross section, with the pencil along the doughnut axis.

Q24.8. In the preceding question, define an E plane and an H plane of the radiation pattern. — *Hint: recall the definitions of these two patterns and have in mind the directions of the two vectors.*

Q24.9. What is an isotropic antenna? Can it be made? If you think it cannot, explain why. — *It radiates equally in all directions. (a) It can be made easily. (c) It is difficult to make. (c) It cannot be made.*

Q24.10. Why is the directivity of an isotropic antenna equal to unity, or zero dB? — *Hint: inspect the defining formula for directivity.*

S **Q24.11.** What are the conditions implicit in the definition of the effective antenna area? — *Hint: inspect carefully the defining formula for the effective area and the text following it.*

Answer. (1) The receiver is matched to the antenna. (2) The wave incident on the antenna from the direction considered is of the same polarization as that of the wave the antenna would radiate in that direction when in transmitting mode.

Q24.12. Can the Friis transmission formula be used for the analysis of a radio communication channel if the transmitting antenna is not matched? Or if the receiving antenna is not matched? How would you modify the formula? — *(a) Only the receiving antenna must be matched. (b) Both transmitting and receiving antennas must be matched. (c) Only the transmitting antenna must be matched.*

PROBLEMS

S **P24.1.** Prove that the impedance of any antenna above a perfectly conducting ground, with the generator driving the antenna connected between the ground and the antenna terminal, is one half that of the symmetrical antenna obtained with the image of the antenna. — *Hint: note that the monopole antenna is connected to half the voltage of the equivalent dipole (monopole plus image), and that the currents in the two antennas are the same (see Fig. Q24.3).*

Solution. In the real system, the radiation field exists only in the upper half space. The power delivered to the symmetrical antenna in the equivalent system with image is therefore twice that of the original antenna. The current being the same, this means that the impedance of the antenna above ground is one half that of the equivalent symmetrical antenna.

P24.2. A thin two-wire transmission line with conductors of radius $a = 1\,\text{mm}$ and distance between them $d = 5\,\text{cm}$ is driven at one end by a generator with a rms value of the emf $E = 10\,\text{V}$ and frequency $f = 100\,\text{MHz}$. The line length is $b = 50\,\text{cm}$, and the other end of the line is open-circuited. Assuming that the line conductors do not radiate, but that the short segment with the generator does, determine approximately the electric field strength at a distance $r = 1\,\text{km}$ from the antenna. (*Hint: consider the short segment with the generator as a Hertzian dipole.*) — *(a) $E = 1.16 \cdot 10^{-4}\,V/m$. (b) $E = 2.16 \cdot 10^{-4}\,V/m$. (c) $E = 4.16 \cdot 10^{-4}\,V/m$.*

P24.3. A Hertzian dipole of length $l = 1\,\text{m}$ is fed with a current of rms value $I = 1\,\text{A}$ and of frequency $f = 1\,\text{MHz}$. Find the rms values of E_θ and H_ϕ in the equatorial plane (plane $\theta = \pi/2$) of the dipole at a distance of $r = 10\,\text{km}$. — (a) $E_\theta = 3.28 \cdot 10^{-5}\ V/m$, $H_\phi = 0.67 \cdot 10^{-7}\ A/m$. (b) $E_\theta = 6.28 \cdot 10^{-5}\ V/m$, $H_\phi = 1.67 \cdot 10^{-7}\ A/m$. (c) $E_\theta = 9.28 \cdot 10^{-5}\ V/m$, $H_\phi = 2.67 \cdot 10^{-7}\ A/m$.

P24.4. Using a system of two half-wave dipoles, construct an antenna system that radiates a circularly polarized wave in one direction. State clearly how you would make the feed. — *Hint: referring to Fig. P24.4, let the currents at the terminals of the dipoles be I_m and jI_m.*

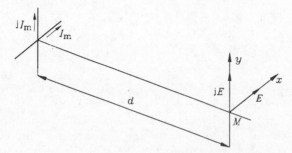

Fig. P24.4. An antenna system radiating a circularly polarized wave in one direction.

S   **P24.5.** A short vertical transmitting antenna of height h has a conducting plate at the top, so that the current along the antenna is practically uniform, of rms value I. At the receiving point, at a distance d from the antenna, only the wave reflected by the ionosphere arrives, as shown in Fig. P24.5. The ionosphere can be approximated by a perfectly conducting plane at a height H above the surface of the ground. Assuming that the ground at both the transmitting and receiving point is perfectly conducting, and neglecting the curvature of the earth, determine the rms value of the electric field intensity at the receiving point. The wavelength of the radiated wave is λ. — (a) $E_{\text{total}} = 2\eta_0 I h d^2 / [4\lambda(H^2 + d^2/4)^{3/2}]$. (b) $E_{\text{total}} = \eta_0 I h d^2 / [4\lambda(H^2 + d^2/4)^{3/2}]$. (c) $E_{\text{total}} = \eta_0 I h d^2 / [2\lambda(H^2 + d^2/4)^{3/2}]$.

Solution. The antenna integral with its image in the ground plane is equivalent to a Hertzian dipole of length $l = 2h$. The wave toward the receiving point makes an angle θ with the antenna, and the distance it covers is $r = 2\sqrt{H^2 + (d/2)^2}$. Having in mind the expression for the radiation electric field of a Hertzian dipole, the magnitude of the electric field intensity vector of the incident wave at the receiving point is given by

$$E_i = \frac{\eta_0 I}{2\pi r}\,\frac{\beta l}{2}\,\sin\theta = \frac{\eta_0 I h \sin\theta}{2\lambda\sqrt{H^2 + d^2/4}}.$$

The total wave at the receiving point is the sum of the incident wave and the reflected wave. The direction and the magnitude of the reflected wave are determined from the condition that the tangential component of the resultant electric field intensity vector on the ground surface is zero. Hence, according to Fig. P24.5, we find that

$$E_{\text{total}} = 2E_i \cos\left(\frac{\pi}{2} - \theta\right) = \frac{\eta_0 I h \sin^2\theta}{\lambda\sqrt{H^2 + d^2/4}} = \frac{\eta_0 I h d^2}{4\lambda\,(H^2 + d^2/4)^{3/2}}.$$

P24.6. In a radio link at $f = 900\,\text{MHz}$, with two half-wave dipoles a distance $r = 100\,\text{m}$ apart, one dipole (e.g., dipole 1) is replaced by a more directional antenna (for a cellular phone, this would be the base-station antenna). Calculate the ratio of the powers received by the second dipole in the two cases, for a directivity of (1) 6 dB, (2) 10 dB and (3) 20 dB of antenna 1. — *(a) 2.43, 6.1, 61. (b) 3.38, 8.3, 83. (c) 1.65, 5.3, 53.*

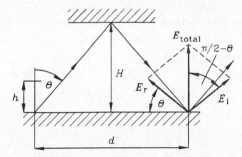

Fig. P24.5. A vertical antenna above ground.

P24.7. Assume that in a communications link two matched lossless antennas, A and B, are r apart in each other's far fields. The antenna directivities and effective areas in the line-of-sight direction are D_A, A_A, and D_B, A_B, respectively. First antenna A transmits a power P_{A1}, while antenna B receives a power P_{B1}. Then antenna B transmits a power P_{B2}, while antenna A receives a power P_{A2}. Using the reciprocity condition, which says that $P_{A1}/P_{B1} = P_{B2}/P_{A2}$ (think about what this means), show that the ratio of the directivity to the effective area is a constant for any antenna. —*Hint: use Eqs. (24.9) and (24.10) to obtain the Friis formula with the directivity of the transmitting antenna and the effective area of the receiving antenna; combine two such formulas, one for antenna A as the transmitting antenna and the other for antenna B as the transmitting antenna, with the reciprocity conditions.*

P24.8. Derive the Friis formula in terms of effective area only. In a microwave relay system for TV each antenna is a reflector with an effective area of $1\,\text{m}^2$, independent of frequency. The antennas are 10 km apart. If the required received power is $P_r = 1$ nW, what is the minimum transmitted power P_t required for transmission at $1\,\text{GHz}$, $3\,\text{GHz}$, and $10\,\text{GHz}$? — *Hint: use Eqs. (24.9) and (24.10). (a) 9 mW, 1 mW, and 0.09 mW. (b) 90 mW, 10 mW, and 0.9 mW. (c) 9 W, 1 W, and 0.09 W.*

P24.9. Derive the Friis formula in terms of directivities only. — *Hint: use Eqs. (24.9) and (24.10).*

25. Some Practical Aspects of Electromagnetic Waves

• Applications of electromagnetic waves are numerous, ranging from cooking food to controlling a faraway spacecraft and receiving information from it. In these applications, a number of practical problems need to be solved. For example, when one sends or receives information using electromagnetic (radio) waves, the maximum distance at which this can be done is limited by the amount of power available at the sending end, and the loss of the wave energy by the time it gets to the receiving end, assuming a certain receiver sensitivity. The path loss varies with the medium through which the wave is propagating, as well as the frequency (wavelength) of the wave. For example, for a coaxial cable the attenuation constant is given by

$$\alpha = \frac{R'}{2Z_0} = \frac{R'}{2}\sqrt{\frac{C'}{L'}}. \tag{25.1}$$

Note that, due to skin effect, R' increases with frequency.

- For long two-conductor transmission lines, the attenuation can be reduced drastically by introducing additional series inductances along the line. This is seen from the expression for the attenuation constant,

$$\alpha \simeq \frac{1}{2}\left(R'\sqrt{\frac{C'}{L'}} + G'\sqrt{\frac{L'}{C'}} \right), \tag{25.2}$$

noting that, in practice, the first term is much greater (several orders of magnitude) than the second, due to relatively large value of R'.

- Attenuation is also present in waveguides, although they have relatively large surfaces for current flow. For example, the attenuation constant for the dominant mode (TE_{10}) in a rectangular waveguide is given by

$$\alpha = R_s \sqrt{\frac{\epsilon}{\mu}} \frac{a/b + 2f_c^2/f^2}{a\sqrt{1 - f_c^2/f^2}} \qquad (\text{TE}_{10}). \tag{25.3}$$

Since R_s is related directly to skin depth, the attenuation constant depends on the metal conductivity and frequency. For example, a $10\,\text{GHz}$ waveguide has an attenuation constant of about $\alpha = 0.0883\,\text{Np/m} = 0.767\,\text{dB/m}$.

- Optical fibers are used as waveguides for electromagnetic waves in the visible an infrared part of the frequency spectrum, with wavelengths between roughly $300\,\text{nm}$ and $10\,\mu\text{m}$. Fibers are so-called *dielectric waveguides*. The simplest dielectric waveguide is a flat dielectric slab, where the total reflection at the slab faces is used to guide the wave.

- In all of the above *guided wave* cases, the ration of the received and transmitted power is of the form $P_{\text{rec}}(r)/P_{\text{tr}} = \text{e}^{-2\alpha r}$.

- In a line-of-sight radio link (where a radio wave travels between two antennas directly, with no reflections), the power loss is given by the Friis transmission formula. If the two antennas are equal, $n\lambda^2$ large, and we assume they are well designed so that the effective areas are roughly equal to their geometric areas, the Friis formula becomes

$$\frac{P_{\text{rec.}}(r)}{P_{\text{transm.}}} = \frac{n^2\lambda^2}{r^2}. \tag{25.4}$$

This means that the larger the antennas are (measured in wavelengths), the lower the loss of power between the transmitter and receiver.

• AM broadcasting systems rely on surface wave transmission between two points on the earth's surface. Short wave radio systems use bounces off of the ionosphere and the earth's surface. UHF and VHF radio waves used for communications by airplanes, as well as microwaves in radio relay links, use the line-of-sight propagation, in which case the range is limited by the curvature of the earth. Therefore almost all radio relay stations are put up on high peaks. If the heights of the two antennas are h_{trans} and h_{rec}, and the radius of the earth is R, the approximate range of the line-of-sight link obtained in practice is

$$r = \sqrt{\frac{8R}{3} h_{\text{trans}}} + \sqrt{\frac{8R}{3} h_{\text{rec}}}. \tag{25.5}$$

In this formula, no attenuation due to the atmosphere is taken into account.

• The upper layer of the atmosphere, approximately between 50 km and 500 km above the earth's surface, is a highly rarefied ionized gas, known as the *ionosphere*. It has a very pronounced influence on the propagation of electromagnetic waves in a wide frequency range, from about 10 kHz up to about 30 MHz, but also at higher frequencies than these. The presence of the ions is equivalent to a *reduction* in permittivity. The equivalent (or effective) permittivity of an ionized gas is defined as

$$\epsilon' = \epsilon_0 \left(1 - \frac{NQ^2}{\omega^2 \epsilon_0 m} \right) = \epsilon_0 \left(1 - \frac{\omega_c^2}{\omega^2} \right), \tag{25.6}$$

where

$$f_c = \frac{\omega_c}{2\pi} = \frac{1}{2\pi} \sqrt{\frac{NQ^2}{\epsilon_0 m}} \tag{25.7}$$

is the *critical frequency* of the ionized gas. Thus the propagation coefficient and the phase velocity of the wave are

$$\beta = \omega \sqrt{\epsilon' \mu_0} = \frac{\omega}{c_0} \sqrt{1 - \frac{\omega_c^2}{\omega^2}}, \qquad v_{\text{ph}} = \frac{\omega}{\beta} = \frac{c_0}{\sqrt{1 - \omega_c^2/\omega^2}}. \tag{25.8}$$

It is seen that waves of angular frequencies $\omega < \omega_c$ cannot propagate in the ionized gas. Ionized gases also introduce a wave attenuation, due to the collisions of ions accelerated by the electric field of the wave with neutral molecules.

• Let a plane wave be emitted from the earth's surface towards the ionosphere at an arbitrary angle θ_0 with respect to the vertical. The wave is either reflected back, or passes through the ionosphere, depending on the wave frequency, f, the highest critical frequency in the ionosphere, $f_{c \text{ max}}$, and the angle θ_0. If the wave is reflected, it bounces back at a height corresponding to a critical frequency given by the equation

$$f_c = f \cos \theta_0. \tag{25.9}$$

Waves of higher frequencies emitted at the same angle will pass through the ionosphere.

- From this summary, it is seen that the optimal frequency range for different applications is very different. For example, the communications via satellites require frequencies not influenced by the ionosphere, while short-wave communications use the ionosphere as a mirror. At lower frequencies and when the two ends can be physically connected, coaxial cables are often used (e.g., cable TV). TV signals can also be received from satellites, in which case frequencies on the order of 10 GHz are used. For cellular telephony, mostly frequencies around 900 MHz and 2 GHz are used, to avoid manmade and atmospheric noise. For radio communication with submarines, frequencies as low as about 10 kHz are used, since these waves have relatively large skin depth (penetration) in sea water.

- Radars are essentially a radio link, where the transmitter and receiver are located at the same place. The transmitter sends a wave of power P_T, which partly reflects off of the target. The power density at the target is $P_T D/(4\pi r^2)$, where r is the distance to the target, and D is the radar antenna directivity. The target scatters the wave proportionally to a quantity called the *radar scattering cross section*, usually denoted by $\sigma(\theta, \phi)$, which is essentially an effective area of the target acting as a receiving antenna. When it reflects the wave, the target acts as a transmitting antenna with a directivity of $4\pi\sigma/\lambda^2$. The reflected wave is received at the transmitting point and conclusions are then made about the target. The applications of radars are very diverse, ranging from many military purposes, to weather radars in meteorology, anticollision radars for cars, etc. The *radar equation* tells us how much power is received back for a given transmitted power. If the same antenna is used for transmission and reception, it reads

$$P_{\text{rec}} = P_T \frac{D^2(\theta, \phi)\sigma^2(\theta, \phi)}{16\pi^2 r^4}. \tag{25.10}$$

In a *Doppler radar*, the transmitted signal is frequency modulated, which enables the speed of the target to be measured.

- In computers and other digital systems, radiation and electromagnetic coupling becomes progressively more pronounced as the clocks in these systems become faster. This can result in a high crosstalk between different segments of the system.

- In normal ovens, food is heated mainly by infrared radiation from the heaters. These waves have very small skin depth, so that the food is heated from the surface inwards by thermal conduction. Using microwave frequencies instead, the skin depth is greatly increased, so that direct electromagnetic heating takes place in a large part of the heated food, reducing greatly the cooking time. In microwave ovens, the standard frequency is 2.45 GHz.

QUESTIONS

Q25.1. Explain what the physical origin of loss in coaxial waveguides is. — *Hint: recall the skin effect.*

Q25.2. Explain what the physical origin of loss in metallic waveguides is, and why the loss can be smaller than in coaxial cables. — *Hint: recall the skin effect and compare the areas available for current flow in the two cases.*

Q25.3. Explain what the physical origin of loss in optical fiber is, and why the loss can be smaller than in metallic structures. — *Hint: recall the polarization losses.*

Q25.4. Explain what the physical origin of loss in a line-of-sight antenna link is. — *Hint: think of the loss due to the spherical character of the wave, of the influence of the atmosphere, of possible raindrops along the way and of possible multiple waves reaching the receiver.*

Q25.5. What is the range in a line-of-sight link limited by? — *(a) Frequency. (b) Height of transmitter. (c) Height of transmitter and receiver, and the radius of the earth.*

S **Q25.6.** Explain in you own words why there is attenuation in an ionized medium with neutral gas molecules. — *Hint: think what happens when the electric field accelerates a charged particle, and the particle collides with a neutral atom.*

Answer. The electric field of the wave accelerates charged particles, and thus transfers to them some energy. If there are no collisions with neutral particles, this energy is returned to the wave in a later time interval. If, however, a charged particle collides with a neutral gas molecule, the acquired energy of the particle is partly transferred to the molecule. There is no mechanism for an *uncharged* particle to interact with the wave and to return this energy back to it. So the energy of the wave is reduced.

Q25.7. A wave of frequency higher than the highest critical frequency for the ionosphere needs to be used for communication between two points of the earth. Is this possible? Explain. — *(a) No. (b) Yes, if the angle of incidence on the ionosphere is large enough. (c) Yes, if the signal is frequency modulated.*

S **Q25.8.** A wave of extremely low frequency (e.g., below $100\,\mathrm{Hz}$) coming from outer space penetrates through the ionosphere and reaches the earth's surface. Explain. — *(a) The ions in the ionosphere are not affected by such low frequencies. (b) The process of wave reflection takes a certain depth. When the wavelength is long, this depth can be the entire ionosphere, and the wave reaches the earth's surface. (c) Low frequencies carry more energy, and therefore pass the ionosphere.*

Answer. The free-space wavelength of such a wave is on the order of few *thousand* kilometers, i.e., much greater than the ionosphere thickness. So the ionosphere does not have a sufficient shielding effect for such waves.

Q25.9. Imagine a line-of-sight link in a hallway with conducting walls on top and bottom, and absorbing walls on the sides. How many waves can contribute to the received signal? How would you construct antenna images that approximate the influence of the walls? — *Hint: assume no reflections off the absorbing walls, and approximate the top and bottom walls by perfectly conducting planes.*

S **Q25.10.** Derive the radar equation, Eq. (25.10).

Answer. At the target, the received power is $P' = P_T D\sigma/(4\pi r^2)$. This power is reflected in an amount proportional to the "directivity" of the target, $D_t = \sigma 4\pi/(\lambda^2)$, and is received by the radar in proportion to its antenna effective area, $A = D\lambda^2/(4\pi)$. So, we have

$$P_R = P' \frac{1}{4\pi r^2} \frac{\sigma 4\pi}{\lambda^2} \frac{D\lambda^2}{4\pi} = P_T \frac{D^2 \sigma^2}{16\pi^2 r^4}.$$

S **Q25.11.** Consider a Doppler radar at 10 GHz. The received signal from one car is in the audio range, and can be between 300 Hz and 4 kHz. What is the range of speed this radar can detect?

Answer. For velocities much smaller than the speed of light, c, the Doppler shift is given by $f_D/f = v/c$, so $v = cf_d/f$. For $f_d = 300$ Hz, $v = 9$ m/s (corresponding to a car moving at about 30 km/h), and for $f_D = 4$ kHz, $v = 120$ m/s (corresponding to a slow plane, flying at about 220 km/h).

S **Q25.12.** Consider an FM ranging radar in which the frequency varies linearly from $f_1 = 10$ GHz to f_2 in $T = 10\,\mu$s. How would you choose f_2 in order to be able to detect targets 1 km away, if the radar bandwidth is 500 MHz?

Answer. The range is given by $r = [cT(f - f_1)]/[2(f_2 - f_1)]$. For $f_1 = 10$ GHz and $r = 1$ km, we get that $f = (2f_2 + 10)/3$. If we chose f_2 so that the full bandwidth is used, i.e., $f_2 = 10$ GHz $+ 500$ MHz $= 10.5$ GHz, we get $f = 10.33$ GHz for a 1-km target range, which is within the radar bandwidth.

PROBLEMS

P25.1. Calculate how much power is received in England if 1 MW is sent from Boston along a transatlantic 50 Ω cable at 10 kHz. You can assume that the main loss in the cable is due to conductor loss and that $R' = 0.005\,\Omega$/m. — *(a) 102 kW. (b) 4.39 kW. (c) 0.553 kW.*

P25.2. What value of Pupin coils would you choose and how would you place them to reduce the loss in the cable with $C' = 167$ pF/m, $G' = 0.3$ pS/m, $L' = 0.2\,\mu$H/m, and $R' = 0.0055\,\Omega$/m. — *Hint: use coils at every kilometer to artificially increase the inductance per unit length about 25 times.*

P25.3. Calculate the skin depth and attenuation coefficient of a rectangular waveguide with dimensions $a = 23$ mm and $b = 10$ mm, at 10 GHz, if the waveguide is made of (1) copper, (2) aluminum, (3) silver, or (4) gold. What do you think are the engineering problems associated with each metal? Can you think of any combined solution? — *We give possible answers for gold: (a) $\delta = 7.86 \cdot 10^{-7}$ m, $\alpha = 0.0149$ Np/m. (b) $\delta = 4.53 \cdot 10^{-7}$ m, $\alpha = 0.0231$ Np/m. (c) $\delta = 9.75 \cdot 10^{-7}$ m, $\alpha = 0.0189$ Np/m.*

P25.4. Calculate the skin depth of gold in the optical domain, at wavelengths of 500 nm, 830 nm, 1.33 μm, and 1.55 μm. How thin would one need to make a sheet of gold to see through? — *The sheet should be thinner than about (a) 100 nm. (b) 20 nm. (c) 5 nm.*

P25.5. Compare the loss in the inner conductor and outer conductor of a coaxial cable at 1 MHz. Assume the conductors are made of copper, that the cable is filled with a dielectric of permittivity $\epsilon_r = 3$, and that the dimensions are such that the inner conductor radius $a = 0.45$ mm, and inner radius of the outer conductor $b = ae$. — *(a) 3.05. (b) 2.71. (c) 2.30.*

P25.6. Plot the power attenuation in dB versus distance from 1 m to 1,000 km on a logarithmic scale for: coaxial cable at 10 GHz with $\alpha = 0.5$ dB/m, waveguide with $\alpha = 0.1$ dB/m, 1.55-μm single-mode optical fiber with $\alpha = 0.1$ dB/km, and a free space link at 10 GHz with a horn antenna with 20-dB directivity and a 1-m diameter dish antenna. — *Hint: recall the attenuation constants in the four cases.*

P25.7. Calculate the dimensions for a rectangular waveguide with a dominant TE_{10} mode at cable TV frequencies between 100 and 600 MHz. — *Let $a/b = 2$. (a) $a = 30$ cm. (b) $a = 50$ cm. (c) $a = 100$ cm.*

S **P25.8.** A UHF radio system for communication between airplanes uses antennas with a directivity of 2. What is the maximum line-of-signt range between two airplanes at an altitude of 10 km? If the required received power is 10 pW, what is the minimum transmitted power P_t required for successful transmission at 100 mHz, 300 MHz, and 1 GHz?

Solution. The range is $r = 2\sqrt{2R_{eff}h} = 824.6$ km. For this range, the minimal required transmitted power is $P_t = (4\pi r/\lambda)^2 P_r/(D_T D_R) = 268\,(lambda^2)$ W, where λ is in meters. At 100 MHz, 300 MHz and 1 GHz, the respective wavelengths are 3 m, 1 m and 0.3 m, and the transmitter powers are 30 W, 268 W and 3 kW.

P25.9. Calculate the effective area of a dish antenna for TV that requires a 1-degree beamwidth in both θ and ϕ planes, assuming one of the standard cable frequencies (for example, 225 MHz). Is this a practical antenna? (Note: you can use an approximate formula for the maximal directivity given the beamwidths, α_1 and α_2, in the two planes, $D \simeq 32,000/(\alpha_1\alpha_2)$, where the beamwidths are given in degrees.) — *(a) 356 m². (b) 1342 m². (c) 4527 m².*

P25.10. If a satellite is 1000 km above the earth's surface, and has a 0.1-degree beamwidth in both planes, calculate the corresponding directivity using the approximate formula in the previous problem. Find the size of the footprint on the earth's surface, and the effective area of the antenna at a satellite frequency of 4 GHz. — *The radius of the footprint is (a) 37 m, (b) 189 m, (c) 356 m. The antenna effective area is (a) 1432 m², (b) 2271 m², (c) 764 m².*

P25.11. Derive the radar equation (25.10) for a radar that uses two antennas, one for transmitting and another for receiving. — *Hint: what you need to have in this case instead of the directivity squared of the antenna?*

P25.12. Assuming a 10-GHz police radar uses an antenna with a directivity of 20 dB (standard horn), and your car has a scattering cross section of $100\lambda^2$, plot the received power as a function of target distance, for a transmitted power of 1 W. If the receiver sensitivity is 10 nW, how close to the radar would you need to slow down to avoid getting a speeding ticket? — *The radar starts detecting the speed of a car at approximately (a) 354 m, (b) 489 m, (c) 1.213 m.*

P25.13. How large is the dynamic range of the radar from problem P25.12? (The dynamic range is the ratio of the largest to smallest signal power detected, expressed in decibels.) — *(a) 50 dB. (b) 29 dB. (c) 64 dB.*

Simple Electromagnetic Labs

Lab 1. Simple circuit elements

Background: Circuits, Chapter 2 in *Introductory Electromagnetics*

Resistors, capacitors and inductors are basic building blocks in linear circuits. Here we look at their real characteristics from the electromagnetic point of view, and the way they can affect circuit performance.

All real passive, linear electrical components behave like combinations of resistors, capacitors, and inductors. For example, a packaged inductor can be represented realistically by an ideal inductor, along with a capacitor that represents capacitance between the windings, and a resistor that represents the loss in the wire. This is then called a *model* of an inductor. Modeling is very important for engineers; without good component models, circuit and systems cannot be designed accurately, or at all.

Purpose: to understand the characteristic behaviour of resistors (R), capacitors (C) and inductors (L), and to learn what makes an object behave like R, C or L. We do this on the simple examples of RL and RC circuits and their response to a step function excitation.

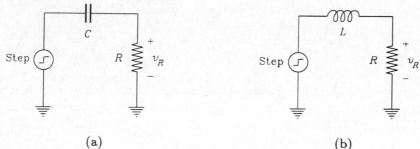

(a)   (b)

Fig. L1.1. (a) A series RC and (b) a series RL circuit with a step function excitation.

Pre-lab problems:

PL1.1. A RC circuit is shown in Fig.L1.1a. Find the voltage across the resistor as a function of time for a step-function generator of amplitude A. (This can be done by solving a differential equation in time domain, or by using a Laplace transform.) How does the value of R influence the waveform? For a given value of resistance, how does the capacitance influence the waveform?

PL1.2. Repeat problem L1.1, but for a series RL circuit, Fig.L1.1b, where the inductor L replaces the capacitor.

Lab:

Equipment and parts:

- an oscilloscope (such as, e.g., Tektronix TDS220);

- a function generator (such as, e.g., HP8114A or HP8116A);

- several resistors, a coaxial Tee-connector, two brass plates, and wire wrapped around a tube.

In your lab report, answer questions L1.1 to L1.10, adding any graphs or sketches that you feel make it clearer. Make sure to label all axes and include scales and units.

Part 1: Understanding the oscilloscope and function generator

An oscilloscope (scope for short) measures a voltage as a function of time. To get familiar with the scope you are using, connect one of its channels to the function generator. Set the function generator output to a square wave with a frequency of 1 MHz and observe the signal. (This involves making sure that the triggering is done properly, either internally in the scope or using an external triggering signal.) Adjust the amplitude on the signal generator so that the scope shows a 2 volt peak-to-peak signal.

L1.1. What output amplitude is the function generator displaying? Is it consistent with the 2 V that you are measuring with the scope?

L1.2. The input impedance of most oscilloscopes is on the order of a 1-MΩ resistance in parallel with a 20-pF capacitance. The model of the signal generator is an ideal voltage source in series with a 50-Ω source resistance. Based on what you measured so far, what is the voltage amplitude of the source?

L1.3. The advantage of the high input impedance of the oscilloscope is that it does not load a circuit you are measuring, because it draws very little current. However, there is also a problem related to this if a coaxial cable is connected to the scope input. The cable is a transmission line that carries a wave, which carries energy. When the wave gets to the high input impedance of the scope (almost an open circuit), the energy is not absorbed by the scope, but is reflected. How can we avoid this reflection? To find out, change the input impedance of the oscilloscope to 50 Ω (note that some scopes have this option built in). Use a Tee and a 50-Ω resistor to do this. Sketch the equivalent circuit of the generator and the input of the scope for this case. Measure the voltage of the generator for this case. Do you get the expected peak-to-peak amplitude?

L1.4. Are you noticing any degradation of the square wave as observed on the scope? What is it due to? (Hint: is there extra inductance and/or capacitance associated with the resistor?) Try to reduce the degradation by changing the way the resistor is attached to the Tee. Try using an rf (high-frequency) resistor.

Part 2: Capacitance effects

Keep the Tee and 50-Ω resistor at the input of the scope. Connect a circuit as shown in Fig.L1.2a.

L1.5. What is the value of the resistance R in this setup?

L1.6. Take the two parallel plates and attach them as shown in Fig.L1.2b. Verify that you get the same thing as before when the plates are touching. Next place a piece of paper between the plates. You have made a capacitor. A capacitor has two parts that store equal but opposite charges, with an electric field (and therefore voltage) between the charges. The electric field stores energy. How does the capacitance affect the signal? Sketch the waveform and explain qualitatively why it has the shape you observed.

L1.7. Calculate the value of the capacitance, using the results from the prelab problem PL1.1.

L1.8. We will later learn that the capacitance of a parallel-plate capacitor is given by $C = \epsilon_0 \epsilon_r A/d$, where $\epsilon_0 = 8.854 \cdot 10^{-12}\,\text{F/m}$, A is the area of each of the plates, and d is the separation of the plates. Using $\epsilon_r = 1.2$ for paper, calculate the capacitance of your capacitor. How does it compare to that obtained by analysing the waveform measured by the scope?

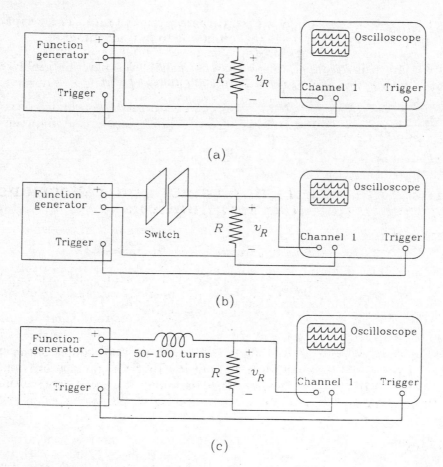

Fig. L1.2. (a) Voltage measured across resistor alone, and (b) across resistor in a series RC circuit and (c) in a series RL circuit.

Part 3: Inductance effects

Repeat Part 2, but using an inductor instead of a capacitor. Inductance is caused by current that produces a magnetic field. The magnetic field stores energy. We will later learn Faraday's law, which tells us that a change in current produces a change in the magnetic field, which causes a voltage.

L1.9. Set up the circuit shown in Fig.L1.2c. Use a plastic tube (about 0.5-1 cm in diameter) wrapped with 50–100 turns of wire as the inductor. Sketch the waveform and explain the shape. Calculate the value of L using the result from prelab problem PL1.2.

L1.10. Do you notice a ripple on the waveform? What you are seeing is not an "error", but a resonance between a capacitor and an inductor. Where is the capacitor? Think of a better model for the wire wrapped around the tube. Should the capacitance be in series or in parallel with the inductance? Explain. Use the measured ripple period (frequency) to calculate the value of this capacitance associated with a wire wound densely around a tube.

Conclusions:

1. Whenever there is charge stored in a circuit, there is capacitance. These capacitances can change the shapes of signals in a circuit, often not to our advantage.

2. Whenever there is current in a circuit, there is a magnetic field and inductance. These inductances also change the performance of the circuit.

3. All real linear, passive electrical components can be modeled by a combination of resistors, capacitors and inductors. Better models lead to better agreement between theory and experimental reality.

Lab 2. Resonant effects in circuits and reflections from transmission-line terminations

Background: Circuits, Chapter 2 in *Introductory Electromagnetics*

In this experiment, we continue the study of basic circuit elements and their effects on circuit behavior. In the first part of the lab, we observe the effects of resonant circuit elements, and in the second part we investigate reflections in coaxial cables.

Purpose: to learn to recognize resonant effects in circuits, on the example of the step response of a simple RLC circuit and a twisted pair resonator. To get a glimpse of transmission line (wave) effects on the example of a pulse propagating along a coaxial cable. An example we look at is the thinLan cable used in ethernet computer networks.

Pre-lab problems:

PL2.1. Solve for the voltage across the resistor in the circuit shown in Fig. L2.1 when the source is a step function of amplitude 1 V. What is the rise/decay time equal to, in terms of the R, L and C? What is the resonant frequency equal to, in terms of the element values? What is the Q factor equal to and how is it influenced by the value of the resistance? (If you do not remember what the Q factor is, look it up in your circuits book.)

PL2.2. Let us go back to the inductor from Lab 1. You should have recognized from experiment that a reasonable model for the real "inductor" is an inductor in parallel with a small capacitor. (What happens to the impedance of the inductor at the low-frequency limit if the capacitor in the model is placed in series instead of in parallel?) Assume the inductance is $5\,\mu H$, and the parasitic capacitance between the windings distributed along its length is $1\,pF$.

Calculate the approximate frequency at which the inductor starts looking more like a capacitor. At what frequency is the inductive part an order of magnitude larger than the capacitive part (at what frequency and below it does the inductor behave nearly as a pure inductance)? At what frequency is the inductive part an order of magnitude smaller than the capacitive part (at what frequency and above it does the inductor behave as a pure capacitance)?

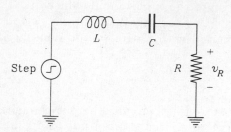

Fig. L2.1. A series RLC resonant circuit excited with a step function voltage source.

PL2.3. The coax usually has a characteristic impedance of $50\,\Omega$ (we have not defined characteristic impedance yet, but this is the parameter you use when you go to the store to purchase a piece of cable). If you put a 50-Ω resistor at the open end of the cable, the pulse will be completely dissipated in the resistor. If the pulse has a width of $20\,\text{ns}$ and an amplitude of $1\,\text{V}$, how much energy is transformed into heat in the resistor?

Lab:

Equipment and parts:

- an oscilloscope and a function generator (such as in Lab 1);

- several resistors, a coaxial Tee-connector, two brass plates, and wire wrapped around a tube, a 10-m long coaxial cable, a potentiometer (variable resistor).

Part 1: Resonant effects in circuits

Use the same setup as in Lab 1 (Fig. L1.2). Connect the resistor, capacitor and inductor in series. For the resistor, use a potentiometer (pot for short). Connect a coaxial cable from the external trigger of the source to the external trigger input of the scope. (This is because oscillating circuits often lead to triggering problems.)

L2.1. Use a Tee at the output of the signal generator. Connect one arm to channel 2 of the scope. Connect the other arm of the Tee to the circuit, and monitor the voltage across the pot on channel 1. Notice that the square wave on channel 2 degrades when we attach the circuit. Explain. Is there a change in channel 1 (voltage across the resistor) if channel 2 is not connected? Explain. Measure the resonant frequency by observing channel 1. Sketch and explain what you see. (The value of the resistor should be small. However, if it is too small, you will not be able to see a voltage drop. On the other hand, if it is too large, the circuit will not oscillate.)

L2.2. Change the capacitance by putting more pieces of paper between the plates. Sketch and explain the waveform on channel 1. Next change the capacitance by reducing the overlap region of the plates, so that the capacitance is roughly half of what you had earlier. Does the voltage observed on the scope show this change in a way you expect? Explain.

L2.3. Measure the decay time and the resonant frequency for the largest capacitance value. Use your prelab problem PL2.1 results to calculate the Q factor of the circuit.

Part 2: More resonant effects in circuits

Next we change the shape of the wire connecting the resistor to the source to be a twisted two-wire shorted cable, about 20 cm long, Fig. L2.2a. Connect the twisted pair in series with the resistor. Change the frequency of the square wave to 5 MHz.

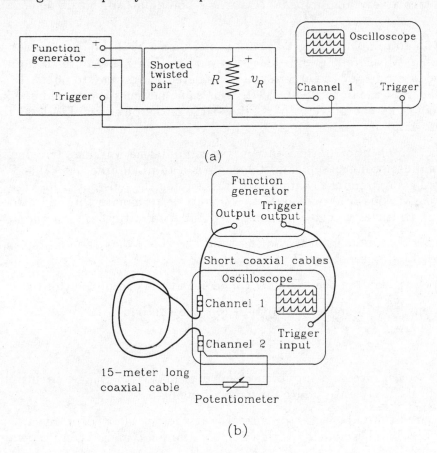

(a)

(b)

Fig. L2.2. (a) Measuring the voltage across the resistor when the connecting wire is in the shape of a twisted pair of wires. (b) Measurements on a piece of a coaxial cable.

L2.4. Sketch the voltage you are measuring across the resistor. Based on the waveform, what combination of R, L and C is a good model for this twisted wire? The twisted pair is shorted at the end (point B). Based on the waveform, what conclusion can you make about short circuits?

L2.5. Now clip the shorted end of the twisted pair of wires (at point B). Sketch the waveform. Based on the waveform, what conclusion can you make about an open circuit?

Part 3: Reflections from coaxial-cable terminations

The coaxial cable that you are familiar with is a transmission line. This means that a wave of voltage (and current) travels along the cable carrying energy. To see some wave effects, set up the experiment as shown in Fig. L2.2b, so that the voltage at the input of the long piece of cable is seen on channel 1, and the voltage at the output is measured on channel 2. The load at the end of the cable is changed by varying the pot connected to the Tee at channel 2.

L2.6. To begin, disconnect the pot, so that the end of the cable is practically open. Set the signal generator output to a narrow pulse (about 20 ns) with a long duty cycle. Set the amplitude of the pulse to 1 V. Sketch the waveforms you observe on channels 1 and 2. What is surprising about the pulse observed on channel 2? Try to explain the effect by considering that the pulse sees a very high impedance at the scope input, so that it reflects back towards the source.

L2.7. When you go to a store to get a coax, you ask for a 50 Ω (or 75 Ω) cable. What do you think this means? Measure the impedance between the inner and outer conductor when the cable is shorted and open at the other end. Is it related in any way to 50 Ω? To see what this impedance means, set the pot value to 50 Ω. This value matches the cable impedance. Repeat the measurement from the previous part. How large is the pulse amplitude at the end of the cable this time? Explain.

L2.8. A practical application of this concept is in computer networking. Imagine that channels 1 and 2 are computer connections, and the cable is a thinLan ethernet cable. What you are seeing are pulses going from the computer that is sending the pulses (in this case, the function generator) to the various other computers. Why do we terminate thinLan cable in 50 Ω and what happens if we leave the cable dangling? To understand this, answer the following:

- if you have a computer attached to the thinLab cable, what should you make the input impedance of its ethernet card?

- what happens if you remove the 50-Ω connector at the end of the thinLan?

- what happens if you remove a computer from the thinLan?

Conclusions:

1. Whenever capacitance and inductance are present together in a circuit, there will be resonant effects that can produce unwanted "ringing".

2. Any two wires that are close to each other, whether shorted or open, have resonant properties and can be modeled with a combination of L and C (and R, which is usually small).

3. When connecting coaxial cables to equipment, we need to be careful with impedance values of connections at both ends of the cable. If the terminating impedance is not right, there will be reflections.

Lab 3. Resistivity and four-point probe

Background: Circuits, Chapters 10 and 11 in *Introductory Electromagnetics*

In this lab, we learn how to measure resistivity (conductivity) of a homogeneous piece of material. We also learn how to extend the measurement technique to a thin resistive layer, and how to apply it to a circuit consisting of a mesh of resistors. We use the four-point method by measuring the current and voltage on four wires that come to sharp points at the ends. This method is used often in the semiconductor industry for measuring properties of silicon, gallium arsenide and other materials. You would need much finer probes for these materials, however, so we will not measure them in this lab.

Purpose: to get an intuitive and qualitative understanding of the resistivity (conductivity) of a material, the way it is measured, and the limitations of the measurement technique.

Pre-lab problems:

PL3.1. Two metal probes, which are d apart, are connected to a current source I and to a flat thick homogeneous piece of material of unknown conductivity σ, Fig. L3.1. Assume that the probes are sharp and that the resistance between the sharp points of the probes and the material is small compared to the resistance between the probe tips through the sample. The current flows into the material from one probe, spreads through the material and flows out through the other probe. Using superposition of the two currents, find the expression for the current density along the line connecting the two probe tips. Knowing that $\boldsymbol{J} = \sigma\boldsymbol{E}$, find the expression for the voltage V between the two probes. Can the conductivity be found by measuring V? What problem might you run into when trying to measure V?

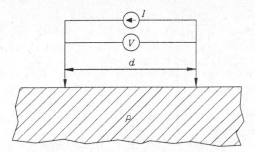

Fig. L3.1. A two-point probe.

PL3.2. Repeat problem P11.11 (refer also to Fig. P11.11). How does a probe with four points help improve the measurement accuracy over the probe with two points?

PL3.3. Repeat problem P11.14. The surface conductivity $\sigma_s = 1/\rho_s$ of a thin layer is related to the surface current density, $\boldsymbol{J}_s$ (in A/m) by $\boldsymbol{J}_s = \rho_s\boldsymbol{E}$. What is the unit for surface resistivity?

PL3.4. Repeat problem P11.15.

Lab:

Equipment and parts:

- a dc power supply; 2 multimeters;

- a power resistor, a four-point connector, sand paper, a beaker with water, salt, beef jerky, a piece of aluminum, a piece of paper and a graphite pencil, many identical resistors soldered in a mesh.

Part 1: Four-point probe measurements of volume resistivity

In this lab, we will use a four-point probe to measure conductivities (resistivities) of several materials. Connect the setup as shown in Fig. L3.2a. The probes need to be clean and need to make a good contact. Use sand paper to clean them after each measurement. Press down firmly to make a good connection between the tips and the material, but be careful not to break the probes.

L3.1. After cleaning the probes, place them in tap water you poured into a container, so that the probe tips just touch the water surface. Based on the measured current and voltage and your prelab results, calculate the conductivity of water.

L3.2. Next dissolve some salt in the water, so that it starts tasting salty. Clean the probe tips and repeat the measurement. What happens to the current reading as a function of time? Explain. Think of a way to measure the current so that the measurement is consistent with the fact that the current changes in time. Calculate the conductivity of the salt water.

L3.3. Put the probes on the top of beef jerky (press into the material a little bit) and calculate the conductivity. Does the reading change with time?

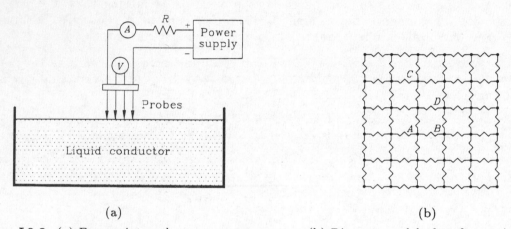

(a)   (b)

Fig. L3.2. (a) Four-point probe measurement setup. (b) Discrete model of surface resistivity.

L3.4. Now we try the same procedure to measure the conductivity of aluminum. Use and aluminum block and clean probes. Aluminum is a good conductor (its conductivity is $38 \cdot 10^6$ S/m). What do you expect the voltage drop across the sample will be per unit of current? Measure the conductivity (press hard on the probes).

L3.5. Measure the resistance of each probe. How large is the probe resistance R_p? Should have the voltage drop across R_p been taken into account in any of your measurements?

Part 2: Four-point probe measurements of surface resistivity

Make a graphite sheet with a pencil on paper, as continuous as you can, and large enough so that it is many probe distances across.

L3.6. Using the results from prelab problem PL3.3, calculate the surface resistivity of graphite. What would you need to know about the graphite layer in order to find the resistivity of graphite?

L3.7. Suppose you measured material samples of different thicknesses, say a very thin sheet, and 1-mm, 5-mm and 5-cm thick samples. Would the calculated conductivities differ for the different samples? Which measurement would you trust most and why?

Part 3: Circuit model of surface resistivity

Solder at least 25 identical resistors in a square mesh as shown in Fig.L3.2b. (A lab technician or the instructor may do this ahead of time for you.) This is a circuit model of a thin layer of resistive material. Imagine you insert a current at a point in the middle of the mesh. Draw the current paths. What do they look like as the mesh becomes finer and larger?

L3.8. Using an ohmmeter, measure the resistance between adjacent points, such as A and B, at different places across the mesh (i.e. in the center and closer to the edges). What resistance are you expecting to approximately measure in the middle of the mesh based on your prelab problem PL3.4?

L3.9. How do your measurements compare with the calculation for different pairs of adjacent points across the mesh?

L3.10. If you have time and you are interested, repeat measurements and calculations for problem P11.16. How would you make a discrete model for volume, not surface, resistivity?

Conclusions:

1. Resistivity and surface resistivity can be measured using a four-point probe voltage and current measurement. The calculation of the resistivity from this measurement is based on superposition of two currents.

2. The method of measurement is slightly different for thick samples and thin layers, as well as for high-resistivity and low-resistivity (high conductivity) materials.

3. We can make a circuit model (mesh of identical resistors) of homogeneous resistive layers as well as resistive blocks of material. To calculate the value of resistance between any two points in the circuit model, we can use the two-point probe analysis based on superposition instead of a relatively complicated circuit analysis.

Lab 4. Magnetic field and currents: an ammeter

Background: Circuits, Chapter 12 in *Introductory Electromagnetics*

A current produces a magnetic field, which can in turn produce a force on either a magnet or another wire with current flowing through it. In this lab, we will produce a static magnetic field with a dc current flowing through a dense spiral winding, usually called a solenoid. The magnetic force will act on a small magnet (a compass needle). By measuring the force on the small magnet, we can measure the intensity of the current flowing through the solenoid.

Purpose: to see how current produces a magnetic field, and to learn how to measure the current by measuring magnetic force. The purpose of this lab is to make a simple instrument, to learn how to calibrate it and extend its operating range, and to understand its limitations.

Pre-lab problems:

PL4.1. A circular current loop of radius a is positioned in the xy plane, centered on the z axis, Fig. L4.1. Use the Biot-Savart law to find the expression for the magnetic flux density vector along the z axis.

PL4.2. Find the magnetic force that acts on a small current element positioned on the z axis in the yz plane, making an angle of 45° with the z-axis (Fig. L4.1).

Lab:

Equipment and parts:

 - a dc power supply; a multimeter;

 - a compass, insulated wire, two potentiometers.

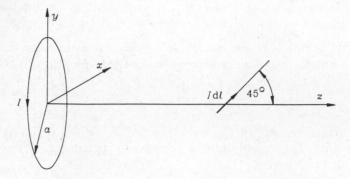

Fig. L4.1. A circular current loop in the xy plane and a current element on the axis of the current loop.

Part 1: Calibrating the ammeter

You can make an ammeter by winding insulated wire (about 30-40 turns) around a compass. Connect the ammeter as shown in Fig. L4.2a. The pot resistor is used in series with the power supply and ammeter in order to limit the current. Set it to 50 Ω. You can measure the current either by inserting a commercial ammeter in series in the circuit, or by measuring the voltage across the pot.

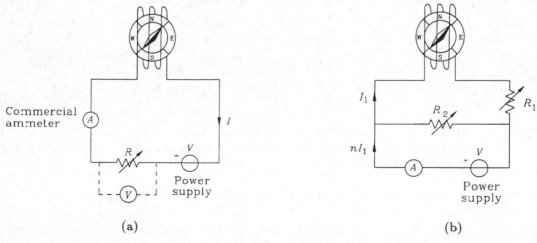

Fig. L4.2. (a) Setup for calibrating the ammeter. (b) Setup for increasing the current range of the ammeter.

L4.1. Start with a low voltage (about 1 V) on the power supply, and slowly increase it. You should see the compass needle deflect as you increase the current through the solenoid up to a certain point. How do you orient the compass initially so that you measure the largest current range?

L4.2. In order to be able to use your ammeter, you need a scale. Draw the scale on a piece of paper by measuring known currents (this is what the commercial ammeter is for) and recording the compass needle deflection. What is the largest value of the current you can measure? What is the smallest nonzero current you can measure? What is the sensitivity of your ammeter, i.e., how accurately can you measure small changes? Is your scale linear (it is linear if the deflection angle is linearly proportional to the current with some multiplication constant)?

L4.3. Compare your scale with that of your classmates. How do they compare? How practical is your ammeter for manufacturing?

Part 2: Measurement of the earth's magnetic field

The earth is a large magnet and the initial position of the needle is determined by the earth's magnetic field. (You can read more about this in Chapter 17 in *Introductory Electromagnetics*.)

L4.4. Determine the position of the compass that would allow you to have the needle deflect a known amount when the magnetic field generated by the current is equal to the earths' magnetic field. Measure the required current for this case.

L4.5. From the value of current obtained in L4.4 and your prelab homework PL4.1 results, determine the magnetic flux density that the needle is in. This is the value of the earth's magnetic flux density at the place where the needle is located. What approximations do you have to make when answering this question?

Part 3: Measuring larger currents (changing the current range)

L4.6. You have found that after the current is increased beyond a certain point, there is no further deflection of the needle (your instrument is saturated). Also, after some current level, you will burn the resistor that regulates the current. The setup shown in Fig. L4.2b can be used to measure larger currents. Explain how it works. Find the current I_1 as a function of the resistor values R_1 and R_2.

L4.7. Determine the values of R_1 and R_2 that allow you to measure a current range 3 times larger than the one you had in Part 1. Set the values of the pots to those you calculated and calibrate a new scale. What is the smallest nonzero current you can measure? What is the sensitivity of your ammeter?

Conclusions:

1. A current produces a magnetic field. By measuring the torque of this field on a small magnet, we can measure the current.

2. Every instrument needs to be calibrated. The calibration is different for different ranges of the instrument.

3. The simple ammeter you made can be used to indirectly measure the intensity of the magnetic field of the earth at the place where the ammeter is located.

Lab 5. Coupling between signal lines

Background: Circuits, Chapters 6, 8, 15 and 16 in *Introductory Electromagnetics*

A printed circuit board has a large number of signal lines that often run parallel to each other and have a common ground plane at some other layer of the board. When there is a voltage on one of the lines, there are charges on it; which induce charges on other lines (especially neighboring ones). In other words, there are mutual capacitances between these lines and the signal on one line will couple to the other lines through these capacitances. Similarly, if currents are flowing in two adjacent lines, their mutual inductance makes a signal in one of them couple to the other. Unwanted coupling of signals is a common problem, and in this lab we learn how to recognize it from measured waveforms.

Purpose: to measure and quantify capacitive (electric) and inductive (magnetic) coupling on the example of a simple 3-trace model of a printed circuit board. We will do the experiments so that we observe capacitive and inductive coupling separately, with a goal of being able to recognize them in other more complex circuits.

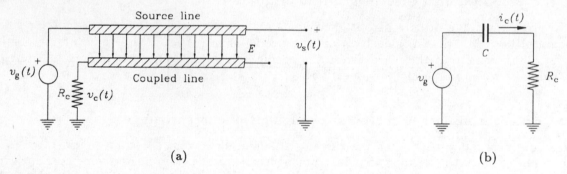

Fig. L5.1. (a) Electric (capacitive) coupling between two lines on a printed circuit board, and (b) equivalent circuit model.

Pre-lab problems:

PL5.1. A simple model of two lines on a printed circuit board, with a common ground plane underneath, is shown in Fig. L5.1a. A voltage generator of voltage $v_s(t)$ is connected to one of the lines, called the source line. The source line produces an electric filed, which induces charges and current in the other line. A voltage can therefore be measured across the resistor connected to the other line, called the coupled line. A circuit model is shown in Fig. L5.1b. Find an approximate expression for the coupled current $i_c(t)$ as a function of the source voltage $v_s(t)$, assuming that the capacitance is small so that most of the voltage drop occurs across the capacitor that models the coupling. For a 2.5-MHz triangular voltage $v_s(t)$ with $10\,\text{V}$ peak-to-peak, a capacitance of $2\,\text{pF}$ between the lines and the coupled-line load $R_c = 50\,\Omega$, find the coupled voltage $v_c(t)$. Sketch the voltage waveforms of $v_s(t)$ and $v_c(t)$. Is the assumption that the voltage drop across the coupling capacitor is much larger than the voltage drop across R_c a good one in this case?

PL5.2. The signal line is next connected to the ground through a resistor, so a current can flow through it, Fig. L5.2. In addition, we ground the end of the coupled line that was floating in PL5.1. The current in the signal line causes a magnetic field and an induced electric field, which will induce a voltage on the coupled line, dictated by Faraday's law. In an equivalent circuit this type of coupling is represented by a mutual inductance, M, between the lines. The induced voltage is given by $v_m(t) = M\,di_s(t)/dt$, where $i_s(t)$ is the source line current. Assume the same triangular wave for the generator voltage as in PL5.1, a mutual inductance of $M = 50\,\text{nH}$, source load $R_s = 50\,\Omega$, and find the inductively coupled voltage $v_m(t)$.

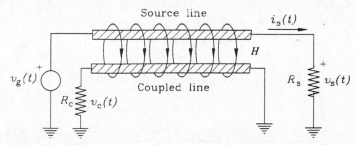

Fig. L5.2. Magnetic (inductive) coupling between two lines on a printed circuit board.

Lab:

Equipment and parts:

- a signal generator; a 2-channel scope (as in Labs 1 and 2);

- 3 metal traces on a board, or a standard breadboard and 3 wires that model the signal, coupled and ground traces;

- 2 Tees with 50-Ω resistors that can change the input impedance of the scope, an additional 50-Ω resistor.

Part 1: Capacitive coupling

Set up the experiment as in Fig. L5.3, with 50-Ω terminations on both scope channels. If you do not have the capability to etch a board with traces with mounted connectors, use a breadboard and parallel pieces of wire. In order to measure just capacitive coupling, leave the coupled line open at one end, so that there can be no inductively induced current through resistor R_c. Externally trigger the scope, and use a 10-V peak-to-peak 2.5-MHz triangular generator waveform.

L5.1. Sketch the waveform of the capacitively coupled voltage, $v_c(t)$, along with the waveform of the source voltage, $v_s(t)$. Based on results from your prelab problem PL5.1, calculate the value C of the coupling (mutual) capacitance between the lines. If your are using a breadboard, measure also the internal capacitance between the breadboard lines.

Part 2: Inductive coupling

Next observe inductive coupling alone by shorting the end of the coupled line. The current is now allowed to flow, so we see inductive (magnetic) coupling.

L5.2. Are we still observing the capacitively coupled voltage across R_c? What happened to the capacitively-coupled current?

L5.3. Sketch the magnetically coupled voltage, along with the signal voltage. Based on the results from your prelab problem PL5.2, calculate the value M of the mutual inductance between the lines.

Part 3: Capacitive and inductive coupling when loading is varied

Now observe the effect of different resistor loadings at the ends of the two lines. First, change the coupled line resistance (channel 2 of the scope) to $1\,\mathrm{M\Omega}$, and leave the source resistance (channel 1) at $50\,\Omega$.

L5.4. Is the capacitively coupled signal larger or smaller than in Part 1? What about the shape of the waveform? Explain. (Hint: is the assumption you made in PL5.1 valid in this case?)

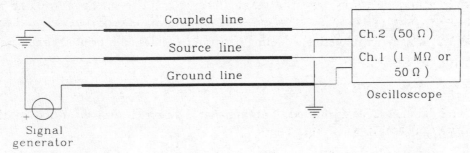

Fig. L5.3. Setup for measuring capacitive and inductive coupling.

L5.5. Now observe the inductive coupling (short the end of the coupled line). Sketch the resulting coupled waveform. What do you think is going on? (Hint: there is also a self-inductance in the loop, which changes the equivalent circuit.)

In the last part, change the coupled line resistance (channel 2 of the scope) to $50\,\Omega$, and the source resistance (channel 1) to $1\,\mathrm{M\Omega}$.

L5.6. Sketch the capacitively coupled voltage. Explain.

L5.7. Sketch the inductively coupled voltage. This one will be confusing. To explain it, think of what the current along the length of the source line looks like.

L5.8. Imagine now you had a digital signal on the signal line, consisting of a train of rectangular pulses. What would the capacitively and inductively coupled signals look like in this case? Sketch the digital pulse and the two coupled signals.

Conclusions:

1. Voltages (charges) give rise to capacitive coupling, and currents give rise to inductive coupling, for example, between signal lines on a pc board. This coupling can dramatically change the signal waveforms.

2. A signal can appear on a line even if the line is not connected to a source or a load.

3. In general, capacitively and inductively coupled signals have the forms of derivatives of the original signal that they are taking energy out of.

Lab 6. Reducing coupling between signal lines (shielding)

Background: Circuits, Chapters 6, 8 (especially section 8.3), 15 and 16 in *Introductory Electromagnetics*

This lab is a follow-up on the previous one. We will learn more about coupling and how to reduce it in some cases. This purposely done reduction of coupling is often called shielding.

Purpose: to learn the physical basis of shielding on the simple example of reducing coupling between two signal lines.

Pre-lab problems:

PL6.1. Two long parallel wires are situated at a height h above the ground plane which is at potential $V_R = 0$. Assume that the left wire is charged with some charge per unit length, and sketch the lines of vector E in the following three cases:

(a) When wire labeled 2 in Fig. L6.1a is uncharged.

(b) When wire labeled 2 in Fig. L6.1a is connected to ground.

(c) When wire labeled 2 is grounded, but another grounded wire is added in the middle between wires 1 and 2, as in Fig. L6.1b.

Explain your sketches of the field plots. If your reasoning is correct, in this simple way you will be able to understand the physics of shielding.

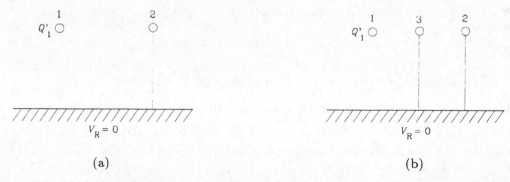

Fig. L6.1. (a) Two parallel wires above a ground plane, and (b) a third parallel grounded wire inserted midway between them at the same height.

PL6.2. Find the expression and value of the source-line resistance R_s in Fig. L5.2 (Lab 5) which makes the electric (capacitive) and magnetic (inductive) voltages, v_c and v_m, equal in amplitude.

Lab:

Equipment and parts: same as in Lab 5, and a potentiometer.

Part 1:

Since we will need to do some comparisons, repeat the measurement of the coupling capacitance from Lab 5. Use the same settings for the signal generator as in Lab 5.

L6.1. Sketch the capacitively coupled voltage and calculate the capacitance. How does it compare to the value you calculated in the previous lab?

L6.2. Connect the setup as shown in Fig. L6.2, with the pot resistor in the source line. Change the value of the pot resistance and monitor the coupled waveforms. What value of the source resistance do you need to have to get just capacitive coupling? What value do you need to observe just inductive coupling? Explain why the waveform for the coupled signal looks the way it does in these limiting cases. Next, adjust the value of R_s so that the capacitively-coupled and inductively-coupled voltages are equal. What waveform do you expect to see for this case?

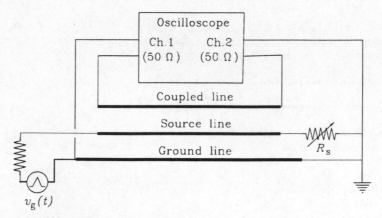

Fig. L6.2. Setup for measuring capacitive and inductive coupling.

Part 2: Capacitive coupling with a grounded shield present

Set up the experiment shown in Fig.L6.3 with a fourth wire (or printed trace) between the source and coupled lines. We will call this line the shield line.

L6.3. Measure the mutual capacitance between the source and coupled lines if the shield line is floating (open at both ends). How different is it from the measured capacitance when there was no shield line in between? Is that what you expect?

L6.4. Now ground the shield line at both ends and measure the mutual capacitance from the coupled signal on the coupled line. Sketch the coupled signal. Explain your result. What happens when only one end of the shield line is grounded? Explain.

Part 3: Inductive coupling with a grounded shield present

L6.5. Again leave the shield line open at both ends and measure the inductive coupling with the shield line present. Does the shield line help reduce the coupling? Explain.

L6.6. Now ground the shield line at both ends. Sketch the coupled signal. Does this help reduce the coupling? Explain. (Hint: there is a current on the shield line that produces its own magnetic field. How large do you expect this field to be compared to the magnetic field of the original source?)

L6.7. Next change the role of the shield line, so that it acts as the ground line, by not using the ground line and instead connecting the shield line to all the ground points. Is the coupling reduced? Explain. (Hint: think of the currents on the ground and source lines.)

Conclusions:

1. A part of unwanted electromagnetic coupling between objects can be reduced by shielding. (Which part?)

2. In order to be effective, shields need to be grounded.

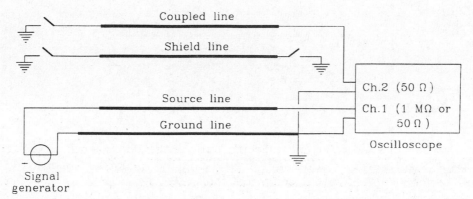

Fig. L6.3. Setup for measuring shielding effect.

Lab 7. Motors and generators

Background: Circuits, Chapters 14 and 17 (section 17.6) in *Introductory Electromagnetics*

Motors and generators are among the most widely used applications of electromagnetic fields. Their principle of operation is based on the law of magnetic force on a current element and Faraday's law of electromagnetic induction. In generators, a changing magnetic field induces a voltage that drives a current through a load. The change in magnetic field can be made by either changing the field in time, or by changing in time the position of a conducting loop with respect to the field. Both methods are used in practice. In motors, electrical energy is used to produce a changing electromagnetic field, resulting in torque on the rotating motor part.

Purpose: to learn about the fundamental properties of motors and generators by making a simple small dc motor with a commutator.

Pre-lab problems:

PL7.1. A wire loop of radius a is spinning around the z axis at an angular velocity ω, in a uniform time-constant magnetic field with the magnetic flux density vector B parallel to the

y axis, Fig. L7.1. Find the expression for the magnetic flux Φ as a function of time, assuming that at $t = 0$ the loop had the minimum flux through it (as in Fig. L7.1.) Find the expression for the *emf* induced around the loop.

PL7.2. Find the maximum induced *emf* in a simple ac generator with a single winding consisting of 7 loops. The loops have an area of 10^{-3} m^2 each, the magnitude of the magnetic flux density is 0.1 T, and the winding makes $n = 60$ revolutions per second.

PL7.3. In large motors, the magnetic flux density, B, is often supplied by electromagnets. What would happen if the current in the electromagnet unintentionally dropped to zero?

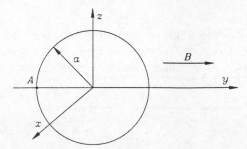

Fig. L7.1. A wire loop rotating in a uniform constant magnetic field.

PL7.4. Now assume that a dc current is flowing through the loop in Fig. L7.1, so that there is a force $d\boldsymbol{F} = I d\boldsymbol{l} \times \boldsymbol{B}$ on every section $d\boldsymbol{l}$ of the loop. Sketch the forces on the loop for different positions of the loop and show that the forces will cause the loop to rotate. Determine the maximum torque on the loop. What is the maximum force one would have to apply at point A to keep the loop from rotating?

Lab:

Equipment and parts:

- A magnet, a battery (or a power supply), a pot resistor (for higher power), some thin insulated wire, paper clips.

- A strobe light; an ohmmeter; a multimeter;

- a commercial dc motor (for example, RadioShack motors such as: 273-223, 1.5-3 V, \$0.99; 273-255, 12 V, \$2.99, or the higher speed 273-256, 9-18 V, 24000 rpm, \$3.49).

Part 1: A simple dc motor with a commutator

You will first put together a very simple dc motor, as shown in Fig. L7.2. Two paper clips, labeled 1 and 2, are bent and used to support the ends of a wire coil. Make the coil circular using thin insulated wire, and about 10-20 turns, as symmetric as you can. Use another paper clip to fasten the setup, and improvize the motor stands. You can use a battery or a dc power supply connected to paper clips 1 and 2 so that there is a closed loop for the current to flow through. Be careful not to connect a high voltage immediately, as the resistance of the loop is very low and the current will be large. Off course, you still have insulation over the wire, so no current will flow.

L7.1. Assume you strip all the insulation at the two ends of the coil, and that you connect it to a power supply so that a constant dc current is flowing through it. Plot the torque on the coil as a function of time. What is the position of the coil for maximum torque, and what is the problem with this torque curve if you wish to make a motor?

If you plotted the torque correctly, you will understand that we need to do something with the two contacts in order for the coil to keep rotating. On the two ends, you need to strip off the insulation as follows. On one end, labeled "negative" in the figure, strip off all the insulation so that full contact can be made at all times as the coil is rotating. On the other end, labeled "positive", carefully strip off only half of the insulation, as shown in the inset in Fig. L7.2. The current will now flow during only half of the coil's rotation cycle.

L7.2. What does the torque on the coil like now as a function of time? Will the motor keep rotating once it is started?

L7.3. What you have made is called a one-section commutator. The next simplest commutator is a two-section device, in which current is reversed every half cycle. Sketch the torque of such a motor as a function of time. (Commercial grade motors have commutators with many sections, so that the torque is almost constant.)

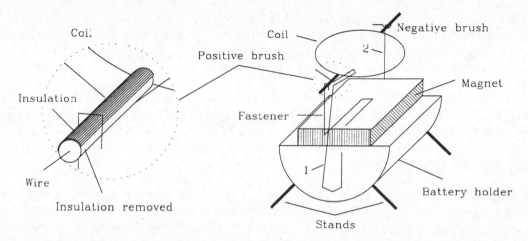

Fig. L7.2. A simple dc motor with a commutator. The inset shows how the wire at the positive terminal is stripped of insulation to make a commutator.

L7.4. Now connect the power supply (or battery) to the motor contacts, keeping the voltage at 0 V. Slowly turn up the voltage, and observe the current increase rapidly if you have the commutator in a position that allows current to flow. Start the coil spinning. (With enough voltage, it should continue to spin.) Measure the rotation speed with a strobe light. What could you make your motor do (e.g., can you think of a way for it to lift a paper clip)?

Part 2: A commercial dc motor

In this part of the lab, we test a small commercial dc motor.

L7.5. Measure the resistance of the motor before it starts turning, using an ohmmeter. (You should get a small resistance, since the motor is essentially a wound wire made of a good conductor.)

L7.6. Connect the motor to the power supply, in series with a pot resistor set to about 50 Ω. The resistor limits the current, and also allows us to measure the voltage across the motor (if we did not have the resistor, we would always measure just the voltage of the power supply).

Measure the voltage across the motor as you turn up the voltage on the supply and run the motor. What happens to the voltage across the motor? Does this make sense, taking into account the result from the previous question that it is essentially a short circuit? Explain what is happening recalling Faraday's law.

L7.7. How large is the power supplied to the motor by the power supply? If all this power is converted to mechanical power, we have a perfect motor with 100% efficiency. The mechanical power that the motor generates is given by $P_{mech} = \tau \cdot \omega$, where τ is the torque, and ω the angular velocity. What happens to the current supplied to the motor as you load down the motor, e.g., by holding the drive shaft slightly? Use a conservation of power argument to explain this.

Part 3: A dc generator

Now we will use the commercial motor and a portion of the motor you made in Part 1 to make a generator. Connect the commercial motor to the power supply (this is simulating a river turning a turbine, for example). Carefully connect the shaft of the motor to one of the ends of your coil that the motor will turn (it is convenient to use heat-shrink tubing for this). Connect an ammeter to the coil so you can measure the current it generates as it is turned by the motor. As you turn up the voltage on the motor, you will need to prevent the other end of the coil that is not connected to the motor shaft from flying off (hold it down gently with a pencil).

L7.8. Measure the power supplied to the motor. If we had 100% efficiency, we would expect the power generated by the coil to be equal to this number. What efficiency are you measuring?

Conclusions:

1. A simple dc motor can be made with a power supply (battery), a magnet, and a wire coil free to rotate in the dc magnetic field. Once the coil is started, the torque produced by the magnetic force acting on the coil with current flowing through it keeps the coil rotating.

2. A generator can be made by rotating a wire loop in a dc magnetic field with some other (usually not electromagnetic) force.

3. Motors and generators do not convert power with 100% efficiency.

Lab 8. Transmission lines with resistive loads, in time domain

Background: Circuits, Chapters 18 and 25 in *Introductory Electromagnetics*

Transmission lines are commonly used to carry power and/or information between two points. For example, the coaxial cable that delivers TV channels to your home is a transmission line. Transmission lines can be analyzed in time and frequency domain. In this lab we observe them in time domain.

Purpose: to learn basics of transmission-line measurements by using time-domain reflectometer (TDR) to measure cable length and reflections from different resistive loads.

Pre-lab problems:

PL8.1. A transmission line is 10 m long, has a characteristic impedance of 50 Ω, and is filled with air as the dielectric, so that the waves propagating along the line travel at the speed of light, Fig. L8.1. (1) How long does it take a pulse to get from one end of the line to the other? (2) Suppose that the pulse is a triangular wave with a duration of 1 ns, with a linear rise from 0 V to 1 V in 1 ns, and then instantaneously falling to zero. At 0 ns the wave is just starting at the left end of the line. Sketch the voltage on the line at the following times: 0 ns; 20 ns; and 40 ns. Repeat this for three cases: when the line is terminated in a 50-Ω load; when the line is shorted at the end; and when the line is open at the end.

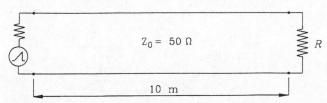

Fig. L8.1. A transmission line terminated in a load.

PL8.2. A voltage pulse has unity amplitude at the generator, $z = 0$. After propagating a distance $z = \ell$ during a time t_1, the voltage pulse is reflected off an open circuit and gets back to the generator after a time t_2. Find the expression for the velocity of propagation along the line. Assume your cable is $\ell/20$ longer than you think. What is the percentage error in your calculated velocity in this case?

PL8.3. Find the reflection and transmission coefficients at the end of a transmission line of characteristic impedance Z_0 for load impedances of $Z_0/2$ and $2Z_0$. If the incident voltage is a rectangular pulse, narrow compared to the round trip time along the line, sketch the reflected and transmitted pulses at the load for these two cases.

Lab:

Equipment and parts:

- function generator (such as, e.g., HP8114A or HP8116A);

- a 2-channel oscilloscope;

- an approximately 10-m long 50-Ω coaxial cable; 2 Tee connectors, two pot resistors, an ohmmeter.

Part 1: A simple Time-Domain Reflectometer (TDR) – measuring cable length

One of the most important techniques for measuring transmission line circuits is time domain reflectometry (TDR), or "cable radar." Usually, one would buy an instrument designed for this purpose, but we will put one together using a function generator, scope, some short coaxial cables, and a 10-m long coaxial cable. Connect the output of the signal generator to channel 1 of the scope. Trigger the scope externally with the signal generator trigger signal. Make sure the input impedance of channel 1 is high.

L8.1. Now use a Tee connector at channel 1 and connect a long (roughly 10 m) piece of coaxial cable from channel 1 to channel 2 of the scope, Fig. L8.2. Set the function generator to a

1-MHz square wave. Sketch the scope display. If the relative permittivity of the dielectric in the line is $\epsilon_r = 2.5$, calculate the length of the cable from the measured reflected pulse. (You can also use a very short, e.g., 20-ns, pulse if your signal generator has this capability.)

L8.2. The input impedance of the scope is $1\,\text{M}\Omega$ in parallel with a 20 pF capacitor (read off the exact values from your scope input). This capacitance is connected in parallel at the end of the cable. What does it do to our length measurement? (Hint: sketch the distributed L and C model for the cable. What is the value of the capacitance per unit length for this cable?)

L8.3. Measure the rise time of the pulse. Based on that, what is the shortest length of cable you can reliably measure?

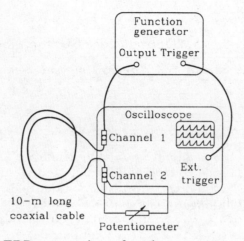

Fig. L8.2. TDR setup using a function generator and a scope.

Part 2: A simple Time-Domain Reflectometer (TDR) – measuring load impedance

Set the signal generator to pulse mode. We want the pulse length to be long, so that any reflected pulses dye out before the next pulse is generated. For example, you can set the pulse to be $5\,\mu s$ long with a 50% duty cycle and 1 V amplitude. Set the horizontal scale on the scope so that you observe one cycle and you can see the rising edge of the pulse well.

L8.4. Set the input impedance of channel 2 to $1\,\text{M}\Omega$. Explain the amplitudes and delays of the pulses observed on both channels.

L8.5. Now connect a Tee to channel 2 and connect a pot resistor in parallel with the scope input using a Tee connector. Change the value of the pot resistance until you get no reflection. Measure the value of the resistance with an ohmmeter. What does this tell you?

L8.6. Next set the pot to the smallest resistance value. Sketch the reflected pulse. Based on the reflected pulse, what is the value of resistance? How does it compare to the value you measure with an ohmmeter? Repeat for the case when channel 2 is shorted with a short wire.

L8.7. Then set the pot to the largest resistance value. Sketch the reflected pulse, and find the value of the resistance based on it.

L8.8. Finally, set the pot resistance to some value larger than $50\,\Omega$ (e.g., $75\,\Omega$) and a value smaller than $50\,\Omega$ (e.g., $25\,\Omega$). Calculate the value of the resistance from measurements of the reflected pulse amplitude. How close are the TDR and ohmmeter resistance measurements?

Conclusions:

1. Time domain reflectometry can be used to measure the length of transmission lines, their characteristic impedance, as well as the impedance of the load that is connected at the end.

2. A simple TDR instrument can be made with a function generator and an oscilloscope.

Lab 9. Transmission lines with RLC loads, in time domain

Background: Circuits, Chapter 18 in *Introductory Electromagnetics*

In this lab, we continue time domain relfectometry measurements on cables with capacitive, inductive and resistive loads.

Purpose: to learn how to recognize *RLC* loads by observing reflected waveforms.

Pre-lab problems:

PL9.1. Find and sketch the reflected, transmitted and total waves at the end of a transmission line with characteristic impedance Z_0 terminated in a purely inductive load of inductance L. The incident waveform is a step function with $1\,\text{V}$ amplitude.

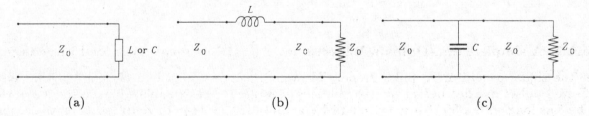

Fig. L9.1. (a) Transmission line with L or C load. (b) L, and (c) C load in the middle of a transmission line.

PL9.2. Repeat prelab problem PL9.1 for a purely capacitive load C.

PL9.3. A generator with $50\,\Omega$ internal resistance produces a 2-V step function at $t = 0$. A 50-Ω, 5-m long cable is connected to the generator. At the end of the cable, an inductor of inductance L is connected in series, and then another $50\,\Omega$ 5-m long cable, terminated in a matched load, is connected to the inductor, Fig. L9.1b. Determine the voltage and current as a function of time at the input of the first line, assuming the waves on the lines travel at the speed of light in a vacuum.

PL9.4. Repeat prelab problem PL9.3 if a capacitor C is connected between the two lines in parallel (instead of the series inductor), Fig. 9.1c.

Lab:

Equipment and parts: same as in Lab 8, and an inductor and a capacitor.

Part 1: Measuring an inductive load

Set up the TDR experiment as in Lab 8. Use again a $5\,\mu s$ period 50% duty cycle pulse. This makes the pulse look like a step function, since the pulse is long compared to the rise time, and all reflections would have died out by the time the pulse is over.

L9.1. Connect an inductor as the load at the end of the long cable. You can use the same inductor you used in Labs 1 and 2. Sketch the reflected waveform observed on channel 1. Based on this waveform and results from prelab problem PL9.1, calculate the value of the inductance.

L9.2. Now connect the inductor in series with a 50-Ω load. (This is essentially the same circuit as your prelab problem PL9.3. Why?) Sketch the waveform observed on channel 1 and explain it.

Part 2: Measuring a capacitive load

L9.3. Next connect a capacitor at the end of the long cable. You can use the plates you used in Labs 1 and 2, with a piece of paper between them. Sketch the waveform observed on channel 1. Calculate the value of the capacitance using the results from prelab problem PL9.2.

L9.4. Now connect a 50-Ω resistor in parallel with the capacitor. (This is essentially the same circuit as your prelab problem PL9.4. Why?) Sketch the waveform observed on channel 1 and explain it. Using results from your prelab problem PL9.4, calculate the value of the capacitance. How does it compare to the value you calculated in L9.3?

Part 3: Measuring an *LC* resonant load

L9.5. Connect the inductor and capacitor in series at the end of the cable. Sketch the resulting waveform observed on channel 1. How does the measured resonant frequency compare to the one obtained by calculating it using previously determined values of L and C?

Part 4: Smallest inductance that can be measured

L9.6. The scope you are using does not have an infinitely fast rise time. What is the rise time of your scope? Based on the rise time as the limit, what is, approximately, the minimal inductance that can be measured using your scope?

Conclusions:

1. TDR can be used to measure inductive and capacitive loads.

2. The instrument speed (rise and fall times) limits the smallest values of inductance and capacitance that can be measured in this fashion.

Lab 10. Transmission lines in frequency domain

Background: Circuits, Chapters 18 and 25 in *Introductory Electromagnetics*

In this lab, we investigate transmission lines in frequency domain. This means that the waves travelling down the lines are sinusoidal in time. This is important for ac power transfer, as well as in high frequency communication signals. In contrast, time domain analysis is more useful in digital transmission in computer networks, but it is also useful for determining the position and type of breaks in cables.

Purpose: to understand the fundamentals of ac analysis of waves on transmission lines.

Pre-lab problems:

PL10.1. A voltage pulse has unity amplitude at the generator, $V(z = 0) = 1\,\mathrm{V}$. After propagating a distance $z = \ell = 10\,\mathrm{m}$, the voltage pulse has a reduced magnitude due to losses and it is equal to $V(\ell) = 0.9\,\mathrm{V}$. Find the expression and numerical value of the attenuation constant α in Np/m and dB/m. Find an approximate expression for α assuming $(\alpha\ell)$ is small.

PL10.2. A 10-m long 50-Ω cable is filled with a dielectric with a relative permittivity of $\epsilon_r = 2.2$. A 50-Ω generator is connected at one end of the line and launches a sinusoidal wave with a 2-V amplitude. Find the input impedance of the line at $f = 100\,\mathrm{MHz}$ if the load is (1) a 150-Ω resistor and (2) a 100-Ω resistor in parallel with a 1-pF capacitor.

PL10.3. At what frequency is the cable from prelab problem PL10.1 a quarter of a wavelength long? At that frequency, what is the input impedance of the cable for: (1) open-circuited other end; (2) short-circuited other end; (3) 200-Ω load; and (4) 20-Ω load?

Lab:

Equipment and parts: same as in Labs 8 and 9.

Part 1: Matched cables in frequency domain

Set up the experiment as in Labs 8 and 9, but now set the function generator to give a sinusoidal output at a 30-MHz frequency and 0.5 V amplitude, and terminate both scope channels in 50 Ω (using a Tee and a resistor if you need to).

L10.1. First disconnect the long cable from channel 1. Sketch the waveform you see on channel 1 and record carefully the position (in time) of the peak of the sine wave (for example, set it to be in the middle of the screen). We need this information, as well as external triggering, in order to be able to track phases throughout the lab.

L10.2. Now attach the long piece of cable back to channel 1, keeping both channels terminated in 50 Ω. What does the waveform on channel 1 look like compared to the one in question L10.1? Explain. Sketch the voltage observed on channel 2 and compare it to that on channel 1. You should see two differences; the amplitude and the phase. Measure the change in phase noting what portion of the cycle it has changed from channel 1. How many wavelengths is the cable long (remember the relative dielectric constant of the cable filling is 2.5)? Based on that, what phase change should you get knowing that a one-wavelength long cable produces 360° of phase change?

L10.3. Now look carefully at the amplitude on channel 2. How large is it? Explain. Remembering that voltage loss in a cable goes like $\exp(-\alpha z)$, where α is the attenuation coefficient in Np/m, and calculate α knowing the length z of your cable. Does the approximate expression for α from your prelab problem PL10.1 result hold in this case?

Part 2: Mismatched loads in frequency domain

In the previous part of the lab, the cables were all terminated in matched 50-Ω scope impedances, and there were no reflections. In this part of the lab, we will look at mismatched cables. So, the voltage (and current) along the cable will be a result of forward and backward (reflected) waves, which add destructively and constructively down the cable. It would be nice if we could measure the voltage all along the cable, but we cannot do that with our setup. However, we can do something that is actually equivalent: we can change the frequency, so that the cable looks like it is changing length. (Why? Recall that the phase of the forward wave goes like $e^{-j\beta z}$, that of the backward wave as $e^{+j\beta z}$, and that $\beta = \omega/c = (2\pi f)/c$. So, if we change f for a constant z, the result is the same as if we changed z.)

L10.4. Terminate the long piece of cable in an open circuit. At what frequency is the cable one wavelength long? Set the signal generator frequency to 1/20 of this frequency. Compare the waveforms observed on channels 1 and 2 and explain. Now set the frequency so that the cable is $\lambda/4$ long. Sketch the waveform on channel 1 and explain. Repeat for frequencies where the cable is $\lambda/2$, $3\lambda/4$ and 1λ long and explain your result each time.

L10.5. Now attach a 75 Ω resistor to the end of the long cable. What is the reflection coefficient at the load? Write down the expression for the total phasor (complex) voltage, noting that there is no value of z for which the expression goes to zero. Now change the frequency, and thereby the electrical length. You will see the voltage go up and down in amplitude, but never to zero. What is the SWR equal to for this load (the standing wave ratio, or SWR, is the ratio of the largest to smallest amplitude along a mismatched line)? What are the smallest and largest possible values of the SWR and for what kinds of loads are they obtained?

Conclusions:

1. Frequency domain analysis of transmission lines can be used to measure cable length, as well as loads terminating the line. Changing frequency of the sinusoidal wave is equivalent to changing the length of the line.

2. The standing wave ratio, or SWR, is the ratio of the maximum to minimum voltage along a mismatched line. It is a very useful parameter for describing the mismatch of a load to a line.

Lab 11. Antennas: microwave oven leak detector

Background: Circuits, Chapters 24 and 25 in *Introductory Electromagnetics*

In this lab, we will build a simple antenna and use it to detect leakage out of a microwave oven. A microwave oven when it leaks (and all of them leak at least a little bit) is an antenna that radiates a wave at 2.45 GHz (this is one of the allowed frequencies for heating). So, when we measure the leakage, we actually make a wireless link between the oven and the receiving antenna, and we can calculate the power radiated by the oven by measuring the received power at a certain distance and knowing the properties of the two antennas (the oven and our detector antenna).

Purpose: to understand the fundamentals of radiation and propagation of electromagnetic waves on the simple example of a home appliance that radiates in the microwave frequency region.

Pre-lab problems:

PL11.1. If a microwave oven has a power of 1 kW, and is leaking -30 dB of that power, what power level will a 50-Ω resistor connected to a 73-Ω half-wave dipole antenna receive at 5 m away from the oven (Fig. L11.1a)? Assume that the oven window is radiating as an aperture that is 0.5 m by 0.5 m large, and that the heating frequency is $f = 2.45$ GHz. What is the wavelength of the leaking wave in air, and how long do you expect the dipole to be? (Recall that a half-wave dipole has a directivity of about 1.6.)

(a)   (b)

Fig. L11.1. (a) The power leaking out of a microwave oven is measured by a dipole antenna terminated in a 50-Ω detector. (b) A quarter-wave monopole antenna above a perfect ground plane.

PL11.2. Knowing that the impedance of a half-wave dipole antenna is approximately $Z = 73\,\Omega$, find the impedance of a quarter-wave long monopole antenna above a perfectly conducting ground, Fig. L11.1b. You may need to shield the LED from external lights.

Lab:

Equipment and parts:

- A microwave oven; some wire; a fast (Schottky) diode (such as the low-cost Motorola MBD101 M652); an LED.

- Not necessary, but could be useful: an operational amplifier with some pots to adjust the gain.

Part 1: Making a dipole antenna

Cut the two pieces of straight wire so that they are about 15 cm long each. Solder the fast Schottky diode between them. (A Schottky diode has a special low-capacitance junction.) Solder a LED back-to-back with the Schottky diode, as shown in Fig. L11.2.

L11.1. Explain what happens to the two diodes when the dipole antenna is receiving enough power to provide on the order of 1 V across its terminals.

Part 2: Measuring leakage out of an (almost) empty oven

Before you turn on the oven, make sure you put a glass or plastic container with a little water in it. The oven should never be completely empty (if "unloaded", it may burn out).

L11.2. Now turn on the oven and move the antenna around. You may see the LED light up, but you probably will not. One reason is that the dipole is most likely not resonant at the microwave oven frequency. Calculate the length of a half-wave dipole at 2.45 GHz. Cut the dipole arms to slightly longer than what you calculated, and then trim it bit by bit until you observe a stronger light from the LED.

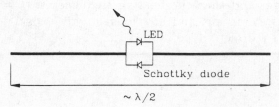

Fig. L11.2. A dipole antenna as a microwave oven leak detector.

L11.3. Once you have observed light from the LED, check at which points your oven leaks mostly and sketch your observations. How far can you move away from the oven and still detect leakage?

L11.4. Assuming you are receiving 0.7 V into the 73-Ω dipole impedance when the LED lights up, how much power are your receiving with your dipole? Measure how far away the dipole is from the oven (roughly). Assuming the oven radiates equally in all directions (this is not quite right, but we are only after orders of magnitude), calculate the power radiated by the microwave oven knowing your dipole has a directivity of 1.6. (This is similar to your prelab problem PL11.1.)

L11.5. Now take a large sheet of metal (e.g., aluminum foil taped to cardboard), and place it at different planes around the oven. Can you get larger or smaller signals out of your detector in this way, depending on where the metal reflector is? Explain.

You may have trouble lighting up the LED. Some other approaches are as follows: (1) try soldering two diodes in series to get enough voltage across the LED; (2) connect the diode with high-impedance lines (inductors) to an operational amplifier circuit that amplifies the voltage and measure it with a voltmeter (you should be able to get much more sensitivity out of this detector, but the cost is higher); (3) make another type of antenna that produces more voltage across the diode, for example a helix above a ground plane, as sketched in Fig.L11.3.

Part 3: Measuring leakage out of a full microwave oven

L11.6. Now place a glass or plastic container filled with water and turn on the microwave. How does the leakage you are measuring compare with that out of an almost empty oven? Explain.

L11.7. Repeat question L11.4 for this case. Does it make sense?

L11.8. *Some other experiments you may try:* Try some other antennas, such as a loop antenna, a monopole above a ground plane, etc. Try also connecting an amplifier and driving a speaker as your leakage sensor, or multiple color LEDs where different colors light up depending on the distance from the oven. Another interesting thing you can do is to connect headphones directly to the antenna and diode and listen to the noise that represents leakage. (The diode rectifies the microwave oven signal that is very noisy and has noise in the audio range, as well as higher.)

Conclusions:

1. Although every microwave oven is shielded to prevent microwave radiation from the oven, there is always some leakage radiation.

2. The leakage radiation from microwave ovens can be detected by means of a simple dipole antenna connected to an appropriate detector and indicator.

Appendices

Appendix 1: Short Survey of Vectors and Vector Calculus

For a summary on the concepts used in this mathematical appendix, the reader is advised to consult the chapter on vectors in his electromagnetics textbook, or a textbook on vector calculus. The exercise provided here are intended as a review relevant to basic electromagnetic field theory.

QUESTIONS

S **QA1.1.** Classify the following quantities as either scalar or vector quantities: mass, time, weight, course of a ship, position of a point with respect to another point, acceleration, power of an engine, current intensity, voltage. — *Hint: recall the definitions of scalars and vectors.*

Answer. Mass, time, weight, power of an engine, current intensity, and voltage are scalars. The other quantities are vectors.

QA1.2. Classify the following quantities as either scalar or vector fields: temperature in a room, mass density of an inhomogeneous body, weight per unit volume of an inhomogeneous body, velocity of air particles in a room, velocity of water particles in a river. — *Hint: recall the definitions of scalar and vector fields.*

QA1.3. Discuss whether the definition of subtraction of vectors follows from the definition of vector addition, or whether an additional definition is indispensable. — *(a) It does follow. (b) The definition of the negative vector quantity is indispensable. (c) The definition of the product of a vector with a scalar is indispensable.*

QA1.4. Sketch the dependence of the dot product of two unit vectors on the angle between them. — *Hint: recall that the magnitude of the unit vectors equals one.*

QA1.5. Why the cross product is not commutative? — *Hint: recall the definition of the cross product.*

S **QA1.6.** Which of the following expressions does not make sense, and why? (1) $A \times (B \cdot C)$, (2) $(A \times B) \cdot C$, (3) $(A \times B) \times (C \cdot D)$, (4) $(A \cdot B)(C \times D)$, (5) $[(A \cdot B) \, C] \times D$, (6) $[(A \times B) \cdot C] \times D$, (7) $[A \times (B \times C)] \cdot D$.

Answer. (1) — the expression in parentheses is a scalar. (3) — the expression in the second parentheses is a scalar. (6) — the expression in the brackets is a scalar.

S **QA1.7.** What is the necessary and sufficient condition for three vectors, A, B, and C, to be in the same plane? — *Hint: construct an appropriate triple scalar product.*

Answer. $(A \times B) \cdot C = 0$, assuming that A and B are not parallel or antiparallel.

QA1.8. Prove that $(A \times B) \cdot A$ and $(A \times B) \cdot B$ are zero. — *Hint: recall the definitions of the dot and cross products.*

QA1.9. If $A \cdot C = B \cdot C$, does it mean that $A = B$? Explain. — *(a) It does. (b) Only if C is negative. (c) It does not.*

QA1.10. If $A \times C = B \times C$, does it mean that $A = B$? Explain. — *The answers are the same as for the preceding question.*

QA1.11. Can the dot product of two vector be negative? Can the magnitude of the cross product of two vectors be negative? — *Hint: recall the definitions of the two products.*

QA1.12. In which cases is (1) $A \cdot B = 0$, (2) $A \times B = 0$, and (3) $(A \times B) \cdot C = 0$? — *Hint: recall the definitions of the dot and cross product.*

S **QA1.13.** Explain the meaning of $(A \cdot B)C$. Is this the same as $C(A \cdot B)$? — *Hint: recall the definition of the dot product.*

Answer. This is the vector C multiplied by the scalar $(A \cdot B)$. The two expressions are the same.

S **QA1.14.** Which of the following sets of coordinates define a point? (1) $x = 2, y = -4, z = 0$, (2) $r = -4, \phi = 0, z = -1$, (3) $r = 3, \theta = -90°, \phi = 0$. — *(a) Only (1). (b) All three. (c) Only (1) and (3).*

Answer. Only (1). In a cylindrical system r cannot be negative, and in a spherical system $0 \leq \theta \leq \pi$.

QA1.15. How do you obtain the components of a vector A in the direction of the three base unit vectors in any coordinate system? — *(a) As the cross product of A with the three base unit vectors. (b) As the dot product of A with the three base unit vectors. (c) As the product of A with the three base unit vectors.*

QA1.16. Define coordinate lines. — *(a) A line along which a coordinate is constant. (b) A line along which two coordinates are constant. (c) A line along which all three coordinates are constant.*

QA1.17. Define differential lengths along coordinate lines. — *(a) A small increase in one coordinate. (b) The length of line segment corresponding to a small increase in one coordinate. (c) The length of line segment corresponding to a small increase in the other two coordinates.*

QA1.18. Define orthogonal coordinate systems. — *(a) At least two base unit vectors need to be orthogonal at all points. (b) The three base unit vectors are orthogonal at all points. (c) All coordinate systems are orthogonal.*

S **QA1.19.** What is a "right-handed coordinate system"? — *(a) A system in which the first axis corresponds to the right hand. (b) A system in which the rotation of the first base unit vector towards the second makes a right-hand screw progress in the direction of the third. (c) All coordinate systems are right-handed.*

Answer. A system is right-handed if the rotation of the first base unit vector towards the second, following the smaller angle between them, makes a right-hand screw progress in the direction of the third base unit vector.

QA1.20. A vector is defined by its three orthogonal components. Determine the vector itself and the unit vector in its direction. — *Convince yourself that* $A = A_1 u_1 + A_2 u_2 + A_3 u_3$, *and* $u_A = A/|A| = A/\sqrt{A_1^2 + A_2^2 + A_3^2}$.

QA1.21. A vector is defined by its starting and end points in the rectangular system. Determine its components, the vector itself, and the unit vector in its direction. — *Hint: adapt the answer to the preceding question.*

QA1.22. A point is defined (a) in a cylindrical coordinate system by its coordinates (r, ϕ, z), and (b) in a spherical coordinate system by its coordinates (r, θ, ϕ). Find the rectangular coordinates of the point in both cases. — *Convince yourself that* $x = r \cos \phi$, $y = r \sin \phi$, $z = z$ *in the first case, and* $x = r \sin \theta \cos \phi$, $y = r \sin \theta \sin \phi$, $z = r \cos \theta$ *in the second case.*

QA1.23. A point is defined in a rectangular coordinate system by its coordinates (x, y, z). Find the coordinates of the point in (a) cylindrical and (b) spherical coordinate systems. — *Convince yourself that* $r = \sqrt{x^2 + y^2}$, $\phi = \tan^{-1}(y/x)$, $z = z$ *in the first case, and* $r = \sqrt{x^2 + y^2 + z^2}$, $\phi = \tan^{-1}(y/x)$, $\theta = \cos^{-1}(z/\sqrt{x^2 + y^2 + z^2})$ *in the second case.*

QA1.24. If u_x, u_y, and u_z are base unit vectors in the rectangular coordinate system, and u_r that in the cylindrical coordinate system, what are the values of the following products? (1) $u_r \cdot u_x$, (2) $u_r \cdot u_y$, (3) $u_r \cdot u_z$, (4) $u_r \times u_x$, (5) $u_r \times u_y$, (6) $u_r \times u_z$. — *Find three errors in the following answers: (1)* $\cos \phi$. *(2)* $-\sin \phi$. *(3) Zero. (4)* $-u_z \sin \phi$. *(5)* $-u_z \cos \phi$. *(6)* $u_x \sin \phi + u_y \cos \phi$.

QA1.25. If u_ϕ is the base unit vector in the cylindrical coordinate system, what the following products are equal to? (1) $u_\phi \cdot u_x$, (2) $u_\phi \cdot u_y$, (3) $u_\phi \cdot u_z$, (4) $u_\phi \times u_x$, (5) $u_\phi \times u_y$, (6) $u_\phi \times u_z$. — *Find three errors in the following answers: (1)* $\sin \phi$. *(2)* $-\cos \phi$. *(3) Zero. (4)* $-u_z \cos \phi$. *(5)* $u_z \sin \phi$. *(6) Zero.*

QA1.26. If u_ζ is the base unit vector in the cylindrical coordinate system in the direction of the z axis, what are the values of the following products? (1) $u_\zeta \cdot u_x$, (2) $u_\zeta \cdot u_y$, (3) $u_\zeta \cdot u_z$, (4) $u_\zeta \times u_x$, (5) $u_\zeta \times u_y$, (6) $u_\zeta \times u_z$. — *Find three errors in the following answers: (1) 1. (2) Zero. (3)* -1. *(4)* u_y. *(5)* u_x. *(6) Zero.*

QA1.27. If u_x, u_y, and u_z are base unit vectors in the rectangular coordinate system and u_r that in the spherical coordinate system, what are the values of the following products? (1) $u_r \cdot u_x$, (2) $u_r \cdot u_y$, (3) $u_r \cdot u_z$, (4) $u_r \times u_x$, (5) $u_r \times u_y$, (6) $u_r \times u_z$. — *Find three errors in the following answers: (1)* $\sin \theta \cos \phi$. *(2)* $-\sin \theta \sin \phi$. *(3)* $\cos \theta$. *(4)* $-\sin \theta \sin \phi u_x + \cos \theta u_y$. *(5)* $\sin \theta \cos \phi u_z + \cos \theta u_x$. *(6)* $-\sin \theta \cos \phi u_y - \sin \theta \sin \phi u_x$.

S **QA1.28.** What is the physical meaning of the vector $-\nabla f$, where f is a scalar function? — *(a) The vector in the direction of the most rapid increase of f. (b) The vector in the direction* *of the most rapid decrease of f. (c) The vector in the direction normal to that of the most rapid change of f.*

Answer. This is a vector in the direction of the most rapid decrease of the function f at the point considered.

QA1.29. Vector r is the position vector in a scalar field described by a function f. What is the directional derivative of the function f in the direction defined by the vector $r \times \nabla f$? — *(a) Cannot be determined. (b) Zero. (c) ∇f.*

QA1.30. What is the unit of the del operator in the cartesian (rectangular) coordinate system? — *(a) Dimensionless. (b) Meter. (c) One over meter.*

QA1.31. What is the divergence of the velocity field of the water flow in a pipe? — *(a) Zero. (b) Constant along the pipe, nonzero over its cross section. (c) Nonzero along the pipe, constant over its cross section.*

QA1.32. A pipe with a liquid flowing through it has a very small hole through which the liquid leaks out of the pipe. If the surface in the definition of the divergence is assumed to be finite, and if it encloses part of the pipe and the hole, is the divergence of the liquid velocity field nonzero at that point? Explain. — *(a) It is nonzero. (b) It is zero. (c) Depends on the size of the hole.*

S **QA1.33.** Propose a model in a time-variable flow of a compressible gas for which the divergence of the velocity field might be nonzero. Explain. — *(a) It does not exist. (b) It is nonzero at all points. (c) It is nonzero at points where a compressed gas is allowed to decompress.*

Answer. If there are regions in which the gas is compressed and then decompressed, or conversely, there will be a flow of gas through any small closed surface in such regions, and the divergence would be nonzero.

QA1.34. A fluid flows through a pipe with a rough wall. Would you expect the curl of the fluid velocity field to be nonzero? Explain. — *(a) No. (c) Yes, even if the pipe has a perfectly smooth wall. (c) Yes, only if the wall is rough.*

QA1.35. A small spherical pressured cloud of gas is suddenly freed to disperse. Assuming completely symmetrical gas dispersion, which of the three functions of the gas velocity (a function of coordinates and time), the gradient, the divergence, and the curl, would be zero, and which nonzero? Explain. — *(a) All three are nonzero. (b) The gradient is nonzero, the other two zero. (c) The gradient and divergence are nonzero, the curl is zero.*

QA1.36. Does the divergence theorem apply to time-dependent fields? Explain. — *(a) Yes. (b) No. (c) Depends on the time variation of the field.*

QA1.37. A volume v is limited by a surface S_0 from outside, but has holes limited by surfaces S_1, S_2 and S_3 from inside. Can the divergence theorem be applied to such a domain? If it can, explain in detail the formulation of the theorem in such a case. — *(a) No. (b) Depends on the sizes and number of the holes. (c) Yes, because this is still a closed surface.*

S **QA1.38.** Is it possible to apply the divergence theorem to a vector function of the form $F \times G$? Explain. — *(a) No, this is not a simple function. (b) No, it is not valid for the cross product of two vectors. (c) Yes, it is valid for any vector function.*

Answer. The divergence theorem applies to any vector function.

QA1.39. Does Stokes's theorem apply to time-dependent fields? — *(a) No. (b) Depends on the time variation. (c) Yes, for any time variation.*

QA1.40. An open surface S is limited by a large contour C_0, but has holes limited by small contours C_1 and C_2. Is it possible to apply Stokes's theorem to such a surface? If so, explain in detail the expression for the theorem in such a case. — *(a) Yes, always. (b) Yes, if the holes are very small. (c) No.*

S **QA1.41.** Is it possible to apply Stokes's theorem to the vector function of the form $(\boldsymbol{F} \cdot \boldsymbol{G})\boldsymbol{F}$? Explain. — *(a) No, it is valid only for simple vector functions. (b) No, it does not apply to a vector function multiplied by a scalar function. (c) Yes it is valid for any vector function.*

Answer. It is possible, since this is a vector function, and Stokes's theorem is valid for any vector function.

PROBLEMS

PA1.1. Prove that the distributive law is valid for vector addition and for the three types of vector products (product of a vector with a scalar, dot product, and cross product). *Hint: recall the definitions of these operations.*

S **PA1.2.** Prove that $\boldsymbol{A} \cdot (\boldsymbol{B} \times \boldsymbol{C}) = \boldsymbol{C} \cdot (\boldsymbol{A} \times \boldsymbol{B}) = \boldsymbol{B} \cdot (\boldsymbol{C} \times \boldsymbol{A})$. — *Hint: either assume the three vectors are length vectors and consider the meaning of the three products, or expand the expression in a rectangular coordinate system and have in mind the properties of a determinant (see PA1.19).*

Solution. Having in mind the properties of a determinant, we find that

$$\begin{vmatrix} A_x & A_y & A_z \\ B_x & B_y & B_z \\ C_x & C_y & C_z \end{vmatrix} = \begin{vmatrix} C_x & C_y & C_z \\ A_x & A_y & A_z \\ B_x & B_y & B_z \end{vmatrix} = \begin{vmatrix} B_x & B_y & B_z \\ C_x & C_y & C_z \\ A_x & A_y & A_z \end{vmatrix}.$$

Alternatively, note that, if $\boldsymbol{A}$, $\boldsymbol{B}$ and $\boldsymbol{C}$ are length vectors, the three triple scalar products represent the volume of the same body.

PA1.3. Prove that $\boldsymbol{A} \times (\boldsymbol{B} \times \boldsymbol{C}) = (\boldsymbol{A} \cdot \boldsymbol{C})\,\boldsymbol{B} - (\boldsymbol{A} \cdot \boldsymbol{B})\,\boldsymbol{C}$. — *Hint: express the two sides of the equation in a rectangular coordinate system.*

PA1.4. Let $\boldsymbol{n}$ be a unit vector in an arbitrary direction, and $\boldsymbol{r}_0$ be the position vector of a point in a plane normal to $\boldsymbol{n}$. What is the equation of the plane (i.e., which equation must be satisfied by the position vector of any point belonging to the plane)? — *Let $\boldsymbol{r}(x, y, z)$ be the position vector of an arbitrary point belonging to the plane. (a) $[\boldsymbol{r}(x, y, z) + \boldsymbol{r}_0] \cdot \boldsymbol{n} = 0$. (b) $[\boldsymbol{r}(x, y, z) - \boldsymbol{r}_0] \cdot \boldsymbol{n} = 0$. (c) $[\boldsymbol{r}(x, y, z) - \boldsymbol{r}_0] \times \boldsymbol{n} = 0$.*

PA1.5. Let $\boldsymbol{n}$ be the unit vector along a ζ axis, and $\boldsymbol{r}_0$ be the position vector of the point $\zeta = 0$ on the axis. What is the expression for the position vector, $\boldsymbol{r}$, of a point with a coordinate ζ on the ζ axis? — *Assume that ζ is a length coordinate. (a) $\boldsymbol{r} = \boldsymbol{r}_0 - \zeta \boldsymbol{n}$. (b) $\boldsymbol{r} = \boldsymbol{n} + \zeta \boldsymbol{r}_0$. (c) $\boldsymbol{r} = \boldsymbol{r}_0 + \zeta \boldsymbol{n}$.*

PA1.6. Two unit vectors, u_1 and u_2, are in the half-plane $x > 0$ of the $z = 0$ plane. The unit vector u_1 makes an angle $\alpha_1 > 0$ with the x axis, and u_2 an angle α_2. Find the dot product of the two unit vectors (1) if $\alpha_1 > \alpha_2 > 0$, and (2) if $\alpha_2 < 0$. Do the results remind you of some trigonometric formulas? — *Hint: note that in both cases $u_1 \cdot u_2 = \cos(\alpha_1 - \alpha_2)$, and express the left-hand side in a rectangular coordinate system.*

S **PA1.7.** Determine the unit vector in the direction of a vector R, with the origin at a point $M'(x', y', z')$, and the tip at a point $M(x, y, z)$. — *Hint: recall that $u_R = R/R$, and write the right-hand side in a rectangular coordinate system.*

Solution. The unit vector is

$$u_R = \frac{R}{R} = \frac{(x - x')\,u_x + (y - y')\,u_y + (z - z')\,u_z}{\sqrt{(x - x')^2 + (y - y')^2 + (z - z')^2}}.$$

PA1.8. Express the base unit vectors of the cylindrical coordinate system in terms of those of the rectangular system, and conversely. — *Hint: project the base unit vectors in one system onto the base unit vectors of the other.*

PA1.9. A point in the cylindrical coordinate system is defined by the coordinates (r, ϕ, z). Determine the rectangular coordinates of the point. — *(a)* $x = r \sin \phi$, $y = r \cos \phi$, $z = z$. *(b)* $x = r \cos \phi$, $y = r \sin \phi$, $z = z$. *(c)* $x = r \cos \phi$, $y = r \sin \phi$, $z = z \tan \phi$.

PA1.10. A point in the spherical coordinate system is defined by the coordinates (r, θ, ϕ). Determine the rectangular coordinates of the point. — *(a)* $x = r \sin \theta \sin \phi$, $y = r \sin \theta \cos \phi$, $z = r \cos \theta$. *(b)* $x = r \cos \theta \cos \phi$, $y = r \sin \theta \sin \phi$, $z = r \sin \theta$. *(c)* $x = r \sin \theta \cos \phi$, $y = r \sin \theta \sin \phi$, $z = r \cos \theta$.

PA1.11. A point in the rectangular system is defined by the coordinates (x, y, z). Determine the cylindrical and spherical coordinates of the point. — *Hint: recall the definitions of the cylindrical and spherical coordinates with respect to a rectangular coordinate system.*

S **PA1.12.** A vector is described at a point $M(r, \phi, z)$ in the cylindrical coordinate system by its rectangular components A_x, A_y, and A_z. Determine the cylindrical components of the vector. — *(a)* $A_r = A_x \cos \phi + A_y \sin \phi$, $A_\phi = -A_x \sin \phi + A_y \cos \phi$, $A_z = A_z$. *(b)* $A_r = A_x \sin \phi + A_y \cos \phi$, $A_\phi = A_x \sin \phi + A_y \cos \phi$, $A_z = A_z$. *(c)* $A_r = -A_x \cos \phi + A_y \sin \phi$, $A_\phi = -A_x \sin \phi - A_y \cos \phi$, $A_z = A_z$.

Solution. The cylindrical components of vector A are

$$A_r = A_x \cos \phi + A_y \sin \phi, \qquad A_\phi = -A_x \sin \phi + A_y \cos \phi, \qquad A_z = A_z.$$

PA1.13. A vector is described at a point $M(r, \theta, \phi)$ in the spherical coordinate system by its rectangular components A_x, A_y, and A_z. Determine the spherical components of the vector. — *Hint: project the components of the vector onto the base unit vector in the spherical system.*

PA1.14. Given that $A = A_x u_x + A_y u_y + A_z u_z$ and $B = B_x u_x + B_y u_y + B_z u_z$, determine the smaller angle between the two vectors. — *Hint: use the dot product of the two vectors.*

PA1.15. The *direction cosines* of a vector are cosines of angles between the vector and the base unit vectors. Determine the sum of the squares of the direction cosines. — *(a) Zero. (b) 1. (c) 3.*

PA1.16. If l_A, m_A, and n_A are direction cosines of a vector A, and l_B, m_B, and n_B are those of a vector B, find the cosine of the smaller angle between the two vectors. — *Hint: make use of the dot product.*

PA1.17. A vector A is given by its components A_x, A_y, and A_z. Find the angles the vector makes with the three rectangular coordinate axes. — *Hint: make use of the dot product of the vector with the three base unit vectors.*

S **PA1.18.** Express the vector product of two vectors given by their rectangular components in the form of a determinant. — *Hint: let the first row in the determinant be the three unit vectors, and find the other terms.*

Solution. The expression for the vector product reads

$$A \times B = \begin{vmatrix} u_x & u_y & u_z \\ A_x & A_y & A_z \\ B_x & B_y & B_z \end{vmatrix}.$$

PA1.19. Express the triple scalar product of three vectors given by their rectangular components in the form of a determinant. — *Convince yourself that*

$$A \cdot (B \times C) = \begin{vmatrix} A_x & A_y & A_z \\ B_x & B_y & B_z \\ C_x & C_y & C_z \end{vmatrix}.$$

PA1.20. Given that $A = 3u_x - 5u_y + 8u_z$, $B = 4u_x + u_y - 3u_z$, and $C = -2u_x + 3u_y - 4u_z$, determine: (1) $A + B - C$, (2) $A \cdot B$, (3) $B \times C$, (4) $A \times C$, (5) $A \cdot (B \times C)$, (6) $(A \cdot B) C$, (7) $(A \times B) \cdot C$, and (8) $(A \times B) \times C$. Find the smaller angle between A and B, and between A and C, the magnitudes of the three vectors, and the unit vectors in their direction. — *Hint: recall the definitions of algebraic operations with vectors.*

PA1.21. If $A = 2u_r - u_\phi + u_z$, with the origin at the point $M(3,0,0)$ of the cylindrical coordinate system, and $B = 2u_r - u_\phi + u_z$, with the origin at the point $N(5, \pi/2, 5)$, determine: (1) $A + B$, (2) $A \cdot B$, (3) $A \times B$, (4) the smaller angle between the two vectors, and (5) their magnitude. — *The results for (4) and (5) are: (a) $\alpha = 60.4°$, $A = 2.1$, $B = 2.45$. (b) $\alpha = 40.4°$, $A = 2.1$, $B = 2.65$. (c) $\alpha = 80.4°$, $A = 2.45$, $B = 2.45$.*

S **PA1.22.** If $A = 3u_r + 2u_\theta - u_\phi$, with the origin at the point $M(1, \pi/2, 0)$ of the spherical coordinate system, and $B = -2u_r - 4u_\theta + 2u_\phi$, with the origin at the point $N(3, \pi/2, \pi)$, determine: (1) $A + B$, (2) $A - B$, (3) $A \cdot B$, (4) $A \times B$, (5) the smaller angle between the two vectors, and (6) their magnitude. — *The results for (5) and (6) are: (a) $\alpha = 90°$, $A = 3.74$, $B = 4.9$. (b) $\alpha = 60°$, $A = 4.74$, $B = 5.9$. (c) $\alpha = 90°$, $A = 4.74$, $B = 5.9$.*

Solution. In order to be able to perform the required vector algebra in a rectangular coordinate system, we first transform the vectors to read $A = 3u_x - u_y - 2u_z$, and $B = 2u_x - 2u_y + 4u_z$.

Then we find that (1) $A + B = 5u_x - 3u_y + 2u_z$, (2) $A - B = u_x + u_y - 6u_z$, (3) $A \cdot B = 0$, (4) $A \times B = -8u_x - 16u_y - 4u_z$, (5) $\angle(A, B) = 90°$, and (6) $A = 3.74$, $B = 4.9$.

PA1.23. If $A = u_x + 4u_y + 3u_z$, and $B = 2u_x + u_y - u_z$, determine the smaller angle between the vectors, the unit vectors along the two vectors, and the ratio of their magnitudes. — *Hint: make use of the dot product to find the angle.*

PA1.24. Determine the differential volume, and three differential areas normal to the three base unit vectors in rectangular, cylindrical, and spherical coordinate systems. — *Hint: use the definitions of the differential volume and differential areas, noting that you need elemental lengths in the directions of the three base unit vectors.*

PA1.25. From questions QA1.9 and QA1.10, we know that the relations $A \cdot B = A \cdot C$ and $A \times B = A \times C$, taken separately, do not mean that $A = B$. Is this still true if *both* these relations are satisfied? — *(a) Yes, still $A \neq B$. (b) No, now the two vectors are equal. (c) The answer depends on the ratio of the two vectors.*

PA1.26. Determine the gradient of the scalar function $f(x, y, z) = x \cos 3y \exp(-4z)$, and the divergence and curl of the vector function $F(x, y, z) = (2x^2 yz)u_x + x \sin y \cos z \, u_y + (x + y + z)u_z$. — *Hint: use the expression for the gradient, divergence and curl in a rectangular coordinate system.*

PA1.27. Prove that the curl of the gradient of the function $f(x, y, z)$ from problem PA1.26 is zero, and that the divergence of the curl of $F(x, y, z)$ is zero. — *Hint: use the final results of the preceding problem and find the required curl of the gradient and divergence of the curl.*

PA1.28. Prove that the identities $\nabla \times [\nabla V(x, y, z)] = 0$ and $\nabla \cdot [\nabla \times A(x, y, z)] = 0$ are satisfied for any twice differentiable functions $V(x, y, z)$ and $A(x, y, z)$. — *Hint: treat the del operator as a simple vector and, for the second case, recall the properties of a triple scalar product.*

PA1.29. Let $A(x, y, z) = xyzu_x - \sin x \cos y \, e^z u_y + xy^2 z^3 u_z$. (1) Evaluate the line integral of $A(x, y, z)$ around a rectangular contour in the plane $z = 0$, with a vertex at the origin, a side a along the x axis, and a side b along the y axis; start from the origin along side a. (2) Evaluate the flux of $\nabla \times A(x, y, a)$ through the contour in the direction of the base unit vector u_z. Can you conclude that the results you obtained are correct? — *(a) Line integral $= -\sin a \sin b$, Flux $= -\sin a \sin b$. (b) Line integral $= -\cos a \cos b$, Flux $= -\sin a \sin b$. (c) Line integral $= -\sin a \sin b$, Flux $= -\cos a \cos b$.*

PA1.30. Let $A(x, y, z) = xyzu_x + x^2 y^2 z^2 u_y + x^3 y^3 z^3 u_z$. (1) Evaluate the flux of $A(x, y, z)$ through a cube with a vertex at the origin, and with sides of length 1 along coordinate lines for $x \geq 0$, $y \geq 0$ and $z \geq 0$. (2) Evaluate the volume integral of $\nabla \cdot A(x, y, z)$ over the cube. Do you have a simple check for the accuracy of the results? — *(a) Flux $= 0.31$, Volume integral $= 0.31$. (b) (a) Flux $= 61/104$, Volume integral $= 41/104$. (c) (a) Flux $= 61/144$, Volume integral $= 61/144$.*

PA1.31. The identity div$(\text{curl} A) = \nabla \cdot (\nabla \times A) = 0$ can be proved by considering the closed surface S having a small hole that shrinks to zero. Prove the identity by considering the closed surface to consist of two arbitrary open surfaces with a common boundary C. — *Hint:*

referring to Fig. PA1.31, apply Stokes's theorem to the two open surfaces and add the results, taking care of the correct reference directions.

PA1.32. The gradient of a scalar function f can alternatively be defined in the form very similar to that of the divergence,

$$\nabla f(c_1, c_2, c_3) = \lim_{\Delta v \to 0} \frac{1}{\Delta v} \oint_{\Delta S} f \, d\mathbf{S}.$$

Using arguments similar to those for deriving the divergence in orthogonal coordinate systems, derive the analogous general expression for the gradient of f. — *Hint: note that the differential volume is in the form of a parallelepiped defined by three differential length elements in a coordinate system, dl_1, dl_2, and dl_3, and that the pairs of the opposite parallelepiped sides are normal to one of the three base unit vectors (Fig. PA1.32).*

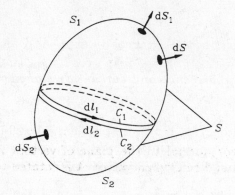
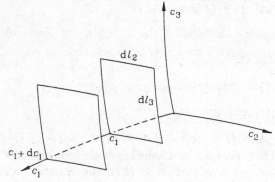

Fig. PA1.31. A closed surface considered as two open surfaces.

Fig. PA1.32. Two elemental surfaces in an orthogonal coordinate system.

PA1.33. From the general formula for the gradient obtained in problem PA1.32, prove that the same expressions for the gradient in rectangular, cylindrical, and spherical coordinate systems are obtained as by using the other formula. — *Hint: use a procedure analogous to that for determining the divergence.*

PA1.34. The curl of a vector function $\mathbf{F}$ can alternatively be defined in the form very similar to that of the divergence, i.e.,

$$\nabla \times \mathbf{F}(c_1, c_2, c_3) = \lim_{\Delta v \to 0} \frac{1}{\Delta v} \oint_{\Delta S} d\mathbf{S} \times \mathbf{F}.$$

Using arguments similar to those for deriving the divergence in orthogonal coordinate systems, derive the analogous general expression for the curl of $\mathbf{F}$. — *Hint: use a procedure analogous to that for determining the divergence.*

PA1.35. From the general formula for the curl obtained in problem PA1.34, prove that the same expressions for the curl in rectangular, cylindrical, and spherical coordinate systems are obtained as by using the other formula. — *Hint: see problem PA1.32, and Fig.PA1.32.*

Appendix 2:
Summary of Vector Identities

In the relationships described in this Appendix, A, B, C, D and F are vector functions, and V, W and f scalar functions of coordinates. It is assumed that they have all necessary derivatives.

A2.1. ALGEBRAIC IDENTITIES

1. $A + B = B + A$

2. $(A + B) + C = A + (B + C)$

3. $A \cdot B = AB \cos(A, B)$

4. $A \cdot A = |A|^2 = A^2$

5. $A \cdot B = B \cdot A$

6. $A \cdot (B + C) = A \cdot B + A \cdot C$

7. $V(A + B) = VA + VB$

8. $A \times B = AB \sin(A, B)n$, where n is the unit vector normal to the plane of vectors A and B, and its direction is determined by the right-hand rule when vector A is rotated to coincide with vector B in the shortest way.

9. $A \times B = -B \times A$

10. $A \times (B + C) = A \times B + A \times C$

11. $A \cdot (B \times C) = C \cdot (A \times B) = B \cdot (C \times A)$

12. $A \times (B \times C) = (A \cdot C)B - (A \cdot B)C$

13. $(A \times B) \cdot (C \times D) = D \cdot [(A \times B) \times C)] = C \cdot [D \times (A \times B)] = \ldots$ (using nos. 11 and 12, several other forms can be obtained)

14. $u_A = A/A$ (unit vector in the direction of vector A)

A2.2. DIFFERENTIAL IDENTITIES

15. $u_x \cdot \nabla V = \partial V / \partial x$ (x — arbitrary axis)

16. $\nabla(V + W) = \nabla V + \nabla W$

17. $\nabla(VW) = V\nabla W + W\nabla V$

18. $\nabla f(V) = f'(V)\nabla V$

19. $\nabla \cdot (A + B) = \nabla \cdot A + \nabla \cdot B$

20. $\nabla \cdot (VA) = V\nabla \cdot A + A \cdot \nabla V$

21. $\nabla \cdot (A \times B) = B \cdot \nabla \times A - A \cdot \nabla \times B$

22. $\nabla \times (A + B) = \nabla \times A + \nabla \times B$

23. $\nabla \times (VA) = (\nabla V) \times A + V\nabla \times A$

24. $\nabla \cdot (\nabla \times A) = 0$

25. $\nabla \cdot (\nabla V) = \nabla^2 V = \Delta V = \text{Laplacian of } V$

26. $\nabla \cdot [\nabla(VW)] = V\nabla \cdot (\nabla W) + 2\nabla V \cdot \nabla W + W\nabla \cdot (\nabla V)$

27. $\nabla \times (\nabla V) = 0$

28. $\nabla \times (\nabla \times A) = \nabla(\nabla \cdot A) - \nabla^2 A$

A2.3. BASIC INTEGRAL IDENTITIES

29. $\displaystyle\int_v \nabla f \, dv = \oint_S f \, dS$

30. $\displaystyle\int_v \nabla \cdot F \, dv = \oint_S F \cdot dS$ (the divergence theorem)

31. $\displaystyle\int_v \nabla \times F \, dv = \oint_S dS \times F$

32. $\displaystyle\int_S (\nabla \times F) \cdot dS = \oint_C F \cdot dl$ (Stokes's theorem)

A2.4. GRADIENT, DIVERGENCE, CURL, AND LAPLACIAN IN ORTHOGONAL COORDINATE SYSTEMS

Rectangular coordinate system

A sketch of the rectangular coordinate system, with indicated coordinates of a point and with the three base unit vectors, is shown in Fig. A2.1.

Notation: $f = f(x,y,z)$, $F = F(x,y,z)$, $F_x = F_x(x,y,z)$, $F_y = F_y(x,y,z)$, $F_z = F_z(x,y,z)$

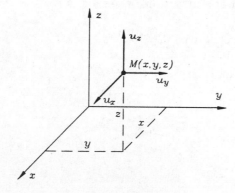

Fig. A2.1. A sketch of the rectangular coordinate system, with indicated coordinates of a point and with the three base unit vectors.

33. $\nabla f = \dfrac{\partial f}{\partial x}\, \boldsymbol{u}_x + \dfrac{\partial f}{\partial y}\, \boldsymbol{u}_y + \dfrac{\partial f}{\partial z}\, \boldsymbol{u}_z$

34. $\nabla \cdot \boldsymbol{F} = \dfrac{\partial F_x}{\partial x} + \dfrac{\partial F_y}{\partial y} + \dfrac{\partial F_z}{\partial z}$

35. $\nabla \times \boldsymbol{F} = \boldsymbol{u}_x \left(\dfrac{\partial F_z}{\partial y} - \dfrac{\partial F_y}{\partial z} \right) + \boldsymbol{u}_y \left(\dfrac{\partial F_x}{\partial z} - \dfrac{\partial F_z}{\partial x} \right) + \boldsymbol{u}_z \left(\dfrac{\partial F_y}{\partial x} - \dfrac{\partial F_x}{\partial y} \right)$

36. $\nabla^2 f \equiv \nabla \cdot (\nabla f) = \dfrac{\partial^2 f}{\partial x^2} + \dfrac{\partial^2 f}{\partial y^2} + \dfrac{\partial^2 f}{\partial z^2}$

37. $\nabla^2 \boldsymbol{F} = (\nabla^2 F_x)\boldsymbol{u}_x + (\nabla^2 F_y)\boldsymbol{u}_y + (\nabla^2 F_z)\boldsymbol{u}_z$

Cylindrical coordinate system

A sketch of the cylindrical coordinate system, with indicated coordinates of a point and with the three base unit vectors, is shown in Fig. A2.2.

Notation: $f = f(r,\phi,z)$, $\boldsymbol{F} = \boldsymbol{F}(r,\phi,z)$, $F_r = F_r(r,\phi,z)$, $F_\phi = F_\phi(r,\phi,z)$, $F_z = F_z(r,\phi,z)$

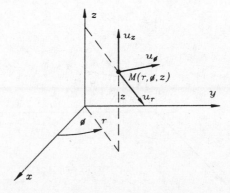

Fig. A2.2. A sketch of the cylindrical coordinate system, with indicated coordinates of a point and with the three base unit vectors.

38. $\nabla f = \dfrac{\partial f}{\partial r}\,\boldsymbol{u}_r + \dfrac{1}{r}\dfrac{\partial f}{\partial \phi}\,\boldsymbol{u}_\phi + \dfrac{\partial f}{\partial z}\,\boldsymbol{u}_z$

39. $\nabla \cdot \boldsymbol{F} = \dfrac{1}{r}\dfrac{\partial(rF_r)}{\partial r} + \dfrac{1}{r}\dfrac{\partial F_\phi}{\partial \phi} + \dfrac{\partial F_z}{\partial z}$

40. $\nabla \times \boldsymbol{F} = \boldsymbol{u}_r\left(\dfrac{1}{r}\dfrac{\partial F_z}{\partial \phi} - \dfrac{\partial F_\phi}{\partial z}\right) + \boldsymbol{u}_\phi\left(\dfrac{\partial F_r}{\partial z} - \dfrac{\partial F_z}{\partial r}\right) + \boldsymbol{u}_z\dfrac{1}{r}\left[\dfrac{\partial(rF_\phi)}{\partial r} - \dfrac{\partial F_r}{\partial \phi}\right]$

41. $\nabla^2 f \equiv \nabla \cdot (\nabla f) = \dfrac{1}{r}\dfrac{\partial}{\partial r}\left(r\dfrac{\partial f}{\partial r}\right) + \dfrac{1}{r^2}\dfrac{\partial^2 f}{\partial \phi^2} + \dfrac{\partial^2 f}{\partial z^2}$

42. $\nabla^2 \boldsymbol{A} = \nabla(\nabla \cdot \boldsymbol{A}) - \nabla \times (\nabla \times \boldsymbol{A})$

Spherical coordinate system

A sketch of the spherical coordinate system, with indicated coordinates of a point and with the three base unit vectors, is shown in Fig. A2.3.

Notation: $f = f(r,\theta,\phi)$, $\boldsymbol{F} = \boldsymbol{F}(r,\theta,\phi)$, $F_r = F_r(r,\theta,\phi)$, $F_\theta = F_\theta(r,\theta,\phi)$, $F_\phi = F_\phi(r,\theta,\phi)$

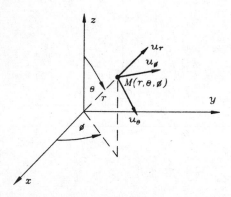

Fig. A2.3. A sketch of the spherical coordinate system, with indicated coordinates of a point and with the three base unit vectors.

43. $\nabla f = \dfrac{\partial f}{\partial r}\, u_r + \dfrac{1}{r}\dfrac{\partial f}{\partial \theta}\, u_\theta + \dfrac{1}{r \sin \theta}\dfrac{\partial f}{\partial \phi}\, u_\phi$

44. $\nabla \cdot F = \dfrac{1}{r^2}\dfrac{\partial (r^2 F_r)}{\partial r} + \dfrac{1}{r \sin \theta}\dfrac{\partial (\sin \theta F_\theta)}{\partial \theta} + \dfrac{1}{r \sin \theta}\dfrac{\partial F_\phi}{\partial \phi}$

45. $\nabla \times F = u_r \dfrac{1}{r \sin \theta}\left[\dfrac{\partial (\sin \theta\ F_\phi)}{\partial \theta} - \dfrac{\partial F_\theta}{\partial \phi}\right] + u_\theta \dfrac{1}{r}\left[\dfrac{1}{\sin \theta}\dfrac{\partial F_r}{\partial \phi} - \dfrac{\partial (r F_\phi)}{\partial r}\right]$

$$+ u_\phi \dfrac{1}{r}\left[\dfrac{\partial (r F_\theta)}{\partial r} - \dfrac{\partial F_r}{\partial \theta}\right]$$

46. $\nabla^2 f \equiv \nabla \cdot (\nabla f) = \dfrac{1}{r^2}\dfrac{\partial}{\partial r}\left(r^2 \dfrac{\partial f}{\partial r}\right) + \dfrac{1}{r^2 \sin \theta}\dfrac{\partial}{\partial \theta}\left(\sin \theta \dfrac{\partial f}{\partial \theta}\right) + \dfrac{1}{r^2 \sin^2 \theta}\dfrac{\partial^2 f}{\partial \phi^2}$

47. $\nabla^2 A = \nabla(\nabla \cdot A) - \nabla \times (\nabla \times A)$

Electrical Properties of Some Materials at Room Temperature and Low Frequencies

Material	ϵ_r	σ (S/m)	Comment
Silver		$6.14 \cdot 10^7$	oxidizes
Copper		$5.65 - 5.8 \cdot 10^7$	oxidizes
Gold		$4.1 \cdot 10^7$	inert
Aluminum		$3.8 \cdot 10^7$	oxidizes
Tungsten		$1.8 \cdot 10^7$	very hard
Zinc		$1.74 \cdot 10^7$	
Brass (30% zinc)		$1.5 \cdot 10^7$	
Nickel		$1.28 \cdot 10^7$	
Bronze		$1 \cdot 10^7$	
Iron		$1 \cdot 10^7$	
Steel		$0.5 - 1 \cdot 10^7$	
Tin		$0.87 \cdot 10^7$	
Nichrome		$0.1 \cdot 10^7$	
Graphite		$7 \cdot 10^4$	
Sea water	70	$3 - 5$	
Wet earth	$5 - 15$	$10^{-2} - 10^{-3}$	
Dry earth	$2 - 6$	$10^{-4} - 10^{-5}$	
Fresh water (lake)	80	10^{-3}	
Distilled water	80	$2 \cdot 10^{-4}$	
Alcohol	25		
Air	1.006		breakdown $3\,\mathrm{kV/m}$
Styrofoam	1.03		
Teflon	2.1		
Polystyrene	2.56		
Rubber	$2.5 - 3$	10^{-15}	
Paper	$2 - 4$		
Quartz	3.8		
Glass	$4 - 10$	10^{-12}	
Mica	5.4		
Porcelain	6	10^{-10}	
Diamond	$5 - 6$	$2 \cdot 10^{-13}$	good heat conductor
Silicon	11		semiconductor, σ depends on doping level
Galium arsenide	13		semiconductor, σ depends on doping level
Barium titanate	$60 - 3600$		anisotropic in crystaline form dependent on mechanical conditions

Standard (IEC) multipliers of fundamental units

Multiple	Prefix	Symbol	Example
10^{12}	tera	T	$1.2\,\text{Tm} = 1.2 \cdot 10^{12}\,\text{m}$
10^9	giga	G	$12\,\text{GW} = 12 \cdot 10^9\,\text{W}$
10^6	mega	M	$5\,\text{MHz} = 5 \cdot 10^6\,\text{Hz}$
10^3	kilo	k	$22\,\text{kV} = 22 \cdot 10^3\,\text{V}$
10^2	hecto	h	$100\,\text{hN} = 100 \cdot 10^2\,\text{N}$
10	deca	da	$32\,\text{dag} = 320\,\text{g}$
10^{-1}	deci	d	$2\,\text{dm} = 2 \cdot 10^{-1}\,\text{m}$
10^{-2}	centi	c	$75\,\text{cm} = 70 \cdot 10^{-2}\,\text{m}$
10^{-3}	milli	m	$56\,\text{m}\Omega = 56 \cdot 10^{-3}\,\Omega$
10^{-6}	micro	μ	$25\,\mu\text{H} = 25 \cdot 10^{-6}\,\text{H}$
10^{-9}	nano	n	$56\,\text{nA} = 56 \cdot 10^{-9}\,\text{A}$
10^{-12}	pico	p	$40\,\text{pF} = 40 \cdot 10^{-12}\,\text{F}$
10^{-15}	femto	f	$1.2\,\text{fm} = 1.2 \cdot 10^{-15}\,\text{m}$
10^{-18}	atto	a	$0.16\,\text{aC} = 0.16 \cdot 10^{-18}\,\text{C}$

Standard (IEC) multipliers of fundamental units

Multiple	Prefix	Symbol	Example
10^{12}	tera	T	$1.2\,\text{Tm} = 1.2 \cdot 10^{12}\,\text{m}$
10^{9}	giga	G	$12\,\text{GW} = 12 \cdot 10^{9}\,\text{W}$
10^{6}	mega	M	$5\,\text{MHz} = 5 \cdot 10^{6}\,\text{Hz}$
10^{3}	kilo	k	$22\,\text{kV} = 22 \cdot 10^{3}\,\text{V}$
10^{2}	hecto	h	$100\,\text{hN} = 100 \cdot 10^{2}\,\text{N}$
10	deca	da	$32\,\text{dag} = 320\,\text{g}$
10^{-1}	deci	d	$2\,\text{dm} = 2 \cdot 10^{-1}\,\text{m}$
10^{-2}	centi	c	$75\,\text{cm} = 70 \cdot 10^{-2}\,\text{m}$
10^{-3}	milli	m	$56\,\text{m}\Omega = 56 \cdot 10^{-3}\,\Omega$
10^{-6}	micro	μ	$25\,\mu\text{H} = 25 \cdot 10^{-6}\,\text{H}$
10^{-9}	nano	n	$56\,\text{nA} = 56 \cdot 10^{-9}\,\text{A}$
10^{-12}	pico	p	$40\,\text{pF} = 40 \cdot 10^{-12}\,\text{F}$
10^{-15}	femto	f	$1.2\,\text{fm} = 1.2 \cdot 10^{-15}\,\text{m}$
10^{-18}	atto	a	$0.16\,\text{aC} = 0.16 \cdot 10^{-18}\,\text{C}$